Studies in Systems, Decision and Control

Volume 428

The series "Studies in Systems, Decision and Control" (SSDC) covers both new developments and advances, as well as the state of the art, in the various areas of broadly perceived systems, decision making and control–quickly, up to date and with a high quality. The intent is to cover the theory, applications, and perspectives on the state of the art and future developments relevant to systems, decision making, control, complex processes and related areas, as embedded in the fields of engineering, computer science, physics, economics, social and life sciences, as well as the paradigms and methodologies behind them. The series contains monographs, textbooks, lecture notes and edited volumes in systems, decision making and control spanning the areas of Cyber-Physical Systems, Autonomous Systems, Sensor Networks, Control Systems, Energy Systems, Automotive Systems, Biological Systems, Vehicular Networking and Connected Vehicles, Aerospace Systems, Automation, Manufacturing, Smart Grids, Nonlinear Systems, Power Systems, Robotics, Social Systems, Economic Systems and other. Of particular value to both the contributors and the readership are the short publication timeframe and the world-wide distribution and exposure which enable both a wide and rapid dissemination of research output.

Indexed by SCOPUS, DBLP, WTI Frankfurt eG, zbMATH, SCImago.

All books published in the series are submitted for consideration in Web of Science.

More information about this series at https://link.springer.com/bookseries/13304

Fei Han · Zidong Wang · Hongli Dong

Distributed Filtering, Control and Synchronization

Local Performance Analysis Methods

Springer

Fei Han
Artificial Intelligence Energy Research Institute
Northeast Petroleum University
Daqing, Heilongjiang, China

Hongli Dong
Artificial Intelligence Energy Research Institute
Northeast Petroleum University
Daqing, Heilongjiang, China

Zidong Wang
Department of Computer Science
Brunel University London
Uxbridge, Middlesex, UK

ISSN 2198-4182 ISSN 2198-4190 (electronic)
Studies in Systems, Decision and Control
ISBN 978-3-030-97074-1 ISBN 978-3-030-97075-8 (eBook)
https://doi.org/10.1007/978-3-030-97075-8

Mathematics Subject Classification: 93-02, 93A14, 93A15, 93E10, 93E11

This Springer imprint is published by the registered company Springer Nature Switzerland AG
The registered company address is: Gewerbestrasse 11, 6330 Cham, Switzerland

This book is dedicated to the Dream Dynasty consisting of a group of bright people who have been to Brunel University of the UK to enjoy memorable research life by distributedly filtering the noises and then driving various lives to the consensus over the finite horizon...

Preface

Over the past decade, thanks to the rapid development of wireless transmission technology, sensor networks, multi-agent systems, and complex networks have become hot research focuses that have attracted an increasing interest. In these emerging systems, the large-scale and the complex couplings are two ubiquitous features that contribute significantly to the complexity of system modeling. Although the analysis and synthesis problems for the large-scale systems have long been the mainstream of research topics and much effort has been made to deal with them, there still exist some distinguished differences between these new systems and the traditional large-scale systems, for example, the local communications among neighboring nodes, energy constraints, cyber-attacks, etc. It's undeniable that these new features will bring about fresh challenges in the related fields.

This book is devoted to dealing with the distributed filtering and control problems and the synchronization problems in the context of sensor networks, multi-agent systems, and complex networks by utilizing the local performance analysis method, respectively. The content of this book covering the applications of the local performance analysis method can be conceptually divided into two parts. In the first part, we focus mainly on the distributed filter design problems for several classes of discrete-time nonlinear stochastic systems subject to complex system dynamics (i.e., random parameter systems, parameter uncertainties, piecewise linear systems, and stochastic nonlinearities), networked-induced phenomena (i.e., missing measurements, censored measurements, quantization, and sensor saturations), scheduling protocols (i.e., Round-Robin protocols and event-triggering protocols), and deception attacks. Some novel distributed filtering algorithms are developed in terms of the solutions to Recursive Linear Matrix Inequalities (RLMIs), which are suitable for online applications. In the second part, distributed controller design problem of multi-agent systems and the synchronization and the filtering problems of complex networks are considered for several classes of nonlinear stochastic systems with random parameters. In this part, the local information from neighboring nodes is employed to present the H_∞-consensus control performance index and H_∞ bounded synchronization performance index to cater for the local performance analysis method, respectively. Some sufficient conditions are established for the existence of distributed controller, and synchronization criteria and filter of complex networks that can be solved easily by using the semi-definite programming method.

The compendious framework and description of this book are outlined as follows. Chapter 1 introduces the recent advances on distributed filter and controller design, the synchronization criterion, the challenges as well as the vector dissipativity theory. By using the globally augmented method, Chapter 2 is concerned with the distributed H_∞ filtering problem for a class of piecewise discrete-time linear systems, which shows the disadvantages of the global augmented method. For addressing the problems caused by the globally augmented method, the local performance analysis method is developed by means of the vector dissipativity theory in the following chapters. The consensus filtering problem is investigated in Chapter 3 for a class of discrete time-varying systems with stochastic nonlinearities and multiple missing measurements. In Chapter 4, the distributed filtering problem is studied for a class of discrete time-varying linear systems with random parameters and event-triggering protocol. The scalable consensus filtering problem is dealt with in Chapter 5 for a class of discrete time-varying systems over sensor networks with the Round-Robin protocol. Chapter 6 investigates the partial-nodes-based scalable H_∞-consensus filtering problem for a class of discrete time-varying systems subject to multiplicative noises and censored measurements over sensor networks. Chapter 7 discusses the distributed H_∞-consensus filtering problem for a class of discrete time-varying systems subject to both multiplicative noises and deception attacks over sensor networks. The distributed resilient filtering problem over the finite horizon is studied in Chapter 8 for a class of discrete time-varying nonlinear systems with stochastic perturbations and time-delays. In addition, Chapter 9 is concerned with the distributed H_∞-consensus control problem over the finite horizon for a class of discrete time-varying multi-agent systems with random parameters, and Chapter 10 deals with the finite horizon bounded H_∞ synchronization and filtering problems for the discrete-time complex networks with missing measurements. Finally, Chapter 11 draws some conclusions on this book and points out some possible research directions related to the work done in this book.

This book is a research monograph whose intended audience is graduate and postgraduate students as well as researchers, serving as both a summary of the recent research results and a source of further research directions.

Daqing, China Fei Han
London, UK Zidong Wang
Daqing, China Hongli Dong

Acknowledgements

We would like to acknowledge the help of many people who have been directly involved in various aspects of the research leading to this book. Special thanks go to Profs. Xiaohui Liu and Lampros Stergioulas from the Brunel University of the U.K., Prof. James Lam from the University of Hong Kong, Profs. Guanrong Chen and Daniel W. C. Ho from the City University of Hong Kong, and Prof. Yugang Niu from East China University of Science and Technology of China. We also extend our thanks to many colleagues who have offered support and encouragement throughout this research effort. In particular, we would like to acknowledge the contributions and friendly support from Guoliang Wei, Bo Shen, Derui Ding, Lifeng Ma, Yurong Liu, Jinling Liang, Xiao He, Jun Hu, Xiu Kan, Sunjie Zhang, Nianyin Zeng, Yang Liu, Lei Zou, Qinyuan Liu, Wangyan Li, Licheng Wang, Shuai Liu, Nan Hou, Hongyu Gao, Yuxuan Shen, Jiahui Li, Bo Jiang, and Xuerong Li. Finally, we are deeply indebted to our families for their never-ending understanding, unfailing encouragement, and constant support when it was most required.

The writing of this book was supported by the National Natural Science Foundation of China under Grants 61873148, 61873058, 61933007, and 62073070; the China Post-Doctoral Science Foundation under Grants 2017M621242 and 2020T130092; Heilongjiang Provincial Key Laboratory of Networking and Intelligent Control (Northeast Petroleum University) of China; the Engineering and Physical Sciences Research Council (EPSRC) of the UK; the Royal Society of the UK; and the Alexander von Humboldt Foundation of Germany. The support of these organizations is much acknowledged.

Contents

List of Notation

$\mathbb{R}^n$	The n-dimensional Euclidean space
$\mathbb{R}^{n\times m}$	The set of all $n \times m$ real matrices
$\mathbb{Z}^-$	The set of all negative integers
$\|A\|$	The norm of matrix A defined by $\|A\| = \sqrt{\text{trace}(A^T A)}$
A^T	The transpose of the matrix A
A^{-1}	The inverse of the matrix A
I	An identity matrix with compatible dimension
0	A zero matrix with compatible dimension
$\text{Prob}(\cdot)$	The occurrence probability of the event "$\cdot$"
$\mathbb{E}\{x\}$	The expectation of the stochastic variable x
$\mathbb{E}\{x\|y\}$	The expectation of the stochastic variable x conditional on y
$(\Omega, \mathscr{F}, \text{Prob})$	The complete probability space
$\lambda_{\max}(A)$	The largest eigenvalue of a square matrix A
$\text{diag}\{\dots\}$	The block-diagonal matrix
$\text{diag}_{p_i}^{j}\{A_j\}$	The block-diagonal matrix $diag\{A_1, A_2, \dots, A_{p_i}\}$, where $j = 1, 2, \dots, p_i$
$\text{vec}\{\dots\}$	A matrix vector with blocks given by the matrices in $\{\dots\}$
$\text{vec}_{p_i}^{j}\{A_j\}$	A matrix vector with blocks $\text{vec}\{A_1, A_2, \dots, A_{p_i}\}$, where $j = 1, 2, \dots, p_i$
$l_2[0, \infty)$	The space of square summable sequences
$\|\cdot\|_2$	The usual l_2 norm
$\text{tr}(A)$	The trace of a matrix A
$\text{col}\{\dots\}$	A matrix column with blocks given by the matrices in $\{\dots\}^T$
$\text{Cov}\{x_k\}$	The covariance of random variable x_k
$\circ$	The Hadamard product being defined as $[A \circ B]_{ij} = A_{ij} \cdot B_{ij}$
$X > Y$	The $X - Y$ is positive definite, where X and Y are real symmetric matrices
$X \geq Y$	The $X - Y$ is positive semi-definite, where X and Y are real symmetric matrices

$\upsilon \leq\leq 0$	The vector υ is nonpositive indicating that every component of υ is nonpositive, where $\upsilon \in \mathbb{R}^q$
$\upsilon \ll 0$	The vector υ is negative indicating that every component of υ is negative, where $\upsilon \in \mathbb{R}^q$

In symmetric block matrices or long matrix expressions, a star "$*$" is used to represent a term that is induced by symmetry. Matrices, if their dimensions are not explicitly stated, are assumed to be compatible for algebraic operations.

Chapter 1
Introduction

In the past decades, special attention was focused on the distributed filtering problem over sensor networks (SNs), the distributed consensus control problem of multi-agent systems (MAs), and the filtering and synchronization problems in complex networks (CNs) due to their extensive applications in a variety of practical domains. The common features of these problems are that the large number of nodes and the complicated coupling relationship among nodes, both of which pose huge challenges on the design of distributed filtering, control and synchronization criteria. More specifically, the challenges come from the following three aspects: (1) the computational burden stemming from the large-scale characteristic; (2) the analysis difficulties due to the complicated dynamic couplings among nodes; and (3) the technical obstacles from various network-induced phenomena as well as the complicated system dynamics. Therefore, the design of the novel distributed controller/filter for nonlinear stochastic systems with network-induced phenomena serve as a very interesting, imperative yet challenging topic. Meanwhile, it is of significant importance to develop new distributed control and filtering algorithms for the nonlinear stochastic systems over the network environment in order to address the above challenges and cover more engineering practice.

1.1 Research Background, Motivations and Research Problems

In this section, we introduce the research background, motivations and research problems of this book.

F. Han et al., *Distributed Filtering, Control and Synchronization*,
Studies in Systems, Decision and Control 428,
https://doi.org/10.1007/978-3-030-97075-8_1

1.1.1 Distributed Filtering

The past decades have witnessed the rapid development of SNs, which are due mainly to their broad applications in the military fields, environmental monitoring, human healthcare and robotics. For more applications, readers can refer to the survey papers [1–3]. With the introduction of SNs, more information from the plant can be collected via a large number of sensor nodes scattered in the interesting region. Then, based on the available measurement information, a new filtering technique, called distributed filtering, has been developed by combining the traditional filtering techniques and distributed algorithms [4]. As discussed in the majority of references, compared with the traditional filtering theory, the advantages of distributed filtering include better robustness, flexibility and reliability because of the distributed filter structures. Accordingly, the distributed filtering has received tremendous attention [3, 5–22].

Actually, only the distributed filter structure cannot entirely show the advantages of the distributed filtering scheme. If this scheme is not implemented in a distributed manner, it results in heavy computational burden due to the well-known large-scale characteristic of SNs. Actually, the distributed implementation implies that every node needs to calculate its own gains. Then, how to achieve such a fundamental objective is attractive for decentralizing the computational burden to each node. It is worthwhile noting that, some insightful results regarding the distributed calculation problems have been published by means of the so-called local performance analysis method stemming from the vector dissipativity theory in recent years [8]. This theory has been proposed in [23] based on the vector Lyapunov functions, which offered a flexible framework for stability analysis in a dynamical system. The basic idea lies in such a realization: each element of vector Lyapunov function can satisfy less rigid requirements as compared to a single Lyapunov function. As such, in the process of designing a distributed algorithm, the filtering error dynamics of each node can be separately analyzed and the filter gains can be separately designed, which is conducive to achieving the local design and distributed calculations of the distributed algorithms. As a result, it is desirable to design locally the distributed filtering algorithm.

1.1.2 Distributed Consensus Control

In the past decade, the distributed consensus control problem of MAs has become an increasingly popular research focus for its potential prospects in variety realms such as automated factories, environmental monitoring, search and rescue operations, target tracking, and formation and coverage control [24, 25]. Essentially, the aim of this problem is to design a protocol such that all agents reach an agreement on some common interesting quantities [24]. This kind of protocol, called distributed protocol, is only allowed the data to be transmitted among neighboring agents. It is noted that such local communications apparently imply the well-known local constraint, i.e., every agent accesses the information only from its neighboring agents, which actually lays local constraints on the protocol design issues. Thus, the distributed consensus control is an indeed local design issue based on the local performance analysis. Practically, an agent itself may be subjected to external disturbances. Therefore, an interesting problem is to study the robustness of consensus protocols to external disturbances, which is known as the distributed H_∞-consensus control problem [26].

Up to now, the distributed control protocol design problem has drawn more and more researchers' attention, see [24–41] and so on. The pioneering work [24] has proposed several kinds of basic distributed protocols: state feedback protocol, state feedback protocol with time-delays and state feedback protocol with dynamic topology. Since the relative states among neighboring agents were often not available, a distributed observer-type consensus protocol has been developed based on relative output measurements in [26]. A discontinuous distributed observer has been designed to estimate the real state with bounded disturbance in [40], and then the state estimate has been employed to design the control inputs to achieve the consensus. The H_∞-consensus control problem with missing measurements has been considered in [27], where the output feedback control protocol was adopted. Considering the relative full state is unavailable, [28] has designed a distributed estimation-based output feedback control protocol with event-based protocol. Although the protocol was distributed in [26, 36, 39], calculating the feedback gain matrix needs global information, that is, each agent has to know the knowledge of the eigenvalues of the asymmetric Laplacian matrix. In order to resolve this problem, a distributed adaptive consensus protocol has been proposed in [34, 37] using only the local information from its own neighbors, which is an indeed distributed manner. Nevertheless, it is applicable to the system only with identical agents. For these above mentioned protocols, a core idea is that the controller structure and the implementation of protocols are all distributed. Unfortunately, the distributed controller gains still need centralized calculations except [34, 37]. It obviously results in losing the scalability, which is an essential design requirement of the distributed algorithm. Therefore, it is an urgent need that the controller gains should be calculated separately on the premise of the local information.

1.1.3 Bounded Synchronization and Filtering

CNs are usually utilized to describe a variety of complicated systems of the real world, such as neural networks, scientific citation web, transportation networks, phone call networks, the World Wide Web, micro grid, metabolic and protein networks, for more details, see [42, 43]. Remarkably, the discoveries of scale-free network and small-world network [44, 45] have stimulated the research interest in CNs. From the existing results, it can be seen that the challenges stem mainly from a large number of dynamical nodes in CNs as well as complicated coupling relations among them.

In particular, the synchronization and filtering problems have attracted a great deal of attention from various fields of science and engineering in the past few decades.

As a basic problem of CNs, the synchronization aims to investigate whether the state of nodes reach the same, the key point of which is to deal with the complicated couplings among massive nodes. So far, a widely-used efficiently idea called the globally augmented method is to package all the node dynamics into a large-scale system without couplings and then employ the existing techniques to investigate the synchronization problems. Based on this idea, a large number of results has been published from a variety of perspectives over the last ten years, for example, different synchronization concepts including exponential synchronization [46–48], projective synchronization [49], finite-time synchronization [50], bounded H_∞ synchronization [51], robust synchronization [52–54], and quasi-synchronization [55]; various control methods to achieve synchronization such as pinning control [56], sampled-data control [57, 58], fuzzy control [52], distributed impulsive control [55], and non-fragile control [59, 60]; different couplings such as stochastic coupling [61, 62] and uncertain coupling [63]; different systems such as coupled semilinear partial differential systems [53] and linearly coupled ordinary differential systems [64]; and different topology structures such as stochastically switching topology [65] and switching disconnected topology [66]. It is worth mentioning that the bounded H_∞ synchronization problem has been discussed in [51] by the recursive linear matrix inequalities (RLMIs) approach, where the finite-horizon bounded H_∞ synchronization performance index was first proposed.

As is well known, the dynamic behavior of CNs is very helpful for achieving certain engineering requirements. Nonetheless, in actual applications, it is commonly unavailable due to various reasons both in physical constraints and technological restrictions [67–71]. In view of the H_∞ filtering theory, the available measurement information is generally utilized to estimate node states in CNs. Unfortunately, the effective information may be missing during the process of measurement or transmission, which could be caused by sensor aging, intermittent sensor failures, limited bandwidth, network congestion and cyber attacks [72–75]. Accordingly, the filtering problem in complex networks by means of various of available measurement information has been received much attention by many scholars in [51, 58, 72, 76–81]. The filtering problem has been addressed in [76] for CNs with sensor saturations, signal quantisation and missing measurements in a unified framework. Liang et al. have designed the state estimator for CNs with missing measurements and time-varying delays in [77]. For an array of discrete time-varying coupled stochastic CNs, Hu et al. have dealt with the recursive state estimation problem in [79] and Shen et al. have attempted to handle both the H_∞ synchronization and state estimation problems in [51]. For the CNs with uncertain inner coupling and incomplete measurements, the H_∞ state estimation problem has been investigated in [72] via the semi-definite programming method.

1.2 Systems, Measurements, Protocols, and Cyber-Attacks

According to our investigations, SNs, MAs and CNs have become very prevalent owing to their extensive applications, such as space and terrestrial exploration, access in hazardous environments, factory automation, remote diagnostics and troubleshooting, domestic robots, experimental facilities, automobiles, aircraft, and manufacturing plant monitoring. Some problems naturally emerge from the introduction of network, for example, the imperfect measurements due to various practical reasons, the scheduling protocols to reduce the consumption of energy and the utilization of network bandwidth. Meanwhile, the complicated plant behaviors, such as random parameters, time-delays, and stochastic nonlinearities, would yield great challenges when investigating the distributed filtering and consensus control problem. On the other hand, the desirable filter structure (the resilient filter, for instance) also increase the design difficulties. These considered factors are non-negligible in practical engineering. In other words, the designed schemes are meaningful only taking the above influence factors into account. Consequently, it is of important significance to tackle the distributed filtering and control problems for complicated systems with imperfect measurements, scheduling protocols, cyber-attacks and so on.

Random Parameter Systems

Discrete-time systems with random parameters arise in many fields, such as radar control, missile track estimation, satellite navigation, digital control of chemical process and economic system, see [82–84] for more details. Some system parameters might be randomly perturbed within certain intervals due probably to the abrupt phenomena, such as random failures and repairs of the components, changes in the interconnections of subsystems, sudden environment changes, modification of the operating point of a linearized model of nonlinear systems, etc. In such a kind of systems, the system matrix and measurement matrix are formulated by random matrices, the elements of which are all random variables. Possibly, these random variables are correlated with each other, and it is easy to present their statistical information in the form of the components [83, 84]. Accordingly, some research efforts have been made on the filter design with random parameter matrices. For example, the recursive optimal estimation problem has been dealt with in [82] for linear discrete-time systems with random parameter matrices in the minimum variance sense. As it is well known, in the process of analysis and synthesis of system, matrix manipulations are very common considering a large number of variables. Thus, it is necessary to present their statistical information in the form of matrix. Additionally, how to handle the quadratic terms of random matrix, called the weighted-covariance matrix, is still a challenging work. Therefore, there are two challenges needed to be solved. One is to seek a matrix to describe the complicated correlations among the random elements in a random matrix. Another is to establish a formula to calculate the weighted-covariance matrix utilizing this matrix. We devote ourself to tackle these challenges associated with the random parameter systems.

Time-Delays
As is well known, time-delays are inherent features of many practical processes such as chemical reactions, nuclear reactors, biological systems, and etc. If not treated properly, they may lead to instability or significantly deteriorate the performances of the corresponding systems [78, 85–89]. To date, the time-delay systems have received considerable attention and various kinds of methods have been developed to obtain less conservative results, for example, constructing new Lyapunov functions [85], free-weighting matrix approach [86], and delay partitioning method [88]. When time-delays come into the local design of distributed estimation algorithms, a critical problem is how to characterize the impact of time-delays, which essentially belongs to a kind of information with respect to time horizon. Differently, the idea of distributed estimation is to utilize the space information collected from nodes distributed in space region at every instant [8]. Taking these into considerations, an interesting problem is put forward on how to combine these two kinds of information within a unified framework. In the existing vector dissipativity theory, the constructed dissipation matrix is employed to formulate the complicated coupling relationships among neighboring nodes, which characterizes the exchange of the space information. As for the time-delays, the corresponding construction method for the dissipation matrix has not been published yet according to our investigations. Inspired by this point, it is desirable to develop a kind of new dissipation matrices to describe the impact of time-delays within the framework of the vector dissipation definition, and then locally design the distributed filtering scheme for the nonlinear time-delayed plant.

Stochastic Nonlinearities
From a practical engineering viewpoint, the stochastic nonlinearities could come from high maneuverability of the tracked targets, intermittent network congestion, random failures and repairs of the components, and sudden environment changes, see [90] for more details. By virtue of the formulation of statistical means via statistical tests, it has been extensively investigated in the literature, see e.g., [84, 90–92]. Additionally, the missing measurements have also received widespread attention, see [90, 93–95] and the references therein. It should be pointed out that, in most existing literatures, when stochastic nonlinearities or missing measurements are taken into account in the distributed filtering issues, the plant and the filtering error systems of all nodes are always augmented to acquire a whole dynamics for performing the analysis based on the Lyapunov stability theory, which is because the autonomous system of filtering errors is unavailable. Obviously, the complexity of the global conditions based on the above procedure is a polynomial function of the size of SN. In consideration of the massive nodes in SN, the computational complexity issue tremendously blocks the practical applications of distributed filtering. As a result, it is quite necessary to find an efficient method to avoid the problems.

As pointed out in Chap. 3, multiplicative noises are a special kind of stochastic nonlinearities. Noting that the working condition may be corrupted by unexpected random disturbances, multiplicative noises have been treated as one of the unavoidable phenomena occurred in real-world systems, such as signal processing, neural networks, heat transfer and population models [96–103]. It should also be pointed out that the evolution of system dynamics probably depends on the time since the system states may suffer from real-time variations. Therefore, much research attention has been paid to time-varying systems with multiplicative noises to better describe the engineering reality. To date, many research results have been reported on the distributed filtering problem for discrete time-varying systems with multiplicative noises [6, 104, 105]. For example, the distributed l_2-l_∞ state estimation problem has been investigated in [104] for periodic time-varying systems subject to multiplicative noises, where the filter was designed to ascertain the globally asymptotic stability of the error system. Very recently, the variance-constrained distributed filtering problem has been considered in [105] for networked systems with multiplicative noises over a finite horizon.

Imperfect Measurements
It is no doubt that the measurements of plant is crucial to investigate the filtering problem or the control problem. Unfortunately, perfect measurements are not generally available because of various practical reasons. The first example is the Partial-Nodes-Based Measurements. In the practical applications of SNs, it is essentially difficult to guarantee that system measurements are available from *all* sensor nodes due to a variety of physical restrictions. For example, certain sensor nodes only have the transmission capability (i.e., without measuring capability) because of the limited resource that prevents the information collection from all nodes [106]. The unavailability of measurement outputs could also due to the sensor failures in certain severe circumstances, e.g., colliery, nuclear plant, and military battlefield. Therefore, in reality, instead of requiring the measurement information from all the sensor nodes, it is more reasonable to assume that the measurement outputs utilized for the filtering tasks are only available from a *fraction* of sensor nodes, which gives rise to the so-called partial-nodes-based (PNB) distributed filtering problem. To be more specific, the main idea of the proposed PNB distributed filtering problem is to make use of the measurement information from *partial* sensor nodes to achieve the desired filtering performance. By now, some pioneering works have been reported on the PNB state estimation topics for complex networks, see e.g., [107, 108]. Nevertheless, the PNB distributed consensus filtering issue has not been fully investigated yet and this motivates us to launch a study on such a problem of clear engineering insights.

The second example is the Censored Measurements. As a kind of typical phenomena encountered in practical engineering, the measurement censoring caused mainly by the inevitable data truncation has recently aroused some research attention with applications ranging from biological and economic systems, computer vision problems, to distributed detection scenarios. Various models have been utilized in the literature to characterize the censored measurements, among which the Tobit model (proposed by economist James Tobin in the middle of last century) has been recognized to be the most popular one. In recent years, the Tobit model has attracted some initial research interest in the area of signal processing, and special effort has been directed towards the Tobit Kalman filtering problems, see e.g., [109–116]. It is worth mentioning that most results concerning Tobit Kalman filtering have been based on the assumption that the probability density function of the system noises obeys the Gaussian distribution. Apparently, these existing results are not applicable to systems suffering from non-Gaussian noises. Very recently, a set-membership filtering approach has been proposed in [117] for a class of time-varying nonlinear systems with censored measurements where the noises reside within certain ellipsoids. Nevertheless, when it comes to the distributed H_∞-consensus filtering problem subject to censoring effects, the corresponding results have been very few due to the lack of appropriate techniques capable of tackling the scalability issue in a large-scale network environment.

The last example of imperfect measurements is caused by cyber-attacks. In the context of networked systems, cyber-attacks become a concern as the network communication is vulnerable to malicious signals from adversaries. Actually, security protections may not be provided or only basic prevention measures are available during the data transmitted through the network, which may lead to the degradation of system performance or even huge economic loss. So far, an ever-growing research interest has been focused on the security filtering or control problem with various cyber-attacks [118–124]. Among different types of attacks, deception attacks are recognized as the most dangerous ones. Particularly, deception attacks are likely to deteriorate the system performance since the adversaries may modify the actual data and inject false data stealthily without being detected. During the past few years, some initial research efforts have been devoted to the filter design problem with deception attacks [6, 120]. For instance, the estimation and control problems has been handled in [120] for linear systems with certain sensors or actuators suffering from attacks, and the system resilience with respect to cyber-attacks has been analyzed as well. In addition, the distributed recursive filtering problem has been addressed in [6] for time-delayed systems with deception attacks and quantization effects. Despite the available literature, the distributed H_∞-consensus filtering problem with deception attacks is still an on-going issue deserving to be investigated further.

Scheduling Protocols

As is well known, protocols play a predominant role in communication systems for regulating the data transmissions to avoid data congestion and/or collision. It is of significant importance to examine the impact of protocols on various filtering issues especially in the context of SNs with respect to their topologies. To date, much effort has been devoted to the investigation of filtering/control problems subject to different types of protocols such as Round-Robin protocol (RRP) [125–130], Weighted Try-Once-Discard protocol (WTODP) [125, 128], and stochastic communication protocol (SCP) [131]. Some representative results can be summarized as follows. The set-membership filtering problem has been discussed in [128], where two typical protocols (i.e., RRP and WTODP) were adopted to determine the transmission order of sensor nodes. A distributed robust estimation algorithm has been proposed in [127], where the RRP was utilized to schedule data transmission in networks. Very recently, a state estimator has been designed in [129] for a class of nonlinear delayed dynamical networks with the RRP. It is worth mentioning that the RRP is perhaps the most frequently applied protocol for its easy deployability in engineering practice.

Another kind of important communication protocols comes from the so-called event-triggering protocols. The original idea of event-triggering protocol is to determine whether the newly estimate (or sampled state, innovation) will be sent out to the controller/filter by using the predefined triggering condition. Only those necessary data, which plays a major role in guaranteeing the system performance, are transmitted. As a result, it can efficiently reduce the consumption of sensor energy and the network resource while retaining the desirable performance. Therefore, it is essential to investigate the distributed filtering problem with event-triggering protocol. The idea from [132] has been generalized to the distributed event-triggered H_∞ filtering with communication delays in [133], where a novel co-design method was developed to determine desired filter parameters and threshold parameters. The recursive filtering method in [134] has been extended to the distributed case with the event-triggering protocol in [28], where the event-triggering conditions were constructed according to the local innovation of sensor node. In this paper, the authors took the event-triggering term as disturbances and employ the traditional robust inequality to design the recursive distributed filtering. The event-triggered distributed filtering problem has been investigated in [135], where the transmitted data may be missed after the event-triggering conditions are violated. No doubt, the event-triggering protocol can help us reduce the number of transmissions, which is desirable in designing the distributed filtering. In these existing references, the globally augmented way has been always adopted to handle the distributed filtering problem with event-triggering protocols. Nevertheless, considering the number of sensor node in sensor network, the obtained global design conditions would yields heavy computational burden, namely, it increases as the network size becomes large. Based on the mentioned observations, we are motivated to seek local analysis method to handle the distributed filtering with event-triggering protocols.

Distributed Resilient Filtering

In the existing references, most of distributed filtering algorithms have acknowledged an implicit assumption that the designed distributed filter can be perfectly implemented by using the obtained filter gains. However, it cannot be ignored that SNs are constituted by a large number of tiny, inexpensive sensor nodes with only limited processing and computing capabilities, which could result in the round-off errors in numerical computation during the implementation. If these considerations are not taken into adequate account, the gain implementation errors would cause severe degradation of the filtering performance. As such, given a target plant to be estimated, a significant issue is how to design a distributed filter for each node which is insensitive to certain errors with respect to its gains, i.e., the so-called distributed resilient filter. To date, a larger number of studies on resilient filtering have been launched by considering different descriptions of resiliency, such as the additive type [19, 22, 136–141] and the interval type [142]. According to our investigations, some results on SNs have been studied in [138, 140], but meaningful investigations on distributed resilient filtering over SNs have not been reported yet, which might result in heavy computational burden due to the large-scale of SNs. Thanks to the local design method, it can design a distributed scheme that is implemented on every node, which is helpful to solve the computational burden problem.

1.3 The Challenges

As discussed in Sect. 1.1.1, due to the intrinsic features (i.e., large scale and complicated couplings) of SNs, MAs and CNs, the study on the distributed filtering and control problem as well as synchronization for these systems is a very important yet challenging research topic.

1.3.1 The Large-Scale Characteristic

In consideration of the complicated coupling relationship among the large number of sensor nods and the sparsity of sensor networks, the essential difficulty lies in how to package the distributed filters of all nodes according to the network topology. In the existing reference, the traditional idea behind the distributed H_{∞}-consensus filtering/control problem is the globally augmented method, which is called "package" by the authors of this book. Specifically, the dynamics of all nodes are augmented into a large-scale system without couplings and then it is analyzed by using the existing methods. Along this line, Shen et al. in [93] have presented a formulation of augment matrix reserving the topology, which has become a fundamental tool for the globally augment method. Liu et al. in [28] have developed a novel matrix simplification technique to handle the challenges resulting from the topology sparseness of SNs and filter structure preserving issues.

Notably, the utilization of the globally augmented method implies that the distributed filtering and control algorithms are based on the global information of all nodes in SN, which implies that the computational complexity is a polynomial function of the network size. It yields a series of questions: (1) although the filter and controller are distributed in the form, the final gain parameters need to be calculated in a centralized way, which lost the robustness of the distributed filtering; (2) for the large-scale network, the computational burden of the globally augmented algorithm becomes heavy, which yields an adverse influence in practical applications; (3) in practical engineering, the topology of SN is constantly varied which implies that the global scheme costs a great deal to adapt the topology variation of SN under the case of the globally augmented design; and (4) since the topology of SN is commonly sparse, the global design yields some sparse matrices with large-scale and thus results in the waste of the computational resources. Taking the above problems into account, it is very necessary to seek some new alternative methods to develop the distributed filtering and control algorithms.

1.3.2 The Dynamic Topology

As for SNs or MAs, the topology is constantly dynamic due to various reasons, such as the uncertain communication environment, the node of running out of charge, newly added nodes, the implementation of the communication protocol and so on. Generally speaking, sensor nodes are deployed in remote widespread areas to carry out the monitoring mission. First of all, the wireless communication channels among these nodes are uncertain due to the harsh nature environment. The connections may be intermittent, which results in the dynamic topology. Next, these nodes are powered by battery. Due to the geography or cost, they are non-replaceable in their service life. If a node is out of charge, then it loses the connections with the other nodes, which also causes the topology changes. Also, some new nodes need to join network to replace the failure node or improve the accuracy. In addition, the nodes are low-cost and they are probably broken down with a certain probability. As a result, a new topology has to be formed. Finally, in order to save node energy and wireless channel resource, some communication protocols are utilized to schedule the information transmission among nodes, such as the event-triggering protocol, the Round-Robin protocol, the random access protocol, and so on. The introduction of these protocols makes the topology of SNs or MAs being probably different at each time instant.

Traditionally, the globally augmented method is employed to address some special dynamic topologies, for example, the switching topologies. Actually, it describes the dynamic topology through switching among a few fixed topologies. In essence, it is a simple extension of fixed topology. Moreover, it is impossible to formulate all possible topologies through the switching. It is noted that, the change of the topology of network is in essence local. In most cases, the variation extent of the topology is relatively small between two adjacent time instants. That is to say, the topology of most nodes are fixed. As a consequence, if we develop the local performance analysis method based on a single node, then only a few nodes need to adjust its own algorithms. Therefore, such a design idea is very suitable for the dynamic topology.

1.3.3 The Scalability Requirement of Distributed Algorithm

Due to the large scale characteristic of the distributed algorithm as well as the dynamic topology of SNs and MAs, it is extremely important to investigate the scalable distributed algorithm. Generally speaking, the scalability refers to the computational complexity of the distributed algorithm depends linearly on the size of the network, which is commonly very large. Unfortunately, the traditional globally augmented method is not scalable, because the computational complexity of the distributed algorithm via this method is a polynomial function of the size of the network, where the degree of polynomial is surely larger than 1. Meanwhile, the scalability of the distributed algorithm implies that the algorithm structure is flexible enough to adapt the dynamic topology. The traditional globally augmented method cannot guarantee this point, because the resulting system to be analyzed and designed includes all the information from all nodes. Accordingly, the resulting algorithm is too cumbersome to adjust the algorithm for the topology changes. On the other hand, the achievement of the scalability of distributed algorithm has the necessary hardware basis, i.e., every node has certain both computational ability and communication ability, which allows every node to implement independently its algorithm. Therefore, the local performance analysis method is urgent needed from the aspect of the practical engineering. This method is helpful to achieve the scalable design of the distributed algorithm.

1.4 The Local Performance Analysis Method

In order to resolve the above problems, the local performance analysis and design method has been proposed by means of the vector dissipativity theory [23, 143, 144]. Let us take SN as an example to explain why we need the vector dissipativity theory.

SN generally constitutes a large number of sensor nodes with local communications. Correspondingly, distributed filter of each node can receive the information from its neighboring nodes. Next, it is commonly desirable to augment the filtering error systems of all sensor nodes into a large-scale system. It is noted that, the dimensionality and complexity of the large-scale systems is very large due to the large size. Then, a single Lyapunov function is calculated along this large-scale system, and the obtained conditions could render the following distributed scheme impractical, which need a huge amount of computing time.

On the other hand, such algorithm architecture is not enough flexible to adapt the dynamic topology, which could cost a certain time to adjust the algorithm. Moreover, a single Lyapunov function implies more conservatism since it cannot consider the different features of nodes. It is noted that the similar problems also arise in MAs and CNs. Therefore, it is an urgent need to seek for an efficient method to analyze such a large-scale system.

1.4.1 The Vector Dissipativity Theory

The vector dissipativity theory has been proposed in [23] by W. Haddad and et al. The original idea is to provide an analysis framework for large-scale systems by utilizing the vector Lyapunov functions to reduce the conservatism of single Lyapunov functions. The vector Lyapunov function in large-scale systems offers a very flexible framework since each component of the vector Lyapunov function can satisfy less-grid requirements as compared to a single scalar Lyapunov function. The vector Lyapunov function was first introduced by Bellman and Matrosov and further developed by Lakshmikantham. Inspired by the idea of the vector Lyapunov functions, the vector dissipativity theory has been first developed for large-scale nonlinear discrete-time dynamical systems.

Similar to the dissipativity theory in [145], the vector dissipativity is characterized by the following vector dissipation inequality:

$$\mathbf{V}(x_{k+1}) \leq\leq \mathbf{V}(x_k) + W\mathbf{S}(z_k, w_k) \tag{1.1}$$

where x_{k+1} is the state of plant, z_k is the output of the plant, w_k is the external noise, $\mathbf{V}(x_k) \triangleq [V_1(x_{1,k}) \quad \ldots \quad V_N(x_{N,k})]^T$ is the vector storage function, $\mathbf{S}(z_k, w_k) \triangleq [S_1(z_{1,k}, w_{1,k}), \ldots, S_N(z_{N,k}, w_{N,k})]^T$ is the vector supply rate function, $V_i(x_{i,k})$ and $S_i(z_{i,k}, w_{i,k})$ respectively are the Lyapunov function and the supply rate function of i-th subsystem, and W is the dissipation matrix reflecting the coupling strength between any two subsystems.

Generally speaking, $S_i(z_{i,k}, w_{i,k})$ is chosen according to the desirable performance index; W is constructed for different plant (i.e., essentially non-negative matrix for continuous time system, semi-stable matrix for discrete time-invariant system). The vector dissipation inequality can be further explained as follows: from the perspective of every subsystem, this inequality formulates a discrete energy flow balance in terms of the stored subsystem energy, the supplied subsystem energy, the subsystem energy from all other subsystems independent of the subsystem coupling strengths, and the subsystem energy dissipated. In consideration of the intrinsic features of SNs, MAs, and CNs, the vector dissipativity theory would play a vital role in investigating the distributed filtering and control and synchronization problems. More specifically, the vector dissipativity theory provides the theoretical foundation of the local performance analysis method.

1.4.2 Local Performance Analysis Method

By utilizing the vector dissipativity theory, V. Ugrinovskii has investigated the distributed robust filtering problem for continuous time-invariant system with fixed or switching topology [127, 146, 147], and gain schedule problem [148], respectively. Inspired by these existing results, this book aims to study the distributed H_∞-consensus filtering and control problem, and the synchronization problem for discrete-time systems via the vector dissipativity theory. For convenience, the developed method based on the vector dissipativity theory is called the local performance analysis method.

Different from the existing results, this method can efficiently solve the heavy burden problem caused by the traditional augmented method for dealing with the complicated couplings. This is because this method directly analyze the dynamics of a certain node, which could avoid the curse of dimensionality for the sparse network. The largest challenge lies in the complicated couplings among neighboring nodes. If we directly analyze the dynamics of a single node, then one needs to address the couplings between this node and its neighboring nodes. In the vector dissipativity theory, the dissipation matrix $W = [W_{ij}]$ formulates the couplings between nodes i and j, where $W_{ij} \geq 0$ denotes the relationship between node i and j at time instant k. More specifically, one firstly obtains the following scalar expression for any node i:

$$V_i(x_{i,k+1}) \leq [WV_i(x_{i,k})]_i + S_i(z_{i,k}, w_{i,k}). \tag{1.2}$$

Next, by augmenting (1.2) with respect to all nodes ($i = 1, 2, \ldots, N$), one then acquires the vector form (1.1).

If one has established the vector dissipation inequality (1.1), then it can be changed into the desirable performance inequality by a simple transformation, i.e., add up all the elements of the vector to obtain a scalar form. For more details, please refer to Chaps. 3–10.

1.5 Outline

In this book, we aim to investigate mainly the distributed filtering and control problems for several classes of discrete-time nonlinear stochastic systems via the local performance analysis method. The organization structure of this book is shown in Fig. 1.1 and the outline of the book is given as follows:

- In Chap. 1, the research background, motivations, and research problems are first introduced, which mainly involve the complicated system behaviors, network-induced phenomena, distributed filtering and control, the vector dissipation theory, and the local design method, then the outline of the book is listed.
- In Chap. 2, the distributed H_∞ filtering problem is investigated for a class of piecewise discrete-time linear systems. Firstly, the modes and its transitions of augmented piecewise linear systems as well as distributed filters are formulated. Also, the structure of augmented distributed filter gains is presented in virtue of the adjacent matrix of sensor networks. Then, a set of sufficient conditions are provided for the distributed filter to ensure that its dynamics is global asymptotically stable while achieving the H_∞ performance constraint. In addition, the distributed filter gains are obtained with the aid of the convex optimization method. At last, an illustrative simulation is presented to demonstrate the effectiveness and applicability of the proposed distributed filtering algorithm.
- In Chap. 3, the distributed H_∞-consensus filtering problem is concerned with for a class of discrete time-varying systems with stochastic nonlinearities and missing measurements. The stochastic nonlinearities are formulated by statistical means and the missing measurements are characterized by a set of random variables obeying the Bernoulli distribution. A novel H_∞-consensus performance index is proposed to measure both the filtering accuracy of every node and the consensus errors among neighbor nodes. Then, a new concept called stochastic vector dissipativity is proposed wherein the dissipation matrix is formulated as a nonsingular

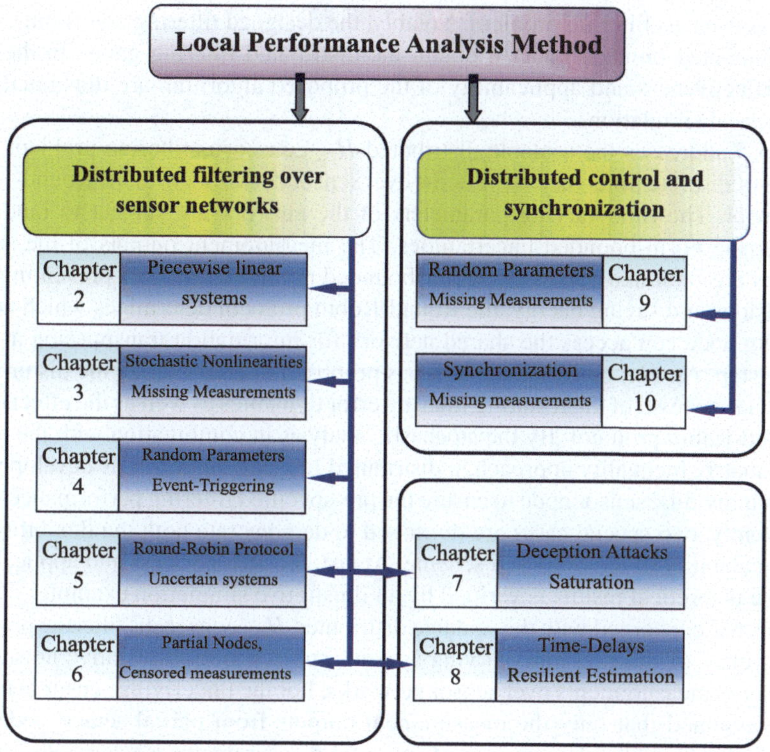

Fig. 1.1 Organization structure of the book

substochastic matrix, which is skillfully constructed by a newly defined interval function on the out-degree. A set of local sufficient conditions in terms of the recursive linear matrix inequalities is presented for each node such that the proposed H_∞-consensus performance can be guaranteed for the locally augmented dynamics over the finite horizon. Furthermore, a novel algorithm proposed in this chapter can be implemented on each node. Finally, an illustrative simulation is presented to demonstrate the effectiveness and applicability of the proposed algorithm.

- Chap. 4 investigates the distributed H_∞-consensus filtering problem for a class of discrete time-varying linear systems with random parameters and event-triggering protocols. An event-triggering protocol for each node is employed to reduce the burden of the network communication. A novel matrix called information matrix is proposed to describe the complicated correlations between any two elements of random matrix. By means of the information matrix, a weighted-covariance formula of random matrix is presented to analyze the system with random parameters. With the aid of constructed dissipation matrix and vector supplied rate functions, a set of local coupled conditions for each node is obtained for establishing the stochastic vector dissipativity-like over the finite-horizon of the filtering error dynamics. Meanwhile, this set of sufficient conditions also solves the

H_∞-consensus filtering problem. Notably, the designed filtering algorithm can be implemented on each node to obtain the distributed filtering gains. In the end, the effectiveness and applicability of the proposed algorithm are illustrated by a numerical simulation.

- Chap. 5 addresses the scalable distributed H_∞-consensus filtering problem for a class of discrete time-varying systems over sensor networks with the Round-Robin protocol. The time-varying parameters of the network are subject to randomly occurring norm-bounded uncertainties. The measurement outputs of the sensor nodes are saturated due to the sector-bounded nonlinearities. For preventing data collisions and saving energy, the Round-Robin protocol determines which neighboring node can access the shared network for information transmission at each time step. An H_∞ performance index is proposed to characterize the disturbance attenuation level of the resulting filtering error dynamics as well as the effect of the Round-Robin protocol. By the stochastic analysis in combination with the recursive matrix inequality approach, a distributed filtering algorithm is developed for each individual sensor node to ensure the pre-specified filtering performance. Subsequently, two special cases are discussed to demonstrate both the flexibility and the scalability of the proposed scheme. At last, the effectiveness and applicability of the theoretical results is verified by an illustrative simulation example.
- Chap. 6 is concerned with the scalable distributed H_∞-consensus filtering problem for a class of discrete time-varying systems subject to multiplicative noises and censored measurements over sensor networks. For the underlying sensor network, it is assumed that only the measurement outputs from partial sensor nodes are available. Also, the phenomenon of censored measurements is taken into account to reflect the limited capability in measuring. A novel H_∞-consensus performance index is put forward to evaluate the disturbance rejection level of the filters against the simultaneous presence of external disturbances, initial conditions as well as censoring effects. By utilizing the vector dissipativity theory and the recursive matrix inequality technique, sufficient conditions are established under which the prescribed H_∞-consensus performance index is achieved. The parameters of the desired distributed filters are calculated via solving certain matrix inequalities, where such a calculation is conducted in a local sense so as to preserve the scalability of the filter design.
- In Chap. 7, the distributed H_∞-consensus filtering problem is concerned with for a class of discrete time-varying systems subject to multiplicative noises and deception attacks over sensor networks. Firstly, the target plant and the measuring sensors are disturbed by multiplicative noises with known statistical information. The malicious signal involved deception attacks is formulated by a specific sector-like bounded condition, which reflects the certain tolerable bound on the difference between the attack signal and the true signal. Attention is paid to the design of filter gains for guaranteeing a desirable filtering performance that simultaneously characterizes the filtering accuracy and the consensus requirement, which need to be considered in the scenario of finite horizon. To handle the proposed filtering problem, the supply rate function is firstly chosen for each node and then the dissipation matrix is constructed as a column sub-stochastic matrix based on

the stochastic vector dissipativity theory. Subsequently, sufficient conditions are presented in terms of the recursive linear matrix inequalities for each node such that the filtering error and the consensus error satisfy the desirable H_∞-consensus performance index over a finite horizon.

- In Chap. 8, the distributed resilient filtering problem is investigated for a class of discrete-time systems with time-delays and stochastic perturbations. Firstly, the plant and the measurements are disturbed by two Gaussian white stochastic processes with known statistical information, respectively. The time-varying delays are considered in the plant to reflect the engineering practice. A resilient filter is designed for each node, where the parameter uncertainties are employed to formulate the gain perturbations during the implementation. Moreover, two Bernoulli-distributed random variables are utilized to describe the randomly occurrence of the parameter uncertainties. By means of the novel vector dissipativity theory, sufficient conditions are established for each node to achieve the exponentially ultimate boundness. Then, based on these sufficient conditions, the upper bound of this performance is optimized for each node by utilizing the robust optimization technique.
- In Chap. 9, the finite-horizon H_∞-consensus control problem is concerned with for a class of multi-agent systems with random parameters. First of all, a newly weighted covariance formula is presented to address the quadratic term associated with random matrix. Then, based on the state average of the neighboring agents, a novel H_∞-consensus performance constraint is presented to cater to the following local performance analysis. By using the definition of the stochastic vector dissipativity-like over the finite horizon, a set of sufficient conditions for each node is obtained such that the controlled output of the closed-loop multi-agent systems satisfy the proposed H_∞-consensus performance index. The obtained consensus control algorithm can be executed on each agent in an indeed distributed manner.
- In Chap. 10, the finite-horizon bounded H_∞ synchronization and H_∞ filteringH_∞ filtering problems are studied for a class of discrete-time complex networks. Firstly, a new bounded H_∞ synchronization performance index is proposed, where only the synchronization errors among neighboring nodes are considered. Next, by establishing the vector dissipativity-like over the finite horizon for the complex network dynamics, the bounded H_∞ synchronization criteria is derived in terms of the local coupled conditions for each node. Similarly, the existence conditions for the filter on each node are also obtained, and then the filter parameters are designed by mean of the recursive linear matrix inequalities. Notably, these algorithms can be implemented on each node in an indeed distributed manner, which reduce efficiently the computational burden arising the massive nodes in the complex networks. Finally, two numerical simulation examples are employed to verify the effectiveness of the schemes proposed in this chapter.
- Chap. 11 draws conclusions on this book and points out some possible research directions related to the work done in this book.

References

1. Akyildiz, I., Su, W., Sankarasubramaniam, Y., Cayirci. E.: A survey on sensor networks. IEEE Commun. Mag. **40**(8), 102–114 (2002)
2. Arampatzis, T., Lygeros, J., Manesis, S.: A Survey of applications of wireless sensors and wireless sensor networks. In: Proceedings of the 2005 IEEE International Symposium on, Mediterrean Conference on Control and Automation, pp. 719–724 (2005)
3. Ding, D., Han, Q., Wang, Z., Ge, X.: A survey on model-based distributed control and filtering for industrial cyber-physical systems. IEEE Trans. Ind. Inf. **15**(5), 2483–2499 (2019)
4. Nancy A.L.: Distributed Algorithms. Morgan Kaufmann Publishers, Inc. (1997)
5. Chen, B., Hu, G., Ho, D., Yu, L.: Distributed covariance intersection fusion estimation for cyber-physical systems with communication constraints. IEEE Trans. Autom. Control **61**(12), 4020–4026 (2016)
6. Ding, D., Wang, Z., Ho, D., Wei, G.: Distributed recursive filtering for stochastic systems under uniform quantizations and deception attacks through sensor networks. Automatica **78**, 231–240 (2017)
7. Dong, H., Wang, Z., Gao, H.: Distributed H_∞ filtering for a class of Markovian jump nonlinear time-delay systems over lossy sensor networks. IEEE Trans. Ind. Electron. **60**(10), 4665–4672 (2013)
8. Han, F., Wei, G., Ding, D., Song, Y.: Local condition-based consensus filtering with stochastic nonlinearities and multiple missing measurements. IEEE Trans. Autom. Control **62**(9), 4784–4790 (2017)
9. Pak, J., Ahn, C., Shi, P., Shmaliy, Y., Lim, M.: Distributed hybrid particle/FIR filtering for mitigating NLOS effects in TOA-based localization using wireless sensor networks. IEEE Trans. Ind. Electron. **64**(6), 5182–5191 (2017)
10. Sun, S., Peng, F., Lin, H.: Distributed asynchronous fusion estimator for stochastic uncertain systems with multiple sensors of different fading measurement rates. IEEE Trans. Signal Process. **66**(3), 641–653 (2018)
11. Zhang, D., Xu, Z., Karimi, H., Wang, Q.: Distributed filtering for switched linear systems with sensor networks in presence of packet dropouts and quantization. IEEE Trans. Circuits Syst.-Part I **64**(10), 2783–2796 (2017)
12. Yang, W., Zhang, Y., Yang, C., Zuo, Z., Wang, X.: Online power scheduling for distributed filtering over an energy-limited sensor network. IEEE Trans. Ind. Electron. **65**(5), 4216–4226 (2018)
13. Yang, W., Yang, C., Shi, H., Shi, L., Chen, G.: Stochastic link activation for distributed filtering under sensor power constraint. Automatica **75**, 109–118 (2017)
14. Wu, Z., Fei, M., Xu, Y., Lu, R.: A distributed Kalman filtering algorithm with fast finite-time convergence for sensor networks. Automatica **95**, 63–72 (2018)
15. Feng, Z., Hu, G., Wen, G.: Distributed consensus tracking for multi-agent systems under two types of attacks. Int. J. Robust Nonlinear Control **26**(5), 896–918 (2016)
16. Yan, H., Yang, Q., Zhang, H., Yang, F., Zhan, X.: Distributed H_∞ state estimation for a class of filtering networks with time-varying switching topologies and packet losses. IEEE Trans. Syst. Man Cybern.: Syst. **48**(12), 2047–2057 (2018)
17. Zhang, H., Zhou, X., Wang, Z., Yan, H., Sun, J.: Adaptive consensus-based distributed target tracking with dynamic cluster in sensor networks. IEEE Trans. Cybern. **49**(5), 1580–1591 (2019)
18. Wang, H., Wu, H.: Design of finite dimensional robust H_∞ distributed consensus filters for dissipative PDE systems with sensor networks. Int. J. Robust Nonlinear Control **25**(10), 1454–1471 (2015)
19. Chen, W., Ding, D., Dong, H., Wei, G.: Distributed resilient filtering for power systems subject to denial-of-service attacks. IEEE Trans. Syst. Man Cybern.: Syst. **49**(8), 1688–1697 (2019)
20. Dong, H., Wang, Z., Lam, J., Gao, H.: Distributed filtering in sensor networks with randomly occurring saturations and successive packet dropouts. Int. J. Robust Nonlinear Control **24**(12), 1743–1759 (2014)

21. Zhu, Y., Zhang, L., Zheng, W.: Distributed H_∞ filtering for a class of discrete-time Markov jump Lur'e systems with redundant channels. IEEE Trans. Ind. Electron. **63**(3), 1876–1885 (2016)
22. Liu, Y., Wang, Z., Ma, L., Alsaadi, F.: A partial-nodes-based information fusion approach to state estimation for discrete-time delayed stochastic complex networks. Inf. Fus. **49**, 240–248 (2019)
23. Haddad, W., Nersesov, S.: Stability and Control of Large-Scale Dynamical Systems. Princeton University Press (2011)
24. Olfati-Saber, R., Murry, R.: Consensus problem in networks of agents with switching topology and time-delays. IEEE Trans. Autom. Control **49**(9), 1520–1533 (2004)
25. Olfati-Saber, R., Fax, J., Murry, R.: Consensus and cooperation in networked multi-agent systems. In: Proceedings of the IEEE: Invited Paper, vol. 95(1), pp. 215–233 (2007)
26. Li, Z., Duan, Z., Chen, G., Huang, L.: Consensus of multiagent systems and synchronization of complex networks: a unified viewpoint. IEEE Trans. Circuits Syst. **57**(1), 213–224 (2010)
27. Wang, Z., Ding, D., Dong, H., Shu, H.: H_∞ consensus control for multi-agent systems with missing measurements: the finite-horizon case. Syst. Control Lett. **62**(10), 827–836 (2013)
28. Liu, Q., Wang, Z., He, X., Zhou, D.: Event-Based recursive distributed filtering over wireless sensor networks. IEEE Trans. Autom. Control **60**(9), 2470–2475 (2015)
29. Liu, Y., Jia, Y.: Consensus problem of high-order multi-agent systems with external disturbances: an H_∞ analysis approach. Int. J. Robust Nonlinear Control **20**(14), 1579–1593 (2010)
30. Lin, P., Jia, Y., Li, L.: Distributed robust H_∞ consensus control in directed networks of agents with time-delay. Syst. Control Lett. **57**(8), 643–653 (2008)
31. Tuna, S.: Conditions for synchronizability in arrays of coupled linear systems. IEEE Trans. Autom. Control **54**(10), 2416–2420 (2009)
32. Wang, P., Jia, Y., Du, J., Zhang, J.: Robust H_∞ consensus control of uncertain multi-agent systems with nonlinear dynamics and time-varying delays. In: Proceedings of 2014 American Control Conference, Portland, Oregon, USA, pp. 2587–2592 (2014)
33. Wen, G., Hu, G., Yu, W., Chen, G.: Distributed H_∞ consensus of higher order multiagent systems with switching topologies. IEEE Trans. Circuits Syst.-II: Express Briefs **61**(5), 359–363 (2014)
34. Yu, W., Ren, W., Zheng, W., Chen, G., Lü, J.: Distributed control gains design for consensus in multi-agent systems with second-order nonlinear dynamics. Automatica **49**(7), 2107–2115 (2013)
35. Saboori, I., Khorasani, K.: H_∞ consensus achievement of multi-agent systems with directed and switching topology networks. IEEE Trans. Autom. Control **59**(11), 199–208 (2014)
36. Zhao, Y., Li, B., Qin, J., Gao, H., Karimi, H.: H_∞ consensus and synchronization of nonlinear systems based on a novel fuzzy model. IEEE Trans. Cybern. **43**(6), 2157–2169 (2013)
37. Li, Z., Wen, G., Duan, Z., Ren, W.: Designing fully distributed consensus protocols for linear multi-agent systems with directed graphs. IEEE Trans. Autom. Control **60**(4), 1152–1157 (2015)
38. Seo, J., Shim, H., Back, J.: Consensus of high-order linear systems using dynamic output feedback compensator: low gain approach. Automatica **45**(11), 2659–2664 (2009)
39. Zhang, H., Lewis, F., Das, A.: Optimal design for synchronization of cooperative systems: state feedback, observer, output feedback. IEEE Trans. Autom. Control **56**(8), 1948–1952 (2011)
40. Liu, X., Ho, D., Cao, J., Yu, W.: Discontinuous observers design for finite-time consensus of multiagent systems with external disturbances. IEEE Trans. Neural Netw. Learn. Syst. **28**(11), 2826–2830 (2017)
41. Liu, X., Cao, J., Yu, W., Song, Q.: Nonsmooth finite-time synchronization of switched coupled neural networks. IEEE Trans. Cybern. **46**(10), 2360–2371 (2016)
42. Boccaletti, S., Latora, V., Moreno, Y.: Complex networks: structure and dynamics. Phys. Rep. **424**, 175–308 (2006)
43. Fu, Z., He, X., Huang, T., Abu-Rub, H.: A distributed continuous time consensus algorithm for maximize social welfare in micro grid. J. Frankl. Inst. **353**, 3966–3984 (2016)

44. Barabáasi, A., Albert, R.: Emergence of scaling in random networks. Science **286**(5439), 509–512 (1999)
45. Strogatz, S.: Exploring complex networks. Nature **410**(6825), 268–276 (2001)
46. Liang, J., Wang, Z., Liu, X.: Exponential synchronization of stochastic delayed discrete-time complex networks. Nonlinear Dyn. **53**(1–2), 153–165 (2008)
47. Lu, J., Ho, D., Cao, J., Kurths, J.: Exponential synchronization of linearly coupled neural networks with impulsive disturbances. IEEE Trans. Neural Netw. **22**(2), 329–335 (2011)
48. Qin, J., Gao, H., Zheng, W.: Exponential synchronization of complex networks of linear systems and nonlinear oscillators: a unified analysis. IEEE Trans. Neural Netw. Learn. Syst. **26**(3), 510–521 (2015)
49. Wu, X., Lu, H.: Generalized projective synchronization between two different general complex dynamical networks with delayed coupling. Phys. Lett. A **374**, 3932–3941 (2010)
50. Ma, Q., Wang, Z., Lu, J.: Finite-time synchronization for complex dynamical networks with time-varying delays. Nonlinear Dyn. **70**, 841–848 (2012)
51. Shen, B., Wang, Z., Liu, X.: A stochastic sampled-data approach to distributed H_∞ filtering in sensor networks. IEEE Trans. Circuits Syst. I-Regular Pap. **58**(9), 2237–2246 (2011)
52. Chen, B., Chiang, C., Nguang, S.: Robust H_∞ synchronization design of nonlinear coupled network via fuzzy interpolation method. IEEE Trans. Circuits Syst.-I: Regul. Pap. **58**, 349–362 (2011)
53. Wu, K., Li, C., Chen, B., Yao, Y.: Robust H_∞ synchronization of coupled partial differential systems with spatial coupling delay. IEEE Trans. Circuits Syst. II: Express Briefs **60**(7), 451–455 (2013)
54. Balasubramaniam, P., Banu, L.: Synchronization criteria of discrete-time complex networks with time-varying delays and parameter uncertainties. Cogn. Neurodynamics **8**(3), 199–215 (2014)
55. He, W., Qian, F., Lam, J., Chen, G., Han, Q.-L., Kurths, J.: Quasi-synchronization of heterogeneous dynamic networks via distributed impulsive control: error estimation, optimization and design. Automatica **62**, 249–262 (2015)
56. Yu, W., Chen, G., *Lü*, J.: On pinning synchronization of complex dynamical networks. Automatica **45**, 429–435 (2009)
57. Rakkiyappan, R., Dharani, S., Zhu, Q.: Stochastic sampled-data H_∞ synchronization of coupled neutral-type delay partial differential systems. J. Frankl. Inst. **352**(10), 4480–4502 (2015)
58. Rakkiyappan, R., Sivaranjani, K.: Sampled-data synchronization and state estimation for nonlinear singularly perturbed complex networks with time-delays. Nonlinear Dyn. **84**(3), 1623–1636 (2016)
59. Anbuvithya, R., Mathiyalagan, K., Sakthivel, R., Prakash, P.: Non-fragile synchronization of memristive BAM networks with random feedback gain fluctuations. Commun. Nonlinear Sci. Numer. Simul. **29**(1–3), 427–440 (2015)
60. Mathiyalagan, K., Anbuvithya, R., Sakthivel, R., Park, J., Prakash, P.: Non-fragile H_∞ synchronization of memristor-based neural networks using passivity theory. Neural Netw. **74**(C) 85–100 (2016)
61. Wang, Z., Wang, Y., Liu, Y.: Global synchronization for discrete-time stochastic complex networks with randomly occurred nonlinearities and mixed time delays. IEEE Trans. Neural Netw. **21**(1), 11–25 (2010)
62. Yang, X., Cao, J., Lu, J.: Synchronization of randomly coupled neural networks with Markovian jumping and time-delay. IEEE Trans. Circuits Syst.-I: Regul. Pap. **60**(2), 363–376 (2013)
63. Kaviarasan, B., Sakthivel, R., Lim, Y.: Synchronization of complex dynamical networks with uncertain inner coupling and successive delays based on passivity theory. Neurocomputing **186**, 127–138 (2016)
64. Lu, W., Chen, T.: New approach to synchronization analysis of linearly coupled ordinary differential systems. Physica D **213**(2), 214–230 (2006)
65. Liu, B., Lu, W., Chen, T.: Synchronization in complex networks with stochastically switching coupling structures. IEEE Trans. Autom. Control **57**(3), 754–760 (2012)

66. Chen, Y., Yu, W., Tan, S., Zhu, H.: Synchronizing nonlinear complex networks via switching disconnected topology. Automatica **70**, 189–194 (2016)
67. Ding, D., Wang, Z., Ho, D., Wei, G.: Observer-based event-triggering consensus control for multi-agent systems with lossy sensors and cyber attacks. IEEE Trans. Cybern. **47**(8), 1936–1947 (2017)
68. Ding, D., Wei, G., Zhang, S., Liu, Y., Alssadi, F.: On scheduling of deception attacks for discrete-time networked systems equipped with attack detectors. Neurocomputing **219**, 99–106 (2017)
69. Dong, H., Wang, Z., Ding, S., Gao, H.: Finite-horizon estimation of randomly occurring faults for a class of nonlinear time-varying systems. Automatica **50**(12), 3182–3189 (2014)
70. Li, Q., Shen, B., Liu, Y., Alsaadi, F.: Event-triggered H_∞ state estimation for discrete-time stochastic genetic regulatory networks with Markovian jumping parameters and time-varying delays. Neurocomputing **174**, 912–920 (2016)
71. Zhang, J., Ma, L., Liu, Y.: Passivity analysis for discrete-time neural networks with mixed time-delays and randomly occurring quantization effects. Neurocomputing **216**, 657–665 (2016)
72. Shen, B., Wang, Z., Ding, D., Shu, H.: H_∞ state estimation for complex networks with uncertain inner coupling and incomplete measurements. IEEE Trans. Neural Netw. Learn. Syst. **24**(12), 2027–2037 (2013)
73. Liu, D., Liu, Y., Alsaadi, F.: A new framework for output feedback controller design for a class of discrete-time stochastic nonlinear system with quantization and missing measurement. Int. J. Gen. Syst. **45**(5), 517–531 (2016)
74. Niu, Y., Ho, D.W.C.: Robust observer design for Itô stochastic time-delay systems via sliding mode control. Syst. Control Lett. **55**(10), 781–793 (2006)
75. Wei, G., Wang, Z., Shu, H.: Robust filtering with stochastic nonlinearities and multiple missing measurements. Automatica **45**(3), 836–841 (2009)
76. Ding, D., Wang, Z., Shen, B., Shu, H.: H_∞ state estimation for discrete-time complex networks with randomly occurring sensor saturations and randomly varying sensor delays. IEEE Trans. Neural Netw. Learn. Syst. **23**(5), 725–736 (2012)
77. Liang, J., Wang, Z., Liu, X.: State estimation for coupled uncertain stochastic networks with missing measurements and time-varying delays: the discrete-time case. IEEE Trans. Neural Netw. **20**(5), 781–793 (2009)
78. Liu, Y., Wang, Z., Liang, J., Liu, X.: Synchronization and state estimation for discrete-time complex networks with distributed delays. IEEE Trans. Syst. Man Cybern. Part B: Cybern. **38**(5), 1314–1325 (2008)
79. Hu, J., Wang, Z., Liu, S., Gao, H.: A variance-constrained approach to recursive state estimation for time-varying complex networks with missing measurements. Automatica **64**, 155–162 (2016)
80. Huang, H., Feng, G., Cao, J.: Robust state estimation for uncertain neural networks with time-varying delay. IEEE Trans. Neural Netw. **19**(8), 1329–1339 (2008)
81. Zou, L., Wang, Z., Gao, H.: Event-Triggered state estimation for complex networks with mixed time delays via sampled data information: the continuous-time case. IEEE Trans. Cybern. **45**(12), 2804–2815 (2015)
82. Koning, W.L.D.: Optimal estimation of linear discrete-time systems with stochastic parameters. Automatica **20**(1), 113–115 (1984)
83. Luo, Y., Zhu, Y., Luo, D., Zhou, J., Song, E., Wang, D.: Globally optimal multisensor distributed random parameter matrices Kalman filtering fusion with applications. Sensors **8**(12), 8086–8103 (2008)
84. Ding, D., Wang, Z., Dong, H., Shu, H.: Distributed H_∞ state estimation with stochastic parameters and nonlinearities through sensor networks: the finite-horizon case. Automatica **48**(8), 1575–1585 (2012)
85. Gao, H., Chen, T.: New results on stability of discrete-time systems with time-varying state delay. IEEE Trans. Autom. Control **52**(2), 328–334 (2007)

86. He, Y., Wu, M., Liu, G., She, J.: Output feedback stabilization for a discrete-time system with a time-varying delay. IEEE Trans. Autom. Control **53**(10), 2372–2377 (2008)
87. Wu, L., Su, X., Shi, P., Qiu, J.: A new approach to stability analysis and stabilization of discrete-time T-S fuzzy time-varying delay systems. IEEE Trans. Syst. Man Cybern.-Part B: Cybern. **41**(1), 273–286 (2011)
88. Li, X., Gao, H.: A new model transformation of discrete-time systems with time-varying delay and its application to stability analysis. IEEE Trans. Autom. Control **56**(9), 2172–2178 (2011)
89. Wang, D., Wang, Z., Shen, B., Alsaadi, E.: Security-guaranteed filtering for discrete-time stochastic delayed systems with randomly occurring sensor saturations and deception attacks. Int. J. Robust Nonlinear Control **27**(7), 1194–1208 (2017)
90. Hu, J., Wang, Z., Gao, H., Stergioulas, L.: Extended Kalman filtering with stochastic nonlinearities and multiple missing measurements. Automatica **48**(9), 2007–2015 (2012)
91. Jacobson, D.: A general result in stochastic optimal control of nonlinear discrete-time systems with quadratic performance criteria. J. Math. Anal. Appl. **47**(1), 153–161 (1974)
92. Yaz, Y., Yaz, E.: On LMI formulations of some problems arising in nonlinear stochastic system analysis. IEEE Trans. Autom. Control **44**(4), 813–816 (1999)
93. Shen, B., Wang, Z., Hung, Y.: Distributed H_∞-consensus filtering in sensor networks with multiple missing measurements: the finite-horizon case. Automatica **46**(10), 1682–1688 (2010)
94. Yu, H., Zhuang, Y., Wang, W.: Distributed H_∞ filtering in sensor networks with randomly occurred missing measurements and communication link failures. Inf. Sci. **222**, 424–438 (2013)
95. Liang, J., Wang, Z., Liu, X.: Distributed state estimation for discrete-time sensor networks with randomly varying nonlinearities and missing measurements. IEEE Trans. Neural Netw. **22**(3), 486–496 (2011)
96. Costa, O., Kubrusly, C.: State-feedback H_∞-control for discrete-time infinite-dimensional stochastic bilinear systems. J. Math. Syst. Estim. Control **6**(2), 1–32 (1996)
97. Crevecoeur, F., Sepulchre, R., Thonnard, J., Lefevre, P.: Improving the state estimation for optimal control of stochastic processes subject to multiplicative noise. Automatica **47**(3), 591–596 (2011)
98. Costa, O., Benites, G.: Linear minimum mean square filter for discrete-time linear systems with Markov jumps and multiplicative noises. Automatica **47**(3), 466–476 (2011)
99. Dragan, V.: Optimal filtering for discrete-time linear systems with multiplicative white noise perturbations and periodic coefficients. IEEE Trans. Autom. Control **58**(4), 1029–1034 (2013)
100. Hu, J., Wang, Z., Shen, B.: Quantised recursive filtering for a class of nonlinear systems with multiplicative noises and missing measurements. Int. J. Control **86**(4), 650–663 (2013)
101. Liu, W., Hu, G.: State estimation for discrete-time Markov jump linear systems with multiplicative noises and delayed mode measurements. Digit. Signal Process. **30**(1), 27–36 (2014)
102. Zhang, L., Zhu, Y., Shi, P., Zhao, Y.: Resilient asynchronous H_∞ filtering for Markov jump neural networks with unideal measurements and multiplicative noises. IEEE Trans. Cybern. **45**(12), 2840–2852 (2015)
103. Dong, H., Wang, Z., Ding, S., Gao, H.: Event-based H_∞ filter design for a class of nonlinear time-varying systems with fading channels and multiplicative noises. IEEE Trans. Signal Process. **63**(13), 3387–3395 (2015)
104. Lu, R., Li, J., Xu, Y.: Distributed l_2-l_∞ state estimation for periodic systems with multiplicative noises. In: Proceedings of 12th World Congress on Intelligent Control and Automation, Guilin, China, pp. 1881–1886 (2016)
105. Ma, L., Wang, Z., Han, Q., Lam, H.: Variance-Constrained distributed filtering for time-varying systems with multiplicative noises and deception attacks over sensor networks. IEEE Sens. J. **17**(7), 2279–2288 (2017)
106. Huang, C., Ho, D., Lu, J.: Partial-information-based distributed filtering in two-targets tracking sensor networks. IEEE Trans. Circuits Syst. I-Regul. Pap. **59**(4), 820–832 (2012)
107. Liu, Y., Wang, Z., Yuan, Y., Alsaadi, F.: Partial-nodes-based state estimation for complex networks with unbounded distributed delays. IEEE Trans. Neural Netw. Learn. Syst. **29**(8), 3906–3912 (2018)

108. Liu, Y., Wang, Z., Yuan, Y., Liu, W.: Event-triggered partial-nodes-based state estimation for delayed complex networks with bounded distributed delays. IEEE Trans. Syst. Man Cybern.: Syst. **49**(6), 1088–1098 (2019)
109. Allik, B., Miller, C., Piovoso, M., Zurakowski, R.: Estimation of saturated data using the Tobit Kalman filter. In: Proceedings of 2014 American Control Conference, Portland, Oregon, USA, pp. 4151–4156 (2014)
110. Allik, B., Miller, C., Piovoso, M., Zurakowski, R.: Nonlinear estimators for censored data: a comparison of the EKF, the UKF and the Tobit Kalman filter. In: Proceedings of the American Control Conference, Chicago, IL, USA, pp. 5146–5151 (2015)
111. Allik, B., Miller, C., Piovoso, M., Zurakowski, R.: The Tobit Kalman filter: an estimator for censored measurements. IEEE Trans. Control Syst. Technol. **24**(1), 365–371 (2016)
112. Arthur, J., Attarian, A., Hamilton, F., Tran, H.: Nonlinear Kalman filtering for censored observations. Appl. Math. Comput. **316**, 155–166 (2018)
113. Han, F., Dong, H., Wang, Z., Li, G., Alsaadi, F.: Improved Tobit Kalman filtering for systems with random parameters via conditional expectation. Signal Process. **147**, 35–45 (2018)
114. Geng, H., Wang, Z., Liang, Y., Cheng, Y., Alsaadi, F.: Tobit Kalman filter with time-correlated multiplicative sensor noises under redundant channel transmission. IEEE Sens. J. **17**(24), 8367–8377 (2017)
115. Geng, H., Wang, Z., Liang, Y., Cheng, Y., Alsaadi, F.: Tobit Kalman filter with fading measurements. Signal Process. **140**, 60–68 (2017)
116. Li, W., Jia, Y., Du, J.: Tobit Kalman filter with time-correlated multiplicative measurement noise. IET Control Theory Appl. **11**(1), 122–128 (2017)
117. Li, J., Wei, G., Ding, D., Li, Y.: Set-membership filtering for discrete time-varying nonlinear systems with censored measurements under Round-Robin protocol. Neurocomputing **281**, 20–26 (2018)
118. Amin, S., Schwartz, G., Sastry, S.: Security of interdependent and identical networked control systems. Automatica **49**(1), 186–192 (2013)
119. Vempaty, A., Ozdemir, O., Agrawal, K., Chen, H., Varshney, P.: Localization in wireless sensor networks: byzantines and mitigation techniques. IEEE Trans. Signal Process. **61**(6), 1495–1508 (2013)
120. Fawzi, H., Tabuada, P., Diggavi, S.: Secure estimation and control for cyber-physical systems under adversarial attacks. IEEE Trans. Autom. Control **59**(6), 1454–1467 (2014)
121. Long, M., Wu, C., Hung, J.: Denial of service attacks on network-based control systems: impact and mitigation. IEEE Trans. Ind. Inf. **1**(2), 85–96 (2005)
122. Marano, S., Matta, V., Tong, L.: Distributed detection in the presence of byzantine attacks. IEEE Trans. Signal Process. **57**(1), 16–29 (2009)
123. Vempaty, A., Tong, L., Varshney, P.: Distributed inference with byzantine data: state-of-the-art review on data falsification attacks. IEEE Signal Process. Mag. **30**(5), 65–75 (2013)
124. Zhang, H., Chen, P., Shi, L., Chen, J.: Optimal denial-of-service attack scheduling with energy constraint. IEEE Trans. Autom. Control **6**(11), 3023–3028 (2015)
125. Donkers, M., Heemels, W., Wouw. N., Hetel, L.: Stability analysis of networked control systems using a switched linear systems approach. IEEE Trans. Autom. Control **56**(9), 2101–2115 (2011)
126. Bauer, N., Donkers, M., Wouw, N., Heemels, W.: Decentralized observer-based control via networked communication. Automatica **49**(7), 2074–2086 (2013)
127. Ugrinovskii, V., Fridman, E.: A round-robin type protocol for distributed estimation with H_∞ consensus. Syst. Control Lett. **69**, 103–110 (2014)
128. Zou, L., Wang, Z., Gao, H.: Set-membership filtering for time-varying systems with mixed time-delays under round-robin and weighted try-once-discard protocols. Automatica **74**, 341–348 (2016)
129. Zou, L., Wang, Z., Gao, H., Liu, X.: State estimation for discrete-time dynamical networks with time-varying delays and stochastic disturbances under the Round-Robin protocol. IEEE Trans. Neural Netw. Learn. Syst. **28**(5), 1139–1151 (2017)

130. Liu, K., Fridman, E., Hetel, L.: Stability and L_2-gain analysis of networked control systems under round-robin scheduling: a time-delay approach. Syst. Control Lett. **61**(5), 666–675 (2012)
131. Kikuchi, K., Cetinkaya, A., Hayakawa, T., Ishii, H.: Stochastic communication protocols for multi-agent consensus under jamming attacks. In: Proceedings of IEEE 56th Annual Conference on Decision and Control (CDC), 12–15 Dec. 2017, Melbourne, Australia
132. Yue, D., Tian, E., Han, Q.: A delay systems method for designing event-triggered controllers of networked control systems. IEEE Trans. Autom. Control **58**(9), 475–481 (2013)
133. Ge, X., Han, Q.: Distributed event-triggered H_∞ filtering over sensor networks with communication delays. Inf. Sci. **291**, 128–142 (2015)
134. Hu, J., Wang, Z., Shen, B., Gao, H.: Gain-constrained recursive filtering with stochastic nonlinearities and probabilistic sensor delays. IEEE Trans. Signal Process. **61**(5), 1230–1238 (2013)
135. Ding, D., Wang, Z., Shen, B., Dong, H.: Event-triggered distributed H_∞ state estimation with packet dropouts through sensor networks. IET Control Theory Appl. **9**(13), 1948–1955 (2015)
136. Mahmoud, M.: Resilient linear filtering of uncertain systems. Automatica **40**(10), 1797–1802 (2004)
137. Zhang, L., Zhu, Y., Ning, Z., Yin, X.: Resilient estimation for networked systems with variable communication capability. IEEE Tran. Autom. Control **61**(12), 4150–4156 (2016)
138. Liu, Q., Wang, Z., He, X., Ghinea, G., Alsaadi, E.: A resilient approach to distributed filter design for time-varying systems under stochastic nonlinearities and sensor degradation. IEEE Trans. Signal Process. **65**(5), 1300–1309 (2017)
139. Wang, F., Wang, Z., Liang, J., Liu, X.: Resilient filtering for linear time-varying repetitive processes under uniform quantizations and Round-Robin protocols. IEEE Trans. Circuits Syst.-I: Regul. Pap. **65**(9), 2992–3004 (2018)
140. Zhang, D., Cai, W., Xie, L., Wang, Q.: Nonfragile distributed filtering for T-S fuzzy systems in sensor networks. IEEE Trans. Fuzzy Syst. **23**(5), 1883–1890 (2015)
141. Liu, Q., Wang, Z., He, X., Zhou, D.: Event-triggered resilient filtering with measurement quantization and random sensor failures: monotonicity and convergence. Automatica **94**, 458–464 (2018)
142. Yang, G., Che, W.: Non-fragile H_∞ filter design for linear continuous-time systems. Automatica **44**(11), 2849–2856 (2008)
143. Haddad, W., Hui, Q., Chellaboina, V., Nersesov, S.: Vector dissipativity theory for discrete-time large-scale nonlinear dynamical systems. Adv. Diff. Equ. **2004**, 37–66 (2004)
144. Haddad, W., Chellabolna, V., Nersesov, S.: Vector dissipativity theory and stability of feedback interconnections for large-scale non-linear dynamical systems. Int. J. Control **77**(10), 907–919 (2004)
145. Willems, J.: Dissipative dynamical systems, Part I: general theory; Part II: linear systems with quadratic supply rates. Arch. Rat. Mech. Anal. **45**, 321–393 (1972)
146. Ugrinovskii, V.: Distributed robust estimation over random switching networks using H_∞ consensus. Automatica **49**(1), 160–168 (2013)
147. Ugrinovskii, V.: Distributed robust filtering with H_∞ consensus of estimates. Automatica **47**(1), 1–13 (2011)
148. Ugrinovskii, V.: Gain-scheduled synchronization of parameter varying systems via relative H_∞ consensus with application to synchronization of uncertain bilinear systems. Automatica **50**(11), 2880–2887 (2014)

Chapter 2
Distributed H_∞ Filtering for Discrete-Time Piecewise Linear Systems

This chapter is concerned with the distributed H_∞ filtering problem for a class of discrete-time piecewise linear (PWL) systems. First of all, both the piecewise linear systems and distributed filters are formulated. Next, in order to deal with the coupling terms, a new large-scale system is derived by augmenting the filtering error dynamics of all nodes. Then, sufficient conditions for the existence of the distributed filter are provided to ensure that the augmented filtering error dynamics is globally asymptotically stable while achieving the H_∞ performance constraint. In addition, the distributed filter gains are calculated with the aid of the convex optimization method. At last, an illustrative simulation is presented to demonstrate the effectiveness and applicability of the proposed distributed filtering algorithm.

2.1 Problem Formulation

First of all, formulate an SN with N nodes distributed in space. Its topology is represented by a digraph $\mathscr{G} = (\mathscr{V}, \mathscr{E}, \mathcal{A})$ with the set of nodes $\mathscr{V} = \{i \mid i = 1, 2, \ldots, N\}$, the set of edges $\mathscr{E} = \{(i, j) \mid (i, j) \in \mathscr{V} \times \mathscr{V}\}$, and the adjacency matrix $\mathcal{A} = [a_{ij}] \in \mathbb{R}^{N\times N}$. If $(i, j) \in \mathscr{E}$ and $i \neq j$, then $a_{ij} = 1$, which means that node j can provide information to node i; otherwise $a_{ij} = 0$. $p_i \triangleq \sum_{j=1}^{N} a_{ij}$ and $q_i \triangleq \sum_{j=1}^{N} a_{ji}$ are, respectively, the in-degree and the out-degree of node i. The set of neighbors of node i with in-degree p_i is denoted by $\mathscr{N}_i = \{j_{i_1}, j_{i_2}, \ldots, j_{i_{p_i}}\}$. The element number of set $\mathscr{N}_i$ is expressed as $|\mathscr{N}_i|$, and $\sum_{i=1}^{N} |\mathscr{N}_i|$ is described as $|\mathscr{N}|$. A digraph is said to be weakly connected if there exists an undirected path from every node to every other node. A graph which is not weakly connected is disconnected [1]. In this chapter, it is assumed that the topology of SN is characterized by a weakly connected digraph. For convenience, the same topology formulation is utilized in the following Chaps. 3–9.

F. Han et al., *Distributed Filtering, Control and Synchronization*,
Studies in Systems, Decision and Control 428,
https://doi.org/10.1007/978-3-030-97075-8_2

Consider a class of discrete-time PWL systems

$$\begin{cases} x_{k+1} = A_{\bar{l}} x_k + B_{\bar{l}} w_k, \\ z_k = E_{\bar{l}} x_k, \quad x_k \in \bar{S}_{\bar{l}} \end{cases} \tag{2.1}$$

where $\bar{l}$ denotes the $\bar{l}$-th local mode, L the index set of local modes: $L = \{1, 2, \ldots, n\}$ with $\bar{l} \in L$. $x_k \in \mathbb{R}^{n_x}$ the state, $w_k \in \mathbb{R}^{n_w}$ the disturbance input belonging to $l_2[0, \infty)$, $z_k \in \mathbb{R}^{n_z}$ the signal to be estimated, x_0 the known initial state, and $\left\{\bar{S}_{\bar{l}}\right\}_{\bar{l} \in L} \subseteq \mathbb{R}^{n_x}$ denotes a number of closed (possibly unbounded) polyhedral subspaces, which is a partition of the state space. System parameters $A_{\bar{l}} \in \mathbb{R}^{n_x \times n_x}$, $B_{\bar{l}} \in \mathbb{R}^{n_x \times n_w}$ and $E_{\bar{l}} \in \mathbb{R}^{n_z \times n_x}$ are all constant matrices.

The measurement model of sensor node i is given as follows:

$$y_{i,k} = C_{\bar{l},i} x_k + D_{\bar{l},i} w_k \tag{2.2}$$

where $y_{i,k} \in \mathbb{R}^{n_y}$ is the measurement of sensor node i, system parameters $C_{\bar{l},i} \in \mathbb{R}^{n_y \times n_x}$ and $D_{\bar{l},i} \in \mathbb{R}^{n_y \times n_w}$ are all constant matrices.

For presentation convenience, a set of ordered pairs $\left(\bar{l}, \bar{m}\right)$ is defined as

$$\Omega_i \triangleq \left\{\left(\bar{l}, \bar{m}\right) | x_k \in \bar{S}_{\bar{l}}, x_{k+1} \in \bar{S}_{\bar{m}}, \bar{l}, \bar{m} \in L\right\}, \tag{2.3}$$

to represent all possible transitions from a region to itself $\left(\bar{l} = \bar{m}\right)$ or another region$\left(\bar{l} \neq \bar{m}\right)$ for sensor node i. When the mode of system (2.1) transits from a mode to another, the dynamics of system (2.1) is governed by that of the former mode at the transition time.

Apart from its own information, sensor node i also receives the estimates and measurements from its neighboring nodes. According to the available information of sensor node i and the typical features of PWL system, the following distributed PWL filter on sensor node i is constructed as follows:

$$\begin{cases} \hat{x}_{i,k+1} = H_{\bar{l}_i,ii} \hat{x}_{i,k} + K_{\bar{l}_i,ii} y_{i,k} + \sum\limits_{j \in \mathcal{N}_i} H_{\bar{l}_i,ij} a_{ij} \hat{x}_{j,k} + \sum\limits_{j \in \mathcal{N}_i} K_{\bar{l}_i,ij} a_{ij} y_{j,k}, \\ \hat{z}_{i,k} = E_{\bar{l}_i} \hat{x}_{i,k}, \quad \hat{x}_{i,k} \in \bar{S}_{\bar{l}} \end{cases} \tag{2.4}$$

where $\hat{x}_{i,k}$ and $\hat{z}_{i,k}$ are the estimated state x_k and signal z_k to be estimated by filter on sensor node i, respectively. $\bar{l}_i$ is the mode of filter on sensor node i. The corresponding definition of $\bar{l}_i$ is similar to $\bar{l}$. Here, $H_{\bar{l}_i,ii}$, $K_{\bar{l}_i,ii}$ $H_{\bar{l}_i,ij}$, and $K_{\bar{l}_i,ij}$ are the matrix-valued filter gains to be determined later. $\hat{x}_{i,0}$ is the known initial value of the filter on sensor node i.

Remark 2.1 The distributed PWL filter (2.4) describes the information exchange structure between sensor node i and its neighboring nodes under the different modes in the PWL system, which can be seen as an extension of the PWL filter under a single sensor case. More specifically, the distributed PWL filters (2.4) is simplified to

$$\begin{cases} \hat{x}_{i,k+1} = H_{\bar{l}_i,ii}\hat{x}_{i,k} + K_{\bar{l}_i,ii}y_{i,k}, \\ \hat{z}_{i,k} = E_{\bar{l}_i}\hat{x}_{i,k}, \quad \hat{x}_{i,k} \in S_{\bar{l}}, \ i \in \mathscr{V}, \end{cases} \tag{2.5}$$

which has been investigated in [2]. Besides, the distributed PWL filter (2.4) can also be seen as another extension of the traditional distributed filter, that is, it is reduced to

$$\begin{cases} \hat{x}_{i,k+1} = H_{ii}\hat{x}_{i,k} + K_{ii}y_{i,k} + \sum\limits_{j\in\mathscr{N}_i} H_{ij}a_{ij}\hat{x}_{j,k} + \sum\limits_{j\in\mathscr{N}_i} K_{ij}a_{ij}y_{j,k}, \\ \hat{z}_{i,k} = E\hat{x}_{i,k}, \quad i, j \in \mathscr{V}, \end{cases} \tag{2.6}$$

which has been studied in [3] and other literatures.

Compared with the traditional filter with a sensor, the distributed filter over sensor networks has a clear distinction: coupling terms, which are commonly addressed by the augmented method. Along this line, a new large-scale system is achieved by augmenting the dynamics of subsystems and distributed filters on all sensor nodes in the SN. Due to the characteristic of PWL system, the augmented system should have the corresponding mode, which is closely dependent on the modes of the distributed filter and the plant. In the following, the mode of the augmented system is presented.

In consideration of the need of augmented way and the features of PWL system, a set of ordered arrays $(\bar{l}, \bar{l}, \ldots, \bar{l})$ is firstly defined as

$$\bar{\mathbb{L}}^N \triangleq \{l_N = (\bar{l}, \bar{l}, \ldots, \bar{l}) | \bar{l} \in L\}$$

to characterize the modes of the following augmented PWL system.

For convenience of the following analysis, denote

$$\begin{aligned} \bar{x}_k &\triangleq \mathrm{vec}_N^T\{x_k^T\}, & \bar{y}_k &\triangleq \mathrm{vec}_N^T\{y_{i,k}^T\}, \\ \bar{z}_k &\triangleq \mathrm{vec}_N^T\{z_{i,k}^T\}, & \hat{x}_k &\triangleq \mathrm{vec}_N^T\{\hat{x}_{i,k}^T\}, \\ \hat{z}_k &\triangleq \mathrm{vec}_N^T\{\hat{z}_{i,k}^T\}, & \bar{A}_{l_N} &\triangleq \mathrm{diag}_N\{A_{\bar{l}}\}, \\ \bar{B}_{l_N} &\triangleq \mathrm{vec}_N^T\{B_{\bar{l}}^T\}, & \bar{C}_{l_N} &\triangleq \mathrm{diag}_N\{C_{\bar{l},i}\}, \\ \bar{D}_{l_N} &\triangleq \mathrm{vec}_N^T\{D_{\bar{l},i}^T\}, & \bar{E}_{l_N} &\triangleq \mathrm{diag}_N\{E_{\bar{l},i}\}. \end{aligned}$$

Then, N identical plants of the form (2.1) are augmented into a new system

$$\begin{cases} \bar{x}_{k+1} = \bar{A}_{l_N}\bar{x}_k + \bar{B}_{l_N}w_k, \\ \bar{y}_k = \bar{C}_{l_N}\bar{x}_k + \bar{D}_{l_N}w_k, \quad \text{if } \bar{x}_k \in S_{l_N}, \\ \bar{z}_k = \bar{E}_{l_N}\bar{x}_k, \quad l_N \in \bar{\mathbb{L}}^N, \\ \bar{x}_0 = \text{vec}^T\{x_{i,0}^T\} \end{cases} \tag{2.7}$$

where $\{S_{l_N}\}_{l_N \in \bar{\mathbb{L}}^N} \subseteq \mathbb{R}^{Nn_x}$ denotes a partition of the state spaces of all the sensor nodes in the SN into a number of closed (possibly unbounded) polyhedral subspaces.

Also, a set of ordered arrays $(\bar{l}_1, \bar{l}_2, \ldots, \bar{l}_N)$ is defined as

$$\mathbb{L}^N \triangleq \left\{ l = (\bar{l}_1, \bar{l}_2, \ldots, \bar{l}_N) \,|\bar{l}_i \in L, i \in \mathscr{V} \right\}$$

to characterize the modes of the following augmented PWL filter. It is worthwhile noting that initial modes of different sensor nodes $\bar{l}_i, i \in \mathscr{V}$ are probably different.

Next, all the distributed PWL filters on sensor node i ($i \in \mathscr{V}$) with the form (2.4) can be written as the following compact form

$$\begin{cases} \hat{x}_{k+1} = \bar{H}_l\hat{x}_k + \bar{K}_l\bar{y}_k, \\ \hat{z}_k = \bar{E}_l\hat{x}_k, \qquad \text{if } \hat{x}_k \in S_l, \quad l \in \mathbb{L}^N, \\ \hat{x}_0 = \text{col}_i^N\left\{\hat{x}_{i,0}\right\} \end{cases} \tag{2.8}$$

where

$$\begin{cases} \bar{H}_l = \left[\bar{H}_{\bar{l}_i,ij}\right]_{N\times N}, \text{ with } \bar{H}_{\bar{l}_i,ij} = H_{\bar{l}_i,ij}a_{ij}, \\ \bar{K}_l = \left[\bar{K}_{\bar{l}_i,ij}\right]_{N\times N}, \text{ with } \bar{K}_{\bar{l}_i,ij} = K_{\bar{l}_i,ij}a_{ij}. \end{cases} \tag{2.9}$$

Furthermore, the parameter matrices $\bar{H}_l$ and $\bar{K}_l$ are expressed as

$$\bar{H}_l \in \mathcal{H}_{l,n_x\times n_x} \triangleq \left[\bar{H}_{\bar{l}_i,ij}\right] \in \mathbb{R}^{Nn_x\times Nn_x}, \quad \bar{K}_l \in \mathcal{K}_{l,n_x\times n_y} \triangleq \left[\bar{K}_{\bar{l}_i,ij}\right] \in \mathbb{R}^{Nn_x\times Nn_y}$$

where

$$\begin{cases} \bar{H}_{\bar{l}_i,ij} \neq 0, \text{ if } j \in \mathscr{N}_i \cup \{i\}; \\ \bar{H}_{\bar{l}_i,ij} = 0, \text{ otherwise}, \end{cases} \tag{2.10}$$

$$\begin{cases} \bar{K}_{\bar{l}_i,ij} \neq 0, \text{ if } j \in \mathscr{N}_i \cup \{i\}; \\ \bar{K}_{\bar{l}_i,ij} = 0, \text{ otherwise}. \end{cases} \tag{2.11}$$

From the above definitions, it follows that $\bar{H}_l$ and $\bar{K}_l$ are closely dependent on the structure of the considered SN. Here, it is worthwhile noting that $\bar{H}_l$ and $\bar{K}_l$ are two sparse matrices due to the sparsity of SN.

Similar to the set (2.3) for sensor node i, define also a product-set:

$$
\begin{aligned}
\Omega &\triangleq \Omega_1 \times \cdots \times \Omega_N \\
&= \left\{ \left(\bar{l}_1, \bar{m}_1\right) \times \cdots \times \left(\bar{l}_N, \bar{m}_N\right) \mid \left(\bar{l}_i, \bar{m}_i\right) \in \Omega_i, i = 1, 2, \ldots, N \right\} \\
&= \left\{ (l, m) \mid l, m \in \mathbb{L}^N \right\}.
\end{aligned}
\tag{2.12}
$$

To represent all possible mode transitions for the augmented system (2.7) and PWL filter (2.8).

Actually, due to utilizing different initial values, there exist probably some differences between the mode of the plant and the filter of sensor node. That is, $\bar{l}$ and l_N may be different from $\bar{l}_i$ and l, respectively.

Next, by letting $\eta_k \triangleq \left[\bar{x}_k^T \;\; \hat{x}_k^T\right]^T$ and $\tilde{z}_k \triangleq \bar{z}_k - \hat{z}_k$, the following augmented system that governs the filtering error dynamics for the SN is obtained:

$$
\begin{cases}
\eta_{k+1} = \mathcal{A}_{l_N l} \eta_k + \mathcal{B}_{l_N l} w_k, \\
\tilde{z}_k = \mathcal{E}_{l_N l} \eta_k, \\
\eta_0 = \left[\bar{x}_0^T \;\; \hat{x}_0^T\right]^T
\end{cases}
\tag{2.13}
$$

where

$$
\mathcal{A}_{l_N l} = \begin{bmatrix} \bar{A}_{l_N} & 0 \\ \bar{K}_l \bar{C}_{l_N} & \bar{H}_l \end{bmatrix}, \quad \mathcal{B}_{l_N l} = \begin{bmatrix} \bar{B}_{l_N} \\ \bar{K}_l \bar{D}_{l_N} \end{bmatrix}, \quad \mathcal{E}_{l_N l} = \begin{bmatrix} \bar{E}_{l_N} & -\bar{E}_l \end{bmatrix}.
$$

Before proceeding further, the following definitions in the case of infinite horizon are ready to present for the filtering error system (2.13).

Definition 2.1 The discrete-time PWL system (2.13) with $w_k \equiv 0$ is globally asymptotically stable, if the following inequality holds:

$$
\|\tilde{z}\|_2^2 < \infty \tag{2.14}
$$

for any initial conditions η_0 and initial mode $l_N l_0 \in \overline{\mathbb{L}}^N \cup \mathbb{L}^N$.

Definition 2.2 The filtering error $\tilde{z}_k$ of the discrete-time PWL system (2.13) satisfies the H_∞ performance constraint with zero initial conditions if the following inequality holds:

$$
\frac{1}{N} \|\tilde{z}\|_2^2 \leq \gamma^2 \|w\|_2^2 \tag{2.15}
$$

for all nonzero $w_k \in l_2[0, \infty)$ and a given disturbance attenuation level $\gamma > 0$.

The distributed PWL filtering problem under consideration in this chapter is to determine the filter gains $\bar{H}_l$ and $\bar{K}_l (l \in \mathbb{L}^N)$ of augmented PWL filter (2.8) such that the filtering error system (2.13) is globally asymptotically stable while achieving the H_∞ performance constraint.

2.2 Main Results

In this section, the analysis and design of the distributed H_∞ filtering scheme are to be implemented based on a piecewise Lyapunov function for the discrete-time PWL system (2.1), respectively.

Theorem 2.1 *Given the filter gains $\bar{H}_l$ and $\bar{K}_l$ and the disturbance attenuation level $\gamma > 0$, the PWL filtering error system* (2.13) *is globally asymptotically stable while achieving H_∞ performance constraint, if there exists a set of symmetric positive definite matrices $P_{l_N l}$, $l_N l \in \left\{\bar{\mathbb{L}}^N \cup \mathbb{L}^N\right\}$ such that the following matrix inequalities are satisfied for any $(l_N l, m_N m) \in \Omega$:*

$$\gamma^2 I - \mathcal{B}_{l_N l}^T P_{m_N m} \mathcal{B}_{l_N l} > 0, \tag{2.16}$$

$$\begin{aligned} &\mathcal{A}_{l_N l}^T P_{m_N m} \mathcal{B}_{l_N l} (\gamma^2 I - \mathcal{B}_{l_N l}^T P_{m_N m} \mathcal{B}_{l_N l})^{-1} \mathcal{B}_{l_N l}^T P_{m_N m} \mathcal{A}_{l_N l} \\ &\quad + \mathcal{A}_{l_N l}^T P_{m_N m} \mathcal{A}_{l_N l} - P_{l_N l} + \frac{1}{N} \mathcal{E}_{l_N l}^T \mathcal{E}_{l_N l} < 0. \end{aligned} \tag{2.17}$$

Proof First of all, choose the Lyapunov function

$$V(\eta_k) = \eta_k^T P_{l_N l} \eta_k.$$

Next, calculate the difference of $V(\eta_k)$ along the system trajectory (2.13) as follows

$$\begin{aligned} &\Delta V_k \\ &\triangleq V(\eta_{k+1}) - V(\eta_k) \\ &= \eta_{k+1}^T P_{m_N m} \eta_{k+1} - \eta_k^T P_{l_N l} \eta_k \\ &= \eta_k^T (\mathcal{A}_l^T P_{m_N m} \mathcal{A}_l - P_{l_N l}) \eta_k + w_k^T \mathcal{B}_{l_N l}^T P_{m_N m} \mathcal{A}_{l_N l} \eta_k + \eta_k^T \mathcal{A}_{l_N l}^T P_{m_N m} \mathcal{B}_{l_N l} w_k \\ &+ w_k^T \mathcal{B}_{l_N l}^T P_{m_N m} \mathcal{B}_{l_N l} w_k + \frac{1}{N} \tilde{z}_k^T \tilde{z}_k - \gamma^2 w_k^T w_k - (\frac{1}{N} \tilde{z}_k^T \tilde{z}_k - \gamma^2 w_k^T w_k) \\ &= \eta_k^T (\mathcal{A}_{l_N l}^T P_{m_N m} \mathcal{A}_{l_N l} - P_{l_N l} + \frac{1}{N} \mathcal{E}_{l_N l}^T \mathcal{E}_{l_N l}) \eta_k \\ &\quad + w_k^T (\mathcal{B}_{l_N l}^T P_{m_N m} \mathcal{B}_{l_N l} - \gamma^2 I) w_k + w_k^T \mathcal{B}_{l_N l}^T P_{m_N m} \mathcal{A}_{l_N l} \eta_k \\ &\quad + \eta_k^T \mathcal{A}_{l_N l}^T P_{m_N m} \mathcal{B}_{l_N l} w_k - (\frac{1}{N} \tilde{z}_k^T \tilde{z}_k - \gamma^2 w_k^T w_k) \\ &= \eta_k^T (\mathcal{A}_{l_N l}^T P_{m_N m} \mathcal{A}_{l_N l} - P_{l_N l} + \frac{1}{N} \mathcal{E}_{l_N l}^T \mathcal{E}_{l_N l} + \mathcal{A}_{l_N l}^T P_{m_N m} \mathcal{B}_{l_N l} \Phi^{-1} \mathcal{B}_{l_N l}^T P_{m_N m} \mathcal{A}_{l_N l}) \eta_k \\ &\quad - (w_k - w_k^*)^T \Phi (w_k - w_k^*) - (\frac{1}{N} \tilde{z}_k^T \tilde{z}_k - \gamma^2 w_k^T w_k) \end{aligned}$$

where $\Phi \triangleq \gamma^2 I - \mathcal{B}_{l_N l}^T P_{m_N m} \mathcal{B}_{l_N l}$, $w_k^* \triangleq \Phi^{-1} \mathcal{B}_{l_N l}^T P_{m_N m} \mathcal{A}_{l_N l} \eta_k$, and $l = m$ when the system stays in the region $S_{l_N l}$ and $l \neq m$ when the system transits from $S_{l_N l}$ to the region $S_{m_N m}$.

Then, it follows from (2.13) and (2.16) that

$$\Delta V_k \leq \gamma^2 w_k^T w_k - \frac{1}{N}\tilde{z}_k^T \tilde{z}_k,$$

which implies that

$$V(\eta_\infty) - V(\eta_0) \leq -\sum_{k=0}^{\infty} \frac{1}{N}\tilde{z}_k^T \tilde{z}_k + \sum_{k=0}^{\infty} \gamma^2 w_k^T w_k.$$

Under the zero initial conditions together with the fact $V(\eta_\infty) \geq 0$, one immediately has

$$\frac{1}{N}\|\tilde{z}\|_2^2 \leq \gamma^2 \|w\|_2^2.$$

In the end, in terms of Definitions 2.1 and 2.2, the proof of this theorem is therefore complete.

The following procedure is to change the matrix inequalities (2.16) and (2.17) into LMIs, which are easily solvable by the Matlab software.

Lemma 2.1 *For given filter gains $\bar{H}_l$ and $\bar{K}_l$ and disturbance attenuation level $\gamma > 0$, consider the PWL filtering error system (2.13), if there exists a set of symmetric positive definite matrices $P_{l_N l}$ and a set of nonsingular matrices $R_{l_N l}$, $l \in \mathbb{L}^N$ such that the following LMIs are satisfied:*

$$\begin{bmatrix} P_{m_N m} - R_{l_N l} - R_{l_N l}^T & * & * \\ \mathcal{A}_{l_N l}^T R_{l_N l} & -P_{l_N l} + \frac{1}{N}\mathcal{E}_{l_N l}^T \mathcal{E}_{l_N l} & * \\ \mathcal{B}_{l_N l}^T R_{l_N l} & 0 & -\gamma^2 I \end{bmatrix} < 0, \quad (l, m) \in \Omega \tag{2.18}$$

then the PWL filtering error system (2.13) *is globally asymptotically stable while achieving the H_∞ performance constraint.*

Proof It follows from Theorem 2.1 that if one shows that (2.18) implies (2.16) and (2.17), then the assertion of Lemma 2.1 will follow.

First and foremost, it is proved that (2.18) implies (2.16). Via the Schur Complement Lemma, (2.18) holds if and only if the following LMIs hold

$$\begin{bmatrix} P_{m_N m} - R_{l_N l} - R_{l_N l}^T & * & * & * \\ \mathcal{A}_{l_N l}^T R_{l_N l} & -P_{l_N l} & * & * \\ \mathcal{B}_{l_N l}^T R_{l_N l} & 0 & -\gamma^2 I & * \\ 0 & \mathcal{E}_{l_N l} & 0 & -NI \end{bmatrix} < 0, \quad (l, m) \in \Omega. \tag{2.19}$$

In light of the concept of principal submatrix, any principal submatrix of a positive definite matrix is also positive definite. It is obvious that matrix

$$\begin{bmatrix} P_{m_N m} - R_{l_N l} - R_{l_N l}^T & * \\ \mathcal{B}_{l_N l}^T R_{l_N l} & -\gamma^2 I \end{bmatrix}$$

is a principal submatrix of the left side of (2.19).

As such, it follows from (2.19) that

$$\begin{bmatrix} P_{m_N m} - R_{l_N l} - R_{l_N l}^T & * \\ \mathcal{B}_{l_N l}^T R_{l_N l} & -\gamma^2 I \end{bmatrix} < 0. \tag{2.20}$$

Subsequently, noting that $R_{l_N l}^T P_{m_N m}^{-1} R_{l_N l} \geq R_{l_N l} + R_{l_N l}^T - P_{m_N m}$, one has

$$P_{m_N m} - R_{l_N l} - R_{l_N l}^T \geq -R_{l_N l}^T P_{m_N m}^{-1} R_{l_N l}. \tag{2.21}$$

By means of (2.21), it follows from (2.20) that

$$\begin{bmatrix} -R_{l_N l}^T P_{m_N m}^{-1} R_{l_N l} & * \\ \mathcal{B}_{l_N l}^T R_{l_N l} & -\gamma^2 I \end{bmatrix} < 0.$$

Based on the Schur Complement Lemma, it leads easily to (2.16).

In the following, one shows that (2.18) implies (2.17). Similarly, it follows from (2.18) and (2.21) that

$$\begin{bmatrix} -R_{l_N l}^T P_{m_N m}^{-1} R_{l_N l} & * & * \\ \mathcal{A}_{l_N l}^T R_{l_N l} & -P_{l_N l} + \frac{1}{N}\mathcal{E}_{l_N l}^T \mathcal{E}_{l_N l} & * \\ \mathcal{B}_{l_N l}^T R_{l_N l} & 0 & -\gamma^2 I \end{bmatrix} < 0. \tag{2.22}$$

Via the Schur Complement Lemma, (2.22) holds if and only if the following inequality holds

$$\begin{bmatrix} -P_{l_N l} + \frac{1}{N}\mathcal{E}_{l_N l}^T \mathcal{E}_{l_N l} & 0 \\ 0 & -\gamma^2 I \end{bmatrix} + \begin{bmatrix} \mathcal{A}_{l_N l}^T R_{l_N l} \\ \mathcal{B}_{l_N l}^T R_{l_N l} \end{bmatrix} R_{l_N l}^{-1} P_{m_N m} R_{l_N l}^{-T} \begin{bmatrix} R_{l_N l}^T \mathcal{A}_{l_N l} & R_{l_N l}^T \mathcal{B}_{l_N l} \end{bmatrix} < 0$$

$$\Leftrightarrow \begin{bmatrix} \mathcal{A}_{l_N l}^T P_{m_N m} \mathcal{A}_{l_N l} - P_{l_N l} + \frac{1}{N}\mathcal{E}_{l_N l}^T \mathcal{E}_{l_N l} & \mathcal{A}_{l_N l}^T P_{m_N m} \mathcal{B}_{l_N l} \\ \mathcal{B}_{l_N l}^T P_{m_N m} \mathcal{A}_{l_N l} & \mathcal{B}_{l_N l}^T P_{m_N m} \mathcal{B}_{l_N l} - \gamma^2 I \end{bmatrix} < 0.$$

By utilizing the Schur Complement Lemma again, the above matrix inequality holds if and only if the following inequality holds

$$\begin{aligned} &\mathcal{A}_{l_N l}^T P_{m_N m} \mathcal{B}_{l_N l} (\gamma^2 I - \mathcal{B}_{l_N l}^T P_{m_N m} \mathcal{B}_{l_N l})^{-1} \mathcal{B}_{l_N l}^T P_{m_N m} \mathcal{A}_{l_N l} \\ &\qquad + \mathcal{A}_{l_N l}^T P_{m_N m} \mathcal{A}_{l_N l} - P_{l_N l} + \frac{1}{N}\mathcal{E}_{l_N l}^T \mathcal{E}_{l_N l} < 0. \end{aligned}$$

Then, it can be concluded that (2.18) implies (2.17). The proof is thus completed by invoking Theorem 2.1 in the end.

Lemma 2.2 ([4]) *Let* $S = \mathrm{diag}\{S_1, S_2, \ldots, S_N\}$ *with* $S_i \in \mathbb{R}^{p\times p}(1 \le i \le N)$ *being invertible matrices. For* $Y \in \mathbb{R}^{Np\times Nq}$*, if* $X = SY$*, then one has* $Y \in \mathcal{Y}_{p\times q} \Leftrightarrow X \in \mathcal{Y}_{p\times q}$.

The role of Lemma 2.2 is to ensure that the solution matrices of the LMIs in the following theorem have the same structure as the desired filter gains. After having conducted the distributed filtering performance analysis in Theorem 2.1 and Lemma 2.1, we are now in a position to design the filter gains pertaining to the distributed PWL filter.

Theorem 2.2 *Let a positive scalar* $\gamma > 0$ *be given, the PWL filtering error system* (2.13) *is globally asymptotically stable while achieving the* H_∞ *performance constraint, if there exist two sets of positive definite matrices* $\tilde{P}_{l_N}$ *and* $\tilde{P}_l$*, two sets of matrices* $\tilde{H}_l$ *and* $\tilde{K}_l$ *satisfying the constraints*

$$\tilde{H}_l \in \mathcal{H}_{l,n_x\times n_x}, \quad \tilde{K}_l \in \mathcal{K}_{l,n_x\times n_y},$$

and sets of matrices $\tilde{P}_{l2}$, S_{l_N}, S_{l2}, *nonsingular matrix* S_l, $l \in \mathbb{L}^N$ *such that the following LMIs are satisfied:*

$$\begin{bmatrix} \tilde{P}_{m_N m} - \tilde{S}_{l_N l} - \tilde{S}_{l_N l}^T & * & * \\ \tilde{\mathcal{A}}_{l_N l}^T & -\tilde{P}_{l_N l} + \frac{1}{N}\tilde{\mathcal{E}}_{l_N l}^T \tilde{\mathcal{E}}_{l_N l} & * \\ \tilde{\mathcal{B}}_{l_N l}^T & 0 & -\gamma^2 I \end{bmatrix} < 0, \quad (l_N l, m_N m) \in \Omega \quad (2.23)$$

where

$$\tilde{\mathcal{A}}_{l_N l}^T = \begin{bmatrix} \bar{A}_{l_N}^T S_{l_N} + \bar{C}_{l_N}^T \tilde{K}_l^T & \bar{A}_{l_N}^T S_{l2} + \bar{C}_{l_N}^T \tilde{K}_l^T \\ \tilde{H}_l^T & \tilde{H}_l^T \end{bmatrix},$$

$$\tilde{\mathcal{B}}_{l_N l}^T = \begin{bmatrix} \bar{B}_{l_N}^T S_{l_N} + \bar{D}_{l_N}^T \tilde{K}_l^T & \bar{B}_{l_N}^T S_{l2} + \bar{D}_{l_N}^T \tilde{K}_l^T \end{bmatrix},$$

$$\tilde{S}_{l_N l} = \begin{bmatrix} S_{l_N} & S_{l2} \\ S_l & S_l \end{bmatrix}, \quad \tilde{P}_{l_N l} = \begin{bmatrix} \tilde{P}_{l_N} & \tilde{P}_{l2}^T \\ \tilde{P}_{l2} & \tilde{P}_l \end{bmatrix},$$

$$\tilde{\mathcal{E}}_{l_N l} = \begin{bmatrix} \bar{E}_{l_N} & -\bar{E}_l \end{bmatrix}.$$

Moreover, the distributed PWL filter gains are given by

$$\bar{H}_l = S_l^{-T}\tilde{H}_l, \quad \bar{K}_l = S_l^{-T}\tilde{K}_l, \quad l \in \mathbb{L}^N.$$

Proof Note that if one shows that (2.23) is an equivalence transformation of (2.18), it follows from Lemma 2.1 that the PWL filtering error systems (2.13) is globally asymptotically stable while achieving H_∞ performance constraint.

First of all, partition the matrix $R_{l_N l}$ in (2.18) as follows:

$$R_{l_N l} = \left[\begin{array}{c|c} R_{l_N} & R_{l2} \\ \hline R_{l3} & R_l \end{array}\right]$$

where R_{l3} and R_l are nonsingular matrices.

Also, the positive definite matrix $P_{l_N l}$ can be partitioned as follows:

$$P_{l_N l} = \left[\begin{array}{c|c} P_{l_N} & P_{l2}^T \\ \hline P_{l2} & P_l \end{array}\right]$$

where P_{l_N} and P_l are positive definite matrices.

Next, introduce the following new matrices:

$$\tilde{K}_l = R_{l3}^T \bar{K}_l, \quad \tilde{H}_l = R_{l3}^T \bar{H}_l R_l^{-1} R_{l3}, \quad \tilde{E}_l = \bar{E}_l R_l^{-1} R_{l3}, \tag{2.24}$$

$$S_{l2} = R_{l2} R_l^{-1} R_{l3}, \quad S_l = R_{l3}^T R_l^{-T} R_{l3}, \tag{2.25}$$

$$J_l = \begin{bmatrix} I & 0 \\ 0 & R_l^{-1} R_{l3} \end{bmatrix}. \tag{2.26}$$

With the aid of (2.9), (2.24), (2.25) and (2.26), the LMIs (2.18) can be turned into LMIs (2.23) by the following procedures: firstly, perform the congruence transformation diag$\{J_l, J_l, I, I\}$ on (2.18); secondly, implement the following transformations:

$$\begin{aligned} \tilde{P}_{l_N l} &= J_l^T P_{l_N l} J_l, \quad \tilde{P}_{m_N m} = J_l^T P_{m_N m} J_l, \\ \tilde{\mathcal{A}}_l^T &= J_l^T \mathcal{A}_l^T R_l J_l, \quad \tilde{\mathcal{B}}_l^T = \mathcal{B}_l^T R_l J_l, \\ \tilde{\mathcal{E}}_l &= \mathcal{E}_l J_l, \quad \tilde{S}_{l_N l} = J_l^T R_{l_N l} J_l. \end{aligned}$$

Since R_{l3} and R_l are nonsingular matrices, it is obvious that the transformation from (2.18) to (2.23) is invertible. Hence, via Lemma 2.1, one has that the PWL filtering error system (2.13) is globally asymptotically stablewhile achieving H_∞ performance constraint.

In the end, the distributed filter gains are to be determined. From (2.24), it firstly needs to compute two nonsingular matrix variables R_{l3} and R_l in order to compute the distributed PWL filter gains $\bar{H}_l$ and $\bar{K}_l$. It is no doubt that two unknown matrix variables could bring more computational burden and conservatism than that of one. In the following, the transfer function is used to reduce the number of unknown variables.

It follows from [5] that the real rational transfer function matrix of augmented PWL filter (2.8) is given by

$$T_{l,\hat{z}\bar{y}} = \bar{E}_l (zI - \bar{H}_l)^{-1} \bar{K}_l. \tag{2.27}$$

In terms of (2.25) and (2.27), the transfer function $T_{l,\hat{z}\bar{y}}$ can be computed as:

$$\begin{aligned} T_{l,\hat{z}\bar{y}} &= \tilde{E}_l R_{l3}^{-1} R_l (zI - R_{l3}^{-T} \tilde{H}_l R_{l3}^{-1} R_l))^{-1} R_{l3}^{-T} \tilde{K}_l \\ &= \tilde{E}_l (z R_{l3}^T R_l^{-1} R_{l3} - \tilde{H}_l)^{-1} \tilde{K}_l \\ &= \tilde{E}_l (zI - S_l^{-T} \tilde{H}_l)^{-1} S_l^{-T} \tilde{K}_l. \end{aligned} \tag{2.28}$$

It can be concluded by comparing (2.27) with (2.28) that

$$\bar{H}_l = S_l^{-T} \tilde{H}_l, \quad \bar{K}_l = S_l^{-T} \tilde{K}_l, \quad \bar{E}_l = \tilde{E}_l.$$

Finally, the proof is completed.

It is noted that the LMIs in (2.23) are coupled for different mode transitions $(l_N l, m_N m) \in \Omega$, so that all the mode transitions $(l_N l, m_N m)$ need to be considered in order to compute the gains $\bar{H}_l$ and $\bar{K}_l$. In a practical application, the parameter gains $\bar{H}_l$ and $\bar{K}_l$ are allowed to be computed off-line and be subsequently implemented in the network, where the coupled LMIs (2.23) can be solved by using the LMI toolbox.

Remark 2.2 Now, attention is focused on the computational complexity of the proposed distributed filter design scheme proposed in this chapter. Note that the PWL system mode size (mode number) is n, the sensor network size (node number) is N, the variable dimensions can be seen from $x_k \in \mathbb{R}^{n_x}$, $\hat{x}_{i,k} \in \mathbb{R}^{n_x}$, $y_{i,k} \in \mathbb{R}^{n_y} (i = 1, \ldots, n)$, $z_k \in \mathbb{R}^{n_z}$, and $w_k \in \mathbb{R}^{n_w}$. For each LMI (2.23), the total row size of the LMI system is $4Nn_x + n_w$, the total number of scalar decision variables is $5N^2 n_x^2 + Nn_x + (n_x^2 + n_x n_y)|\mathcal{N}|$. In addition, n^2 standard LMIs given by (2.23) need to be solved. Thus, the computational complexity of the proposed distributed algorithm can be represented as $O(n^2 N^3 n_x^3 + n^2 N^2 n_x^2 n_w)$. Therefore, it is not difficult to see that the overall complexity of the given algorithm depends polynomially on the PWL system mode size, the sensor network size and the variable dimensions. Note that the computational burden is mainly caused by the basic mathematical operations. Thankfully, in the areas of computational mathematics, optimization, and operational research, it is a very active area on improving the computational efficiency of mathematical operations, and substantial speed-ups can be expected in the future.

2.3 An Illustrative Example

In this section, a simulation example is presented to illustrate the effectiveness of distributed PWL filtering scheme developed in this chapter.

Consider an SN with 10 sensor nodes, whose topology is represented by a directed ring $\mathcal{G} = (\mathcal{V}, \mathcal{E}, \mathcal{A})$ with the set of nodes $\mathcal{V} = \{1, 2, 3, 4, 5, 6, 7, 8, 9, 10\}$, the set of edges $\mathcal{E} = \{(1, 1), (1, 2), (2, 2), (2, 3), (3, 3), (3, 4), (4, 4), (4, 5), (5, 5), (5, 6), (6, 6), (6, 7), (7, 7), (7, 8), (8, 8), (8, 9), (9, 9), (9, 10), (10, 10), (10, 1)\}$, and the adjacent matrix $\mathcal{A}$ is given as follows:

$$\mathcal{A} = \begin{bmatrix} 1&1&0&0&0&0&0&0&0&0 \\ 0&1&1&0&0&0&0&0&0&0 \\ 0&0&1&1&0&0&0&0&0&0 \\ 0&0&0&1&1&0&0&0&0&0 \\ 0&0&0&0&1&1&0&0&0&0 \\ 0&0&0&0&0&1&1&0&0&0 \\ 0&0&0&0&0&0&1&1&0&0 \\ 0&0&0&0&0&0&0&1&1&0 \\ 0&0&0&0&0&0&0&0&1&1 \\ 1&0&0&0&0&0&0&0&0&1 \end{bmatrix}.$$

Consider the PWL system (2.1) with $L \triangleq \{1, 2\}$. The corresponding mode sets for the plant and the filter on each sensor $i \in \mathcal{V}$ are defined as

$$\bar{l} \triangleq \begin{cases} 1, & x_k^{(1)} \geq 0, \\ 2, & x_k^{(1)} < 0, \end{cases}$$

and

$$\bar{l}_i \triangleq \begin{cases} 1, & \hat{x}_{i,k}^{(1)} \geq 0, \\ 2, & \hat{x}_{i,k}^{(1)} < 0 \end{cases}$$

where $x_k^{(1)}$ and $\hat{x}_{i,k}^{(1)}$ are the first element of x_k and $\hat{x}_{i,k}$, respectively.

The system parameters are given as follows:

$$\begin{aligned}
&A_1 = \begin{bmatrix} 0.39 & 0.7 \\ 0.3 & -0.6 \end{bmatrix}, \quad B_1 = \begin{bmatrix} 1 \\ 0.5 \end{bmatrix}, \quad A_2 = \begin{bmatrix} -0.6 & 0.3 \\ 0.7 & 0.5 \end{bmatrix}, \quad B_2 = \begin{bmatrix} 0.5 \\ 1 \end{bmatrix}, \\
&C_{1,1} = \begin{bmatrix} -0.70 & 0.52 \end{bmatrix}, \quad C_{1,2} = \begin{bmatrix} -0.75 & 0.50 \end{bmatrix}, \quad C_{1,3} = \begin{bmatrix} -0.74 & 0.53 \end{bmatrix}, \\
&C_{1,4} = \begin{bmatrix} -0.75 & 0.52 \end{bmatrix}, \quad C_{1,5} = \begin{bmatrix} -0.75 & 0.65 \end{bmatrix}, \quad C_{1,6} = \begin{bmatrix} -0.82 & 0.62 \end{bmatrix}, \\
&C_{1,7} = \begin{bmatrix} -0.75 & 0.80 \end{bmatrix}, \quad C_{1,8} = \begin{bmatrix} -0.74 & 0.75 \end{bmatrix}, \quad C_{1,9} = \begin{bmatrix} -0.75 & 0.65 \end{bmatrix}, \\
&C_{1,10} = \begin{bmatrix} -0.75 & 0.70 \end{bmatrix}, \quad C_{2,1} = \begin{bmatrix} 0.72 & -0.62 \end{bmatrix}, \quad C_{2,2} = \begin{bmatrix} 0.65 & -0.70 \end{bmatrix}, \\
&C_{2,3} = \begin{bmatrix} 0.65 & -0.65 \end{bmatrix}, \quad C_{2,4} = \begin{bmatrix} 0.65 & -0.60 \end{bmatrix}, \quad C_{2,5} = \begin{bmatrix} 0.75 & -0.65 \end{bmatrix}, \\
&C_{2,6} = \begin{bmatrix} 0.82 & -0.62 \end{bmatrix}, \quad C_{2,7} = \begin{bmatrix} 0.75 & -0.80 \end{bmatrix}, \quad C_{2,8} = \begin{bmatrix} 0.74 & -0.75 \end{bmatrix}, \\
&C_{2,9} = \begin{bmatrix} 0.75 & -0.65 \end{bmatrix}, \quad C_{2,10} = \begin{bmatrix} 0.75 & -0.70 \end{bmatrix}, \\
&D_{1,1} = 0.28, \quad D_{1,2} = 0.27, \quad D_{1,3} = 0.26, \quad D_{1,4} = 0.25, \quad D_{1,5} = 0.24, \\
&D_{1,6} = 0.28, \quad D_{1,7} = 0.27, \quad D_{1,8} = 0.26, \quad D_{1,9} = 0.25, \quad D_{1,10} = 0.24, \\
&D_{2,1} = 0.24, \quad D_{2,2} = 0.25, \quad D_{2,3} = 0.26, \quad D_{2,4} = 0.27, \quad D_{2,5} = 0.28, \\
&D_{2,6} = 0.24, \quad D_{2,7} = 0.25, \quad D_{2,8} = 0.26, \quad D_{2,9} = 0.27, \quad D_{2,10} = 0.28, \\
&E_1 = \begin{bmatrix} 0 & 0.5 \end{bmatrix}, \quad E_2 = \begin{bmatrix} 0 & 0.5 \end{bmatrix}.
\end{aligned}$$

According to Theorem 2.2, by using the LMI toolbox, the distributed PWL filter gains satisfying the constraints are calculated as follows:

$$\bar{H}_{1,11} = \begin{bmatrix} 0.1735 & 0.1028 \\ -0.2302 & -0.1163 \end{bmatrix}, \quad \bar{H}_{1,12} = \begin{bmatrix} 0.0200 & 0.0101 \\ -0.0413 & -0.0223 \end{bmatrix},$$

$$\bar{H}_{1,22} = \begin{bmatrix} 0.1733 & 0.1073 \\ -0.2278 & -0.1217 \end{bmatrix}, \quad \bar{H}_{1,23} = \begin{bmatrix} 0.0200 & 0.0097 \\ -0.0411 & -0.0212 \end{bmatrix},$$

$$\bar{H}_{1,33} = \begin{bmatrix} 0.1735 & 0.1050 \\ -0.2365 & -0.1218 \end{bmatrix}, \quad \bar{H}_{1,34} = \begin{bmatrix} 0.0200 & 0.0099 \\ -0.0420 & -0.0222 \end{bmatrix},$$

$$\bar{H}_{1,44} = \begin{bmatrix} 0.1732 & 0.1049 \\ -0.2290 & -0.1189 \end{bmatrix}, \quad \bar{H}_{1,45} = \begin{bmatrix} 0.0171 & 0.0056 \\ -0.0369 & -0.0102 \end{bmatrix},$$

$$\bar{H}_{1,55} = \begin{bmatrix} 0.1705 & 0.0588 \\ -0.1382 & 0.1743 \end{bmatrix}, \quad \bar{H}_{1,56} = \begin{bmatrix} 0.0155 & 0.0098 \\ -0.0183 & -0.0180 \end{bmatrix},$$

$$\bar{H}_{1,66} = \begin{bmatrix} 0.1734 & 0.1010 \\ -0.2346 & -0.1157 \end{bmatrix}, \quad \bar{H}_{1,67} = \begin{bmatrix} 0.0224 & 0.0081 \\ -0.0500 & -0.0196 \end{bmatrix},$$

$$\bar{H}_{1,77} = \begin{bmatrix} 0.1762 & 0.0907 \\ -0.2795 & -0.1114 \end{bmatrix}, \quad \bar{H}_{1,78} = \begin{bmatrix} 0.0263 & 0.0110 \\ -0.0663 & -0.0290 \end{bmatrix},$$

$$\bar{H}_{1,88} = \begin{bmatrix} 0.1757 & 0.0925 \\ -0.2772 & -0.1147 \end{bmatrix}, \quad \bar{H}_{1,89} = \begin{bmatrix} 0.0244 & 0.0113 \\ -0.0598 & -0.0292 \end{bmatrix},$$

$$\bar{H}_{1,99} = \begin{bmatrix} 0.1744 & 0.0977 \\ -0.2603 & -0.1194 \end{bmatrix}, \quad \bar{H}_{1,90} = \begin{bmatrix} 0.0200 & 0.0101 \\ -0.0413 & -0.0223 \end{bmatrix},$$

$$\bar{H}_{1,00} = \begin{bmatrix} 0.1751 & 0.0940 \\ -0.2707 & -0.1154 \end{bmatrix}, \quad \bar{H}_{1,01} = \begin{bmatrix} -3.8420 * 10^{-10} & -2.7467 * 10^{-10} \\ 1.1377 * 10^{-10} & 7.3350 * 10^{-10} \end{bmatrix},$$

$$\bar{H}_{2,11} = \begin{bmatrix} -0.1237 & -0.0514 \\ 0.0637 & 0.0405 \end{bmatrix}, \quad \bar{H}_{2,12} = \begin{bmatrix} -0.0034 & -0.0027 \\ 0.0032 & 0.0016 \end{bmatrix},$$

$$\bar{H}_{2,22} = \begin{bmatrix} -0.1246 & -0.0547 \\ 0.0612 & 0.0384 \end{bmatrix}, \quad \bar{H}_{2,23} = \begin{bmatrix} -0.0036 & -0.0026 \\ 0.0030 & 0.0014 \end{bmatrix},$$

$$\bar{H}_{2,33} = \begin{bmatrix} -0.1242 & -0.0526 \\ 0.0618 & 0.0387 \end{bmatrix}, \quad \bar{H}_{2,34} = \begin{bmatrix} -0.0034 & -0.0027 \\ 0.0030 & 0.0015 \end{bmatrix},$$

$$\bar{H}_{2,44} = \begin{bmatrix} -0.1239 & -0.0528 \\ 0.0636 & 0.0404 \end{bmatrix}, \quad \bar{H}_{2,45} = \begin{bmatrix} -0.0026 & -0.0027 \\ 0.0027 & 0.0016 \end{bmatrix},$$

$$\bar{H}_{2,55} = \begin{bmatrix} -0.1235 & -0.0504 \\ 0.0641 & 0.0420 \end{bmatrix}, \quad \bar{H}_{2,56} = \begin{bmatrix} -0.0035 & -0.0028 \\ 0.0031 & 0.0016 \end{bmatrix},$$

$$\bar{H}_{2,66} = \begin{bmatrix} -0.1233 & -0.0499 \\ 0.0656 & 0.0420 \end{bmatrix}, \quad \bar{H}_{2,67} = \begin{bmatrix} -0.0043 & -0.0019 \\ 0.0036 & 0.0012 \end{bmatrix},$$

$$\bar{H}_{2,77} = \begin{bmatrix} -0.1236 & -0.0419 \\ 0.0548 & 0.0316 \end{bmatrix}, \quad \bar{H}_{2,78} = \begin{bmatrix} -0.0042 & -0.0022 \\ 0.0030 & 0.0011 \end{bmatrix},$$

$$\bar{H}_{2,88} = \begin{bmatrix} -0.1236 & -0.0431 \\ 0.0565 & 0.0332 \end{bmatrix}, \quad \bar{H}_{2,89} = \begin{bmatrix} -0.0038 & -0.0025 \\ 0.0031 & 0.0013 \end{bmatrix},$$

$$\bar{H}_{2,99} = \begin{bmatrix} -0.1235 & -0.0469 \\ 0.0616 & 0.0380 \end{bmatrix}, \quad \bar{H}_{2,90} = \begin{bmatrix} -0.0040 & -0.0023 \\ 0.0034 & 0.0013 \end{bmatrix},$$

$$\bar{H}_{2,00} = \begin{bmatrix} -0.1236 & -0.0445 \\ 0.0591 & 0.0354 \end{bmatrix}, \quad \bar{H}_{2,01} = \begin{bmatrix} 2.0524 * 10^{-10} & 1.1711 * 10^{-10} \\ -7.3530 * 10^{-11} & -3.9173 * 10^{-11} \end{bmatrix},$$

$$\bar{K}_{1,11} = \begin{bmatrix} -0.0853 \\ 0.3128 \end{bmatrix}, \quad \bar{K}_{1,12} = \begin{bmatrix} -0.0127 \\ 0.0340 \end{bmatrix},$$

$$\bar{K}_{1,22} = \begin{bmatrix} -0.0749 \\ 0.2953 \end{bmatrix}, \quad \bar{K}_{1,23} = \begin{bmatrix} -0.0134 \\ 0.0349 \end{bmatrix},$$

$$\bar{K}_{1,33} = \begin{bmatrix} -0.0799 \\ 0.3053 \end{bmatrix}, \quad \bar{K}_{1,34} = \begin{bmatrix} -0.0129 \\ 0.0346 \end{bmatrix},$$

$$\bar{K}_{1,44} = \begin{bmatrix} -0.0772 \\ 0.2945 \end{bmatrix}, \quad \bar{K}_{1,45} = \begin{bmatrix} -0.0114 \\ 0.0287 \end{bmatrix},$$

$$\bar{K}_{1,55} = \begin{bmatrix} -0.1072 \\ 0.5148 \end{bmatrix}, \quad \bar{K}_{1,56} = \begin{bmatrix} -0.0071 \\ 0.0116 \end{bmatrix},$$

$$\bar{K}_{1,66} = \begin{bmatrix} -0.0751 \\ 0.2715 \end{bmatrix}, \quad \bar{K}_{1,67} = \begin{bmatrix} -0.0159 \\ 0.0449 \end{bmatrix},$$

$$\bar{K}_{1,77} = \begin{bmatrix} -0.0983 \\ 0.3281 \end{bmatrix}, \quad \bar{K}_{1,78} = \begin{bmatrix} -0.0195 \\ 0.0611 \end{bmatrix},$$

$$\bar{K}_{1,88} = \begin{bmatrix} -0.0977 \\ 0.3328 \end{bmatrix}, \quad \bar{K}_{1,89} = \begin{bmatrix} -0.0170 \\ 0.0527 \end{bmatrix},$$

$$\bar{K}_{1,99} = \begin{bmatrix} -0.0895 \\ 0.3182 \end{bmatrix}, \quad \bar{K}_{1,90} = \begin{bmatrix} -0.0164 \\ 0.0493 \end{bmatrix},$$

$$\bar{K}_{1,00} = \begin{bmatrix} -0.0924 \\ 0.3235 \end{bmatrix}, \quad \bar{K}_{1,01} = \begin{bmatrix} 2.0877 * 10^{-10} \\ -7.2431 * 10^{-10} \end{bmatrix},$$

$$\bar{K}_{2,11} = \begin{bmatrix} 0.1167 \\ 0.0052 \end{bmatrix}, \quad \bar{K}_{2,12} = \begin{bmatrix} 0.0015 \\ 0.0011 \end{bmatrix},$$

$$\bar{K}_{2,22} = \begin{bmatrix} 0.1159 \\ 0.0114 \end{bmatrix}, \quad \bar{K}_{2,23} = \begin{bmatrix} 0.0014 \\ 0.0012 \end{bmatrix},$$

$$\bar{K}_{2,33} = \begin{bmatrix} 0.1195 \\ 0.0097 \end{bmatrix}, \quad \bar{K}_{2,34} = \begin{bmatrix} 0.0015 \\ 0.0013 \end{bmatrix},$$

$$\bar{K}_{2,44} = \begin{bmatrix} 0.1251 \\ 0.0077 \end{bmatrix}, \quad \bar{K}_{2,45} = \begin{bmatrix} 0.0014 \\ 0.0010 \end{bmatrix},$$

$$\bar{K}_{2,55} = \begin{bmatrix} 0.1139 \\ -0.0011 \end{bmatrix}, \quad \bar{K}_{2,56} = \begin{bmatrix} 7.4393 * 10^{-4} \\ 0.0010 \end{bmatrix},$$

$$\bar{K}_{2,66} = \begin{bmatrix} 0.1086 \\ 0.0015 \end{bmatrix}, \quad \bar{K}_{2,67} = \begin{bmatrix} 7.5415 * 10^{-4} \\ 0.0012 \end{bmatrix},$$

$$\bar{K}_{2,77} = \begin{bmatrix} 0.0936 \\ 0.0092 \end{bmatrix}, \quad \bar{K}_{2,78} = \begin{bmatrix} 9.2162 * 10^{-4} \\ 0.0092 \end{bmatrix},$$

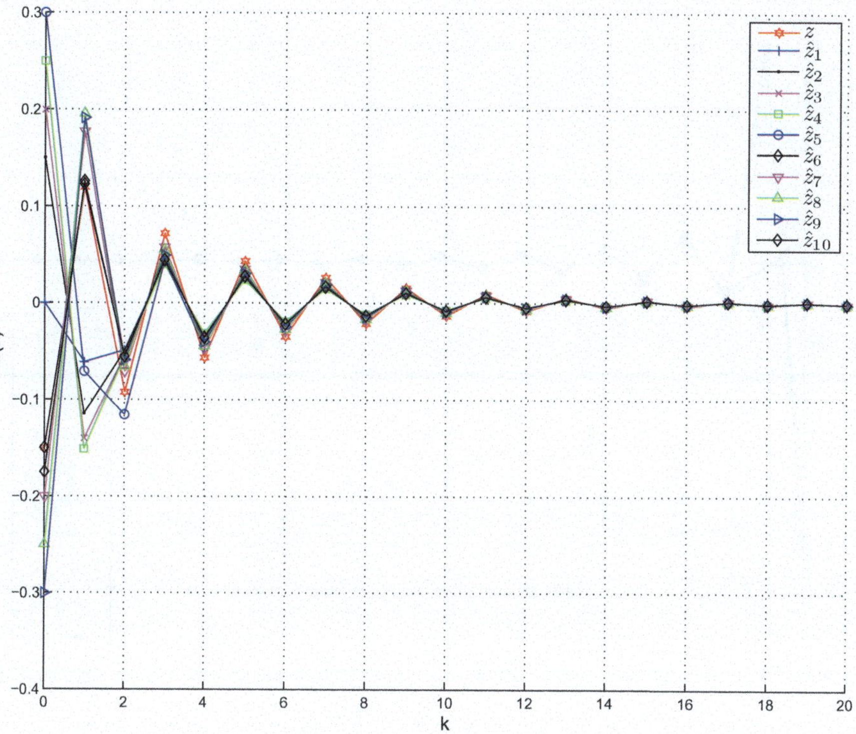

Fig. 2.1 The true state z_k and its estimates $\hat{z}_{i,k}$

$$\bar{K}_{2,88} = \begin{bmatrix} 0.0985 \\ 0.0085 \end{bmatrix}, \quad \bar{K}_{2,89} = \begin{bmatrix} 0.0012 \\ 0.0013 \end{bmatrix},$$

$$\bar{K}_{2,99} = \begin{bmatrix} 0.1085 \\ 0.0052 \end{bmatrix}, \quad \bar{K}_{2,90} = \begin{bmatrix} 0.0010 \\ 0.0013 \end{bmatrix},$$

$$\bar{K}_{2,00} = \begin{bmatrix} 0.1026 \\ 0.0068 \end{bmatrix}, \quad \bar{K}_{2,01} = \begin{bmatrix} -1.1066 * 10^{-10} \\ -7.6333 * 10^{-12} \end{bmatrix},$$

and the optimal performance index γ^* given in (2.15) is 1.57. Moreover, denote γ_i^* as the optimal performance index of the single sensor i with the same parameters in this illustrative example. By using the existing method in [2, 6–8], the corresponding optimal performance index can be obtained as $\gamma_1^* = 1.96$, $\gamma_2^* = 1.95$, $\gamma_3^* = 1.95$, $\gamma_4^* = 1.96$. From the comparison between γ_i^* and γ^*, it shows the distinct advantage of proposed distributed PWL filtering.

In the simulation, choose $w_k = 0.01 \sin(0.01\pi k) \exp(-k/150) \in l_2[0, \infty)$. The initial modes are given as $\bar{l}_0 = 1$, $l_{1,0} = l_{2,0} = l_{3,0} = l_{4,0} = l_{5,0} = 1$ and $l_{6,0} = l_{7,0} = l_{8,0} = l_{9,0} = l_{10,0} = 2$. The initial conditions are given as $x_0 = [0.2 \quad -0.3]^T$, $\hat{x}_{1,0} = [0 \quad 0]^T$, $\hat{x}_{2,0} = [0.3 \quad 0.3]^T$, $\hat{x}_{3,0} = [0.4 \quad 0.4]^T$, $\hat{x}_{4,0} = [0.5 \quad 0.5]^T$,

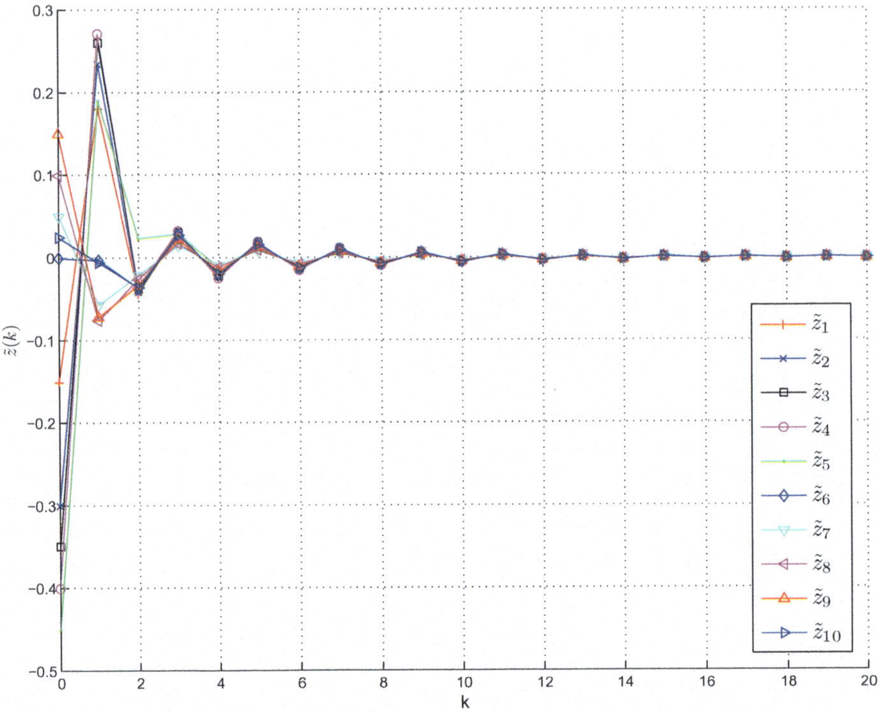

Fig. 2.2 The filtering errors $\tilde{z}_{i,k}$

$\hat{x}_{5,0} = [0.6 \quad 0.6]^T$, $\hat{x}_{6,0} = [-0.3 \quad -0.3]^T$, $\hat{x}_{7,0} = [-0.4 \quad -0.4]^T$, $\hat{x}_{8,0} = [-0.5 \quad -0.5]^T$, $\hat{x}_{9,0} = [-0.6 \quad -0.6]^T$, $\hat{x}_{10,0} = [-0.35 \quad -0.35]^T$. The numerical results are plotted in Figs. 2.1 and 2.2. Figure 2.1 shows the trajectories of true state z_k and its estimate $\hat{z}_{i,k}$ $(i \in \mathscr{V})$ from sensor node i. In addition, Fig. 2.2 describes the filtering errors $\tilde{z}_{i,k} (i = 1, 2, \ldots, 10)$. The simulation results have confirmed the effectiveness of distributed H_∞ filtering scheme proposed in this chapter.

2.4 Summary

In this chapter, the distributed H_∞ filtering problem has been dealt with for a class of discrete-time PWL systems over SNs. By using the globally augmented way, Lyapunov stability theory and convex optimization method, sufficient conditions have been derived such that the augmented PWL filtering error dynamics is globally asymptotically stable while achieving H_∞ performance constraint. Subsequently, distributed H_∞ filter parameters have been calculated by solving a set of LMIs. Finally, an illustrative simulation has been provided to demonstrate the effectiveness of proposed distributed filtering approach. In the future, one of possible research directions would be to develop the efficient methods to reduce the computational burden caused by the globally augmented way. More specifically, the derived sufficient conditions in terms of the LMIs involve all the information from all nodes in an SN. It is no doubt that such algorithm structure is too cumbersome, which results in inconvenience when considering other problems, such as the dynamic topology. Fortunately, the vector dissipativity theory may be a good choice for distributed filtering problem in large-scale network [1], since this method avoids utilizing the globally augmented way.

References

1. Ugrinovskii, V.: Distributed robust filtering with H_∞ consensus of estimates. Automatica **47**(1), 1–13 (2011)
2. Chen, M., Feng, G.: Delay-depent H_∞ filtering of piecewise-linear systems with time-varying delays. IEEE Trans. Circuits Syst.-I: Regul. Papers **55**(7), 2087–2096 (2008)
3. Dong, H., Wang, Z., Lam, J., Gao, H.: Distributed filtering in sensor networks with randomly occurring saturations and successive packet dropouts. Int. J. Robust Nonlinear Control **24**(12), 1743–1759 (2014)
4. Shen, B., Wang, Z., Hung, Y.: Distributed H_∞-consensus filtering in sensor networks with multiple missing measurements: the finite-horizon case. Automatica **46**(10), 1682–1688 (2010)
5. Souza, C., Xie, L.: On the discrete-time bounded real lemma with application in the characterization of static state feedback H_∞ controllers. Syst. Control Lett. **18**(1), 61–71 (1992)
6. Feng, G.: Robust filtering design of piecewise discrete time linear systems. IEEE Trans. Autom. Control **53**(2), 599–605 (2005)
7. Wang, Z., Wei, G., Feng, G.: Reliable H_∞ control for discrete-time piecewise linear systems with infinite distributed delays. Automatica **45**(12), 346–358 (2009)
8. Wei, G., Han, F., Wang, L., Song, Y.: Reliable H_∞ filtering for discrete piecewise linear systems with infinite distributed delays. Int. J. Gen. Syst. **43**(3–4), 346–358 (2014)

Chapter 3
Consensus Filtering with Stochastic Nonlinearities and Multiple Missing Measurements

Chapter 2 investigates the distributed filtering for the piecewise linear system by mean of the so-called globally augmented method. This idea is very simple and the existing results are easily generalized to the distributed filtering case. Nonetheless, the computation burden is related to the size of the network. For addressing this problem, the local performance analysis method will be developed by utilizing the vector dissipativity theory in this chapter.

This chapter addresses the distributed H_∞-consensus filtering problem for a class of discrete time-varying systems with stochastic nonlinearities and missing measurements. First of all, the stochastic nonlinearities are formulated by statistical means and the missing measurements are characterized by a set of random variables obeying the Bernoulli distribution. Next, a novel H_∞-consensus performance index is proposed to measure both the filtering accuracy of every node and the consensus among neighboring nodes. Then, a new concept called stochastic vector dissipativity is presented wherein the dissipation matrix is first formulated by a nonsingular sub-stochastic matrix, which is skillfully constructed by a newly defined interval function on the out-degree. Also, a set of local sufficient conditions in terms of the recursive linear matrix inequalities is constructed for each node such that the proposed H_∞-consensus performance is guaranteed for the local augmented dynamics over the finite horizon. Furthermore, a novel algorithm proposed here is implemented on each node. Finally, an illustrative simulation is presented to demonstrate the effectiveness of the proposed algorithm.

3.1 Problem Formulation

First and foremost, consider a class of discrete time-varying nonlinear stochastic systems defined on $k \in \mathscr{H} \triangleq \{0, 1, \ldots, n-1\}$:

F. Han et al., *Distributed Filtering, Control and Synchronization*,
Studies in Systems, Decision and Control 428,
https://doi.org/10.1007/978-3-030-97075-8_3

$$\begin{cases} x_{k+1} = A_k x_k + f_k(x_k, v_k) + B_k w_k, \\ x_0 = \phi \end{cases} \tag{3.1}$$

with N sensors modeled by

$$y_{i,k} = \zeta_{i,k} C_{i,k} x_k + D_{i,k} \vartheta_{i,k}, \quad i \in \mathscr{V} \tag{3.2}$$

where $x_k \in \mathbb{R}^{n_x}$ is the state, $y_{i,k} \in \mathbb{R}^{n_y}$ is the measurement from node i, $w_k \in \mathbb{R}^{n_w}$ and $\vartheta_{i,k} \in \mathbb{R}^{n_\vartheta}$ are the external disturbances belonging to $l_2[0, n-1]$ (n_y (n_ϑ) is same for each sensor). v_k is a sequence of zero-mean Gaussian noises with unit variance. ϕ is an initial value of the system state. $\zeta_{i,k} \in \mathbb{R}$ is a Bernoulli distributed white sequence taking values of 1 or 0 with $\text{Prob}\{\zeta_{i,k} = 1\} = \beta_i$ and $\text{Prob}\{\zeta_{i,k} = 0\} = 1 - \beta_i$, where $\beta_i \in [0, 1]$ is a known constant. Moreover, it is assumed that all stochastic variables $\zeta_{i,k}$ and v_k are independent with respect to k and i. A_k, B_k, $C_{i,k}$, and $D_{i,k}$ are all known time-varying matrices with compatible dimensions.

Additionally, $f_k(x_k, v_k)$ with $f_k(0, v_k) = 0$ is a stochastic nonlinear function with the first moment for all x_k:

$$\mathbb{E}\{f_k(x_k, v_k)|x_k\} = 0, \tag{3.3}$$

and the covariance matrices

$$\mathbb{E}\left\{f_k(x_k, v_k) f_j^T(x_j, v_j)|x_k\right\} = 0, k \neq j, \tag{3.4}$$

$$\mathbb{E}\left\{f_k(x_k, v_k) f_k^T(x_k, v_k)|x_k\right\} = \sum_{r=1}^{s} \pi_{r,k} \pi_{r,k}^T x_k^T \Gamma_{r,k} x_k \tag{3.5}$$

where s is a known positive integer, $\pi_{r,k} \in \mathbb{R}^{n_x \times 1}$ and $\Gamma_{r,k} \in \mathbb{R}^{n_x \times n_x}$ ($r = 1, 2, \ldots, s$) are known matrices.

Remark 3.1 As discussed in [1–4], the stochastic nonlinearities described by (3.3)–(3.5) are quite general and possess a convincing engineering significance. More specifically, such a function $f_k(x_k, v_k)$ can cover many special cases, for example, 1) the multiplicative noise $G_k x_k v_k$, where $G_k \in \mathbb{R}^{n_x \times n_x}$ and $v_k \in \mathbb{R}$; 2) the random vector $\psi_k(x_k) v_k$ where $\|\psi_k(x_k)\|^2 \leq \tau_k \|x_k\|^2$ and $v_k, \tau_k \in \mathbb{R}$; and 3) the nonlinear function $\text{sgn}[\psi_k(x_k)] G_k x_k v_k$, where sgn denotes the signum function.

In this chapter, the following distributed filter is constructed on node i:

$$\hat{x}_{i,k+1} = A_k \hat{x}_{i,k} + L_{i,k}(y_{i,k} - \beta_i C_{i,k} \hat{x}_{i,k}) + \sum_{j \in \mathscr{N}_i} K_{ij,k}(\hat{x}_{j,k} - \hat{x}_{i,k}) \tag{3.6}$$

where $\hat{x}_{i,k} \in \mathbb{R}^{n_x}$ is the estimate of x_k on node i, $L_{i,k}$ and $K_{ij,k}$ are the filter gains to be determined later.

According to the denotation $\mathscr{N}_i = \left\{j_{i_1}, j_{i_2}, \ldots, j_{i_{p_i}}\right\}$, for convenience, let

$$K_{ij_1,k} \triangleq K_{ij_{i_1},k}, \quad K_{ij_2,k} \triangleq K_{ij_{i_2},k}, \quad \cdots, \quad K_{ij_{p_i},k} \triangleq K_{ij_{ip_i},k}. \tag{3.7}$$

Remark 3.2 Due to taking the missing measurement model into consideration, one has that the random variable $\zeta_{i,k}$ is unknown when the communication channels employ the user data protocol (UDP), which is widely used in SNs. At the same time, from the viewpoint of engineering, all parameters should be known in the implementation process of designed filter. According to this basic rule, $\zeta_{i,k}$ can be replaced with its expected value β_i. In addition, if the transmission control protocol (TCP) is adopted, which implies that $\zeta_{i,k}$ is known, it can be directly utilized to fully leverage the information of missing measurements.

Remark 3.3 The filter (3.6) has the following structure features. More specifically, it can be divided into two parts: 1) $A\hat{x}_{i,k} + L_{i,k}(y_{i,k} - C_{i,k}\hat{x}_{i,k})$ corresponds to the Kalman filter of node i, which is employed to estimate the system state by utilizing the local information of node i; 2) $\sum_{j\in\mathcal{N}_i} K_{ij,k}(\hat{x}_{j,k} - \hat{x}_{i,k})$ corresponds to the consensus role between node i and its neighbors, which is used to promote the consensus by utilizing the consensus errors. Moreover, if the consensus is achieved, then (3.6) can be reduced to the common Kalman filter.

Setting $e_{i,k} \triangleq x_k - \hat{x}_{i,k}$, the filtering error dynamics of node i is described from (3.1), (3.2) and (3.6) as follows:

$$\begin{cases} e_{i,k+1} = (A_k - \beta_i L_{i,k}C_{i,k} - \sum\limits_{j\in\mathcal{N}_i} K_{ij,k})e_{i,k} + f_k(x_k, v_k) + B_k w_k \\ \qquad\qquad - (\zeta_{i,k} - \beta_i)L_{i,k}C_{i,k}x_k - L_{i,k}D_{i,k}\vartheta_{i,k} + \sum\limits_{j\in\mathcal{N}_i} K_{ij,k}e_{j,k}, \\ e_{i,0} = \phi - \hat{x}_{i,0}. \end{cases} \tag{3.8}$$

Denote $e_k \triangleq [e_{1,k}^T \;\; e_{2,k}^T \;\; \cdots \;\; e_{N,k}^T]^T$. Before proceeding further, the following definitions are first proposed.

Definition 3.1 For a given disturbance attenuation level $\gamma > 0$, and some given positive definite matrices U_{1i}, U_{2i}, $R_{i,k}$, $Q_{i,k}$, $T_{1i,k}$, and $T_{2i,k} > 0$, the filtering error dynamics (3.8) is said to satisfy the H_∞-consensus performance constraint over the finite horizon $\mathscr{H}$ if the following inequality holds

$$\mathbb{E}\left\{\sum_{k=0}^{n-1} \Phi_{\mathscr{G}}(e_k)\right\} \le \frac{\gamma^2}{N}\sum_{i=1}^{N}\left(x_0^T U_{1i} x_0 + e_{i,0}^T U_{2i} e_{i,0} + \sum_{k=0}^{n-1}\|w_k\|^2_{T_{1i,k}} + \sum_{k=0}^{n-1}\|\vartheta_{i,k}\|^2_{T_{2i,k}}\right) \tag{3.9}$$

where $\Phi_{\mathscr{G}}(e_k) \triangleq \frac{1}{N}\sum_{i=1}^{N}\left(\sum_{j\in\mathcal{N}_i}\|e_{j,k} - e_{i,k}\|^2_{R_{i,k}} + \|e_{i,k}\|^2_{Q_{i,k}}\right)$.

Remark 3.4 In the existing references, the H_∞-consensus performance index is presented respectively in terms of the consensus errors (i.e., $R_{i,k} \equiv I$ and $Q_{i,k} \equiv 0$ in [5–7]) or the filtering errors (i.e., $R_{i,k} \equiv 0$ and $Q_{i,k} \equiv I$ in [1, 8–10]). As such, it is still unclear what the relationship between the consensus speed and the filter accuracy is. Furthermore, it is also obscure whether the desirable performance behavior with the fastest consensus speed and the highest filtering accuracy can be achieved or not. These considerations motivate us to establish a uniform framework combining these two existing performance indices.

Denote $\eta_{i,k} \triangleq [x_k^T \quad e_{i,k}^T]^T$, $\eta_k \triangleq [\eta_{1,k}^T \quad \cdots \quad \eta_{N,k}^T]^T$, $\xi_{i,k} \triangleq [w_k^T \quad \vartheta_{i,k}^T]^T$ and $\xi_k \triangleq [\xi_{1,k}^T \quad \cdots \quad \xi_{N,k}^T]^T$. Combining (3.1) and (3.8) yields

$$\eta_{i,k+1} = (\mathscr{A}_{i,k} - \sum_{j\in\mathscr{N}_i} \mathscr{K}_{ij,k})\eta_{i,k} - (\zeta_{i,k} - \beta_i)\mathscr{C}_{i,k}\eta_{i,k} + \mathscr{B}_{i,k}\xi_{i,k} + \sum_{j\in\mathscr{N}_i} \mathscr{K}_{ij,k}\eta_{j,k} + \mathscr{F}_k \tag{3.10}$$

where

$$\begin{aligned}
&\mathscr{A}_{i,k} \triangleq \mathrm{diag}\{A_k, A_k - \beta_i L_{i,k}C_{i,k}\}, \ \mathscr{K}_{ij,k} \triangleq \mathrm{diag}\{0, K_{ij,k}\}, \\
&\mathscr{B}_{i,k} \triangleq \begin{bmatrix} B_k & 0 \\ B_k & -L_{i,k}D_{i,k} \end{bmatrix}, \ \mathscr{C}_{i,k} \triangleq \begin{bmatrix} 0 & 0 \\ L_{i,k}C_{i,k} & 0 \end{bmatrix}, \\
&\mathscr{F}_k \triangleq [f_k^T(x_k, v_k) \quad f_k^T(x_k, v_k)]^T.
\end{aligned}$$

Similar to (3.7), define

$$\mathscr{K}_{ij_1,k} \triangleq \mathscr{K}_{ij_{i_1},k}, \quad \mathscr{K}_{ij_2,k} \triangleq \mathscr{K}_{ij_{i_2},k}, \quad \ldots, \quad \mathscr{K}_{ij_{p_i},k} \triangleq \mathscr{K}_{ij_{ip_i},k}.$$

For convenience, Definition 3.1 can be converted into the following form with respect to η_k.

Definition 3.2 Given a disturbance attenuation level $\gamma > 0$, and some positive definite matrices U_{1i}, U_{2i}, $R_{i,k}$, $Q_{i,k}$, $T_{1i,k}$, and $T_{2i,k} > 0$. The locally augmented dynamics (3.10) is said to satisfy the H_∞-consensus performance constraint over the finite horizon $\mathscr{H}$ if the following inequality holds

$$\mathbb{E}\left\{\sum_{k=0}^{n-1} \Phi_{\mathscr{G}}(\eta_k)\right\} \le \frac{\gamma^2}{N}\sum_{i=1}^{N}\left(\eta_{i,0}^T U_i \eta_{i,0} + \sum_{k=0}^{n-1} \|\xi_{i,k}\|_{T_{i,k}}^2\right) \tag{3.11}$$

where $\Phi_{\mathscr{G}}(\eta_k) \triangleq \frac{1}{N}\sum_{i=1}^{N}\left[\sum_{j\in\mathscr{N}_i} \|\eta_{j,k} - \eta_{i,k}\|_{\tilde{R}_{i,k}}^2 + \|\eta_{i,k}\|_{\tilde{Q}_{i,k}}^2\right]$, $U_i \triangleq \mathrm{diag}\{U_{1i}, U_{2i}\}$, $\tilde{R}_{i,k} \triangleq \mathrm{diag}\{0, R_{i,k}\}$, $\tilde{Q}_{i,k} \triangleq \mathrm{diag}\{0, Q_{i,k}\}$ and $T_{i,k} \triangleq \mathrm{diag}\{T_{1i,k}, T_{2i,k}\}$.

The purpose of this chapter is to determine the filter gains $L_{i,k}$ and $K_{ij,k}$ for node i ($i \in \mathscr{V}$) such that the filtering errors satisfy the proposed H_∞-consensus performance constraint over the finite horizon $\mathscr{H}$.

3.2 Main Results

In this section, sufficient conditions of the existence of distributed filter are obtained for the locally augmented dynamics (3.10) to satisfy the H_∞-consensus performance criterion (3.11) via the vector dissipativity theory. It should be pointed out that, due to the existence of stochastic nonlinearities and missing measurements, the concept of stochastic vector dissipativity-like over the finite horizon $\mathscr{H}$ is proposed in the following. The core idea is to establish the vector dissipation inequality in the statistical sense, which is characterized by the constructed vector storage function (i.e., vector Lyapunov function), vector supply rate function and the dissipation matrix.

Denote firstly

$$\begin{aligned} S_i(\eta_k, \xi_{i,k}) &\triangleq \gamma^2 \|\xi_{i,k}\|^2_{T_{i,k}} - \|\eta_{i,k}\|^2_{\tilde{Q}_{i,k}} - \sum_{j \in \mathscr{N}_i} \|\eta_{j,k} - \eta_{i,k}\|^2_{\tilde{R}_{i,k}} \\ &= \gamma^2 \|\xi_{i,k}\|^2_{T_{i,k}} - \eta_k^T \mathscr{R}_{i,k} \eta_k, \quad i \in \mathscr{V} \end{aligned} \tag{3.12}$$

where

$$\mathscr{R}_{i,k} \triangleq \begin{bmatrix} a_{i1}\tilde{R}_{i,k} & \cdots & * & \cdots & * \\ \vdots & \ddots & \vdots & \ddots & \vdots \\ -a_{i1}\tilde{R}_{i,k} & \cdots & \tilde{Q}_{i,k} + p_i \tilde{R}_{i,k} & \cdots & * \\ \vdots & \ddots & \vdots & \ddots & \vdots \\ 0 & \cdots & -a_{iN}\tilde{R}_{i,k} & \cdots & a_{iN}\tilde{R}_{i,k} \end{bmatrix}.$$

Definition 3.3 The discrete time-varying locally augmented dynamics (3.10) is stochastically vector dissipative-like over the finite horizon $\mathscr{H}$ with respect to a vector of supply rate functions

$$\mathbf{S}(\eta_k, \xi_k) \triangleq [S_1(\eta_k, \xi_{1,k}) \quad \cdots \quad S_N(\eta_k, \xi_{N,k})]^T,$$

if there exists a sequence of vector storage functions $\mathbf{V}(\eta_k) \triangleq [V_1(\eta_{1,k}) \ \cdots \ V_N(\eta_{N,k})]^T$ with $\mathbf{V}(0) = 0$ and $V_i(\eta_{i,k})$ being all nonnegative definite functions, and a sequence of nonsingular column substochastic dissipation matrices $W_k \in \mathbb{R}^{N\times N}$ such that the following vector dissipation inequality is satisfied for any $k \in \mathscr{H}$:

$$\mathbb{E}\{\mathbf{V}(\eta_{k+1})\} \leq\leq W_k \mathbb{E}\{\mathbf{V}(\eta_k)\} + \mathbb{E}\{\mathbf{S}(\eta_k, \xi_k)\}. \tag{3.13}$$

Next, a set of local conditions is constructed to guarantee that the locally augmented dynamics (3.10) is stochastically vector dissipative-like over the finite horizon $\mathscr{H}$. For constructing a desirable dissipation matrix W_k, let $\mathbb{J}_{q_i}$ be an interval-valued function with respect to out-degree q_i and define:

$$\mathfrak{I}_{q_i} = \begin{cases} (0, \frac{1+q_i}{2q_i}), & \text{if } q_i \neq 0; \\ (0, 1], & \text{if } q_i = 0. \end{cases} \tag{3.14}$$

Lemma 3.1 *Given the disturbance attenuation level $\gamma > 0$, the scalar sequence $\alpha_{i,k} \in \mathfrak{I}_{q_i}$, the weighted matrix sequences $\tilde{R}_{i,k}$, $\tilde{Q}_{i,k}$, and $T_{i,k}$ defined in Definition 3.2 and the filter gain sequences $L_{i,k}$ and $K_{ij,k}$, the locally augmented dynamics (3.10) is stochastically vector dissipative-like over the finite horizon $\mathscr{H}$ with respect to the vector supply rate function $\mathbf{S}(\eta_k, \xi_k)$, if there exists a vector storage function $\mathbf{V}(\eta_k)$ with the form $V_i(\eta_{i,k}) = \eta_{i,k}^T \mathscr{P}_{i,k} \eta_{i,k}$ where $\{\mathscr{P}_{i,k}\}_{k \in \mathscr{H} \cup \{n\}}$ is a sequence of positive definite matrices, and a scalar sequence $\lambda_{i,k}$ such that the following inequalities for all $k \in \mathscr{H}$, $i \in \mathscr{V}$ hold*

$$\begin{bmatrix} \Psi_{i,k}^{11} & * & * \\ \Psi_{i,k}^{21} & \Psi_{i,k}^{22} & * \\ \Psi_{i,k}^{31} & \Psi_{i,k}^{32} & \Psi_{i,k}^{33} \end{bmatrix} > 0, \tag{3.15}$$

$$\bar{\pi}_{r,k}^T \mathscr{P}_{i,k+1} \bar{\pi}_{r,k} \leq \lambda_{i,k}, \quad (r = 1, \ldots, s) \tag{3.16}$$

where

$$\begin{aligned} \Psi_{i,k}^{11} &= \theta_{i,k} \mathscr{P}_{i,k} - \lambda_{i,k} \sum_{r=1}^{s} \operatorname{diag}\{\Gamma_{r,k}, 0\} - p_i \tilde{R}_{i,k} - \tilde{Q}_{i,k} \\ &\quad - \beta_i (1 - \beta_i) \mathscr{C}_{i,k}^T \mathscr{P}_{i,k+1} \mathscr{C}_{i,k}, \\ \theta_{i,k} &= \frac{1 + q_i (1 - \alpha_{i,k})}{1 + q_i}, \quad \Psi_{i,k}^{21} = \operatorname{col}_{p_i}\{\tilde{R}_{i,k}\}, \\ \Psi_{i,k}^{22} &= \operatorname{diag}\left\{ \frac{\alpha_{j_{i_1},k} \mathscr{P}_{j_{i_1},k}}{1 + q_{j_{i_1}}} - \tilde{R}_{i,k}, \ldots, \frac{\alpha_{j_{i p_i},k} \mathscr{P}_{j_{i p_i},k}}{1 + q_{j_{i p_i}}} - \tilde{R}_{i,k} \right\}, \\ \Psi_{i,k}^{31} &= \mathscr{A}_{i,k} - \sum_{j \in \mathscr{N}_i} \mathscr{K}_{ij,k}, \quad \Psi_{i,k}^{32} = [\mathscr{K}_{ij_1,k} \ \cdots \ \mathscr{K}_{ij_{p_i},k}], \\ \Psi_{i,k}^{33} &= \mathscr{P}_{i,k+1}^{-1} - \frac{1}{\gamma^2} \mathscr{B}_{i,k} T_{i,k}^{-1} \mathscr{B}_{i,k}^T, \quad \bar{\pi}_{r,k} = [\pi_{r,k}^T \ \pi_{r,k}^T]^T. \end{aligned}$$

Here, $\operatorname{col}_{p_i}\{\tilde{R}_{i,k}\} \triangleq [\underbrace{\tilde{R}_{i,k}^T \ \ \tilde{R}_{i,k}^T \ \ \cdots \ \ \tilde{R}_{i,k}^T}_{p_i}]^T$.

Proof The key point of proof is to verify the vector dissipation inequality (3.13). First, calculate the storage function on node i ($i \in \mathscr{V}$) along the system trajectory (3.10) as follows:

$$
\begin{aligned}
&\mathbb{E}\{V_i(\eta_{i,k+1})|\eta_{i,k},\eta_{j,k},j\in\mathcal{N}_i\}\\
&=\mathbb{E}\{\eta_{i,k+1}^T\mathscr{P}_{i,k+1}\eta_{i,k+1}|\eta_{i,k},\eta_{j,k},j\in\mathcal{N}_i\}\\
&=2\xi_{i,k}^T\mathscr{B}_{i,k}^T\mathscr{P}_{i,k+1}\sum_{j\in\mathcal{N}_i}\mathscr{K}_{ij,k}\eta_{j,k}\\
&\quad+\sum_{j\in\mathcal{N}_i}\eta_{j,k}^T\mathscr{K}_{ij,k}^T\mathscr{P}_{i,k+1}\sum_{j\in\mathcal{N}_i}\mathscr{K}_{ij,k}\eta_{j,k}\\
&\quad+2\xi_{i,k}^T\mathscr{B}_{i,k}^T\mathscr{P}_{i,k+1}(\mathscr{A}_{i,k}-\sum_{j\in\mathcal{N}_i}\mathscr{K}_{ij,k})\eta_{i,k}\\
&\quad+2\eta_{i,k}^T(\mathscr{A}_{i,k}-\sum_{j\in\mathcal{N}_i}\mathscr{K}_{ij,k})^T\mathscr{P}_{i,k+1}\sum_{j\in\mathcal{N}_i}\mathscr{K}_{ij,k}\eta_{j,k}\\
&\quad+\xi_{i,k}^T(\mathscr{B}_{i,k}^T\mathscr{P}_{i,k+1}\mathscr{B}_{i,k}-\gamma^2T_{i,k})\xi_{i,k}+\gamma^2\xi_{i,k}^TT_{i,k}\xi_{i,k}\\
&\quad+\eta_{i,k}^T(\mathscr{A}_{i,k}-\sum_{j\in\mathcal{N}_i}\mathscr{K}_{ij,k})^T\mathscr{P}_{i,k+1}(\mathscr{A}_{i,k}-\sum_{j\in\mathcal{N}_i}\mathscr{K}_{ij,k})\eta_{i,k}\\
&\quad+\mathbb{E}\{\mathscr{F}_k^T\mathscr{P}_{i,k+1}\mathscr{F}_k|\eta_{i,k},\eta_{j,k},j\in\mathcal{N}_i\}+\beta_i(1-\beta_i)\eta_{i,k}^T\mathscr{C}_{i,k}^T\mathscr{P}_{i,k+1}\mathscr{C}_{i,k}\eta_{i,k}.
\end{aligned}
$$

Next, it follows from (3.3) to (3.5) and (3.16) that

$$
\begin{aligned}
&\mathbb{E}\{\mathscr{F}_k^T\mathscr{P}_{i,k+1}\mathscr{F}_k|\eta_{i,k},\eta_{j,k},j\in\mathcal{N}_i\}\\
&=\mathbb{E}\{\mathscr{F}_k^T\mathscr{P}_{i,k+1}\mathscr{F}_k|\eta_{i,k}\}\\
&=\sum_{r=1}^{s}x_k^T\Gamma_{r,k}\bar{\pi}_{r,k}^T\mathscr{P}_{i,k+1}\bar{\pi}_{r,k}x_k\\
&\le\lambda_{i,k}\sum_{r=1}^{s}\eta_{i,k}^T\mathrm{diag}\{\Gamma_{r,k},0\}\eta_{i,k}.
\end{aligned}
$$

From (3.15), one has $\Psi_{i,k}^{33}>0$. Then, by using the completing-the-square technique and the matrix inverse formula, one obtains

$$
\begin{aligned}
&\mathbb{E}\{V_i(\eta_{i,k+1})|\eta_{i,k},\eta_{j,k},j\in\mathcal{N}_i\}\\
&\le\gamma^2\xi_{i,k}^TT_{i,k}\xi_{i,k}-\sum_{j\in\mathcal{N}_i}\|\eta_{j,k}-\eta_{i,k}\|_{\tilde{R}_{i,k}}^2-\|\eta_{i,k}\|_{\tilde{Q}_{i,k}}^2\\
&\quad+\sum_{j\in\mathcal{N}_i}\|\eta_{j,k}-\eta_{i,k}\|_{\tilde{R}_{i,k}}^2+\|\eta_{i,k}\|_{\tilde{Q}_{i,k}}^2\\
&\quad+\lambda_{i,k}\sum_{r=1}^{s}\eta_{i,k}^T\mathrm{diag}\{\Gamma_{r,k},0\}\eta_{i,k}\\
&\quad+\beta_i(1-\beta_i)\eta_{i,k}^T\mathscr{C}_{i,k}^T\mathscr{P}_{i,k+1}\mathscr{C}_{i,k}\eta_{i,k}\\
&\quad+\sum_{j\in\mathcal{N}_i}\eta_{j,k}^T\mathscr{K}_{ij,k}^T(\Psi_{i,k}^{33})^{-1}\sum_{j\in\mathcal{N}_i}\mathscr{K}_{ij,k}\eta_{j,k}
\end{aligned}
$$

$$+ 2\eta_{i,k}^T(\mathscr{A}_{i,k} - \sum_{j\in\mathscr{N}_i} \mathscr{K}_{ij,k})^T(\Psi_{i,k}^{33})^{-1}\sum_{j\in\mathscr{N}_i} \mathscr{K}_{ij,k}\eta_{j,k}$$
$$+ \eta_{i,k}^T(\mathscr{A}_{i,k} - \sum_{j\in\mathscr{N}_i} \mathscr{K}_{ij,k})^T(\Psi_{i,k}^{33})^{-1}(\mathscr{A}_{i,k} - \sum_{j\in\mathscr{N}_i} \mathscr{K}_{ij,k})\eta_{i,k}.$$

From the expression of $S_i(\eta_k, \xi_{i,k})$ and (3.15), one has

$$\begin{aligned}
&\mathbb{E}\{V_i(\eta_{i,k+1})|\eta_{i,k}, \eta_{j,k}, j \in \mathscr{N}_i\} \\
\le& S_i(\eta_k, \xi_{i,k}) + \sum_{j\in\mathscr{N}_i} \frac{\alpha_{j,k}}{1+q_j}\eta_{j,k}^T\mathscr{P}_{j,k}\eta_{j,k} + \theta_{i,k}\eta_{i,k}^T\mathscr{P}_{i,k}\eta_{i,k} \\
=& \begin{bmatrix} \frac{a_{i1}\alpha_{1,k}}{1+q_1} & \cdots & \theta_{i,k} & \cdots & \frac{a_{iN}\alpha_{N,k}}{1+q_N} \end{bmatrix} \mathbf{V}(\eta_k) + S_i(\eta_k, \xi_{i,k}).
\end{aligned}$$

Then, taking the vector $\begin{bmatrix} \frac{a_{i1}\alpha_{1,k}}{1+q_1} & \cdots & \theta_{i,k} & \cdots & \frac{a_{iN}\alpha_{N,k}}{1+q_N} \end{bmatrix}$ as its i-th row, W_k can be constructed. Since $\alpha_{i,k} \in \mathbb{J}_{q_i}$, W_k is strictly diagonally dominant, and thus it is nonsingular [11]. Moreover, $\mathbf{1}^T W_k = \mathbf{1}^T$. Consequently, W_k is a dissipation matrix. Hence, it follows from the structure of W_k that

$$\mathbb{E}\{V_i(\eta_{i,k+1})|\eta_{i,k}, \eta_{j,k}, j \in \mathscr{N}_i\} \le [W_k\mathbf{V}(\eta_k)]_i + S_i(\eta_k, \xi_{i,k})$$

where $[W_k\mathbf{V}(\eta_k)]_i$ is the i-th element of vector $W_k\mathbf{V}(\eta_k)$. By using the elementary properties of the conditional expectation, one obtains

$$\mathbb{E}\{V_i(\eta_{i,k+1})\} \le \mathbb{E}\{[W_k\mathbf{V}(\eta_k)]_i\} + \mathbb{E}\{S_i(\eta_k, \xi_{i,k})\}.$$

Noting that the above inequality holds for $\forall\, i \in \mathscr{V}$, one derives a vector form:

$$\mathbb{E}\{\mathbf{V}(\eta_{k+1})\} \le\le W_k\mathbb{E}\{\mathbf{V}(\eta_k)\} + \mathbb{E}\{\mathbf{S}(\eta_k, \xi_k)\}. \tag{3.17}$$

Finally, in light of Definition 3.3, the proof of Lemma 3.1 is therefore complete.

Notably, Lemma 3.1 presents the sufficient conditions to guarantee that the vector dissipation inequality (3.13) holds. The main challenge is to construct a nonsingular column substochastic dissipation matrix W_k. For this purpose, we design a novel matrix $W_k = [W_{ji,k}]$ with $W_{ji,k} = \frac{a_{ji}\alpha_{i,k}}{1+q_i}, i \neq j$ and $W_{ii,k} = \theta_{i,k}$. It follows from the definition of out-degree q_j that the column sum of off-diagonal elements of W_k is equal to $\frac{q_i\alpha_{i,k}}{1+q_i}$. Then, one immediately has that 1) W_k is column substochastic; 2) W_k is nonsingular if $\alpha_{i,k} \in \mathbb{J}_{q_i}$. As a result, W_k is the desirable dissipation matrix.

In what follows, the established vector dissipation inequality is converted into the desired H_∞-consensus performance criterion (3.11) via the following Theorem 3.1. In this procedure, the property of the dissipation matrix plays the leading role.

Theorem 3.1 *Given the scalar sequence $\alpha_{i,k} \in \mathbb{J}_{q_i}$, the initial weighted matrix U_i, the weighted matrix sequences $\tilde{R}_{i,k}$, $\tilde{Q}_{i,k}$, and $T_{i,k}$ defined in Definition 3.2, and the* *filter gain sequences $L_{i,k}$ and $K_{ij,k}$. The locally augmented system* (3.10) *satisfies the H_∞-consensus performance criterion* (3.11) *with the disturbance attenuation level $\gamma > 0$, if there exists a sequence of positive definite matrices $\{\mathscr{P}_{i,k}\}_{k\in\mathscr{H}\cup\{n\}}$ with the initial condition $\mathscr{P}_{i,0} \leq \gamma^2 U_i$ and the scalar sequence $\lambda_{i,k}$ satisfying* (3.15) *and* (3.16) *for all $k \in \mathscr{H}$, $i \in \mathscr{V}$.*

Proof First of all, left-multiplying by $\mathbf{1}^T$ on both sides of (3.13) yields

$$\mathbb{E}\left\{\mathbf{1}^T\mathbf{V}(\eta_{k+1})\right\} \leq \mathbb{E}\left\{\mathbf{1}^T\mathbf{S}(\eta_k, \xi_k)\right\} + \mathbb{E}\left\{\mathbf{1}^T W_k \mathbf{V}(\eta_k)\right\}.$$

Denoting $v(\eta_k) \triangleq \mathbf{1}^T\mathbf{V}(\eta_k)$, it follows from $S_i(\eta_k, \xi_{i,k})$ that

$$\mathbb{E}\{v(\eta_{k+1})\} - \mathbb{E}\{v(\eta_k)\} + N\Phi_{\mathscr{G}}(\eta_k) \leq \gamma^2 \sum_{i=1}^{N} \xi_{i,k}^T T_{i,k}\xi_{i,k}.$$

Next, applying the summation on the above inequality from 0 to $n-1$ leads to

$$\mathbb{E}\{v(\eta_n)\} - \mathbb{E}\{v(\eta_0)\} + \sum_{k=0}^{n-1} N\Phi_{\mathscr{G}}(\eta_k) \leq \gamma^2 \sum_{i=1}^{N}\sum_{k=0}^{n-1} \|\xi_{i,k}\|^2_{T_{i,k}},$$

which implies

$$\sum_{k=0}^{n-1} \Phi_{\mathscr{G}}(\eta_k) \leq \frac{1}{N}\sum_{i=1}^{N}\left(\eta_{i,0}^T\mathscr{P}_{i,0}\eta_{i,0} + \gamma^2 \sum_{k=0}^{n-1} \|\xi_{i,k}\|^2_{T_{i,k}}\right).$$

Finally, considering the initial condition $\mathscr{P}_{i,0} \leq \gamma^2 U_i$, the H_∞-consensus performance constraint (3.9) is satisfied, and thus the proof is complete from Definition 3.2.

Up to now, we have finished the performance analysis of the distributed filtering. In the following, the distributed filter gains $L_{i,k}$ and $K_{ij,k}$ are designed based on Theorem 3.1.

Theorem 3.2 *Given the scalar sequence $\alpha_{i,k} \in \mathfrak{I}_{q_i}$, the initial weighted matrices U_{1i} and U_{2i}, and the weighted matrix sequences $\tilde{R}_{i,k}$, $\tilde{Q}_{i,k}$, and $T_{i,k}$ defined in Definition 3.2, the locally augmented system* (3.10) *satisfies the H_∞-consensus performance constraint* (3.11) *with the disturbance attenuation level $\gamma > 0$, if there exist two matrix sequences $\{E_{ij,k}\}_{k\in\mathscr{H}}$ and $\{F_{i,k}\}_{k\in\mathscr{H}}$, the scalar sequence $\{\lambda_{i,k}\}_{k\in\mathscr{H}}$, and two positive definite matrix sequences $\{P^1_{i,k}\}_{k\in\mathscr{H}\cup\{n\}}$ and $\{P^2_{i,k}\}_{k\in\mathscr{H}\cup\{n\}}$ with the initial conditions $P^1_{i,0} \leq \gamma^2 U_{1i}$ and $P^2_{i,0} \leq \gamma^2 U_{2i}$, such that the following conditions for all $k \in \mathscr{H}$, $i \in \mathscr{V}$ hold:*

$$\begin{bmatrix} \bar{\Psi}_{i,k}^{11} & * & * & * & * \\ \Psi_{i,k}^{21} & \Psi_{i,k}^{22} & * & * & * \\ \bar{\Psi}_{i,k}^{31} & \bar{\Psi}_{i,k}^{32} & \mathscr{P}_{i,k+1} & * & * \\ 0 & 0 & \Psi_{i,k}^{43} & \gamma^2 T_{i,k} & * \\ \Psi_{i,k}^{51} & 0 & 0 & 0 & \mathscr{P}_{i,k+1} \end{bmatrix} > 0, \tag{3.18}$$

$$\begin{bmatrix} -\lambda_{i,k} & * \\ \mathscr{P}_{i,k+1}\bar{\pi}_{r,k} & -\mathscr{P}_{i,k+1} \end{bmatrix} \leq 0, \quad (r = 1, \ldots, s) \tag{3.19}$$

where

$$\begin{aligned}
\bar{\Psi}_{i,k}^{11} &= \theta_{i,k}\mathscr{P}_{i,k} - \lambda_{i,k}\sum_{r=1}^{s} \mathrm{diag}\{\Gamma_{r,k}, 0\} - p_i\tilde{R}_{i,k} - \tilde{Q}_{i,k}, \\
\bar{\Psi}_{i,k}^{31} &= \mathrm{diag}\left\{ P_{i,k+1}^1 A_k,\ P_{i,k+1}^2 A_k - \beta_i F_{i,k} C_{i,k} - \sum_{j\in\mathscr{N}_i} E_{ij,k} \right\}, \\
\bar{\Psi}_{i,k}^{32} &= [\mathscr{E}_{ij_1,k} \ \cdots \ \mathscr{E}_{ij_{p_i},k}], \quad \mathscr{E}_{ij,k} = \mathrm{diag}\{0, E_{ij,k}\}, \\
\Psi_{i,k}^{43} &= \begin{bmatrix} B_k^T P_{i,k+1}^1 & B_k^T P_{i,k+1}^2 \\ 0 & -D_{i,k}^T F_{i,k}^T \end{bmatrix}, \quad \Psi_{i,k}^{51} = \begin{bmatrix} 0 & 0 \\ \sigma_i F_{i,k} C_{i,k} & 0 \end{bmatrix}, \\
\sigma_i &= \sqrt{\beta_i(1-\beta_i)}, \quad \mathscr{P}_{i,k} = \mathrm{diag}\left\{ P_{i,k}^1, P_{i,k}^2 \right\}.
\end{aligned} \tag{3.20}$$

Moreover, the filter gains $L_{i,k}$ and $K_{ij,k}$ are calculated as follows:

$$L_{i,k} = (P_{i,k+1}^2)^{-1} F_{i,k}, \quad K_{ij,k} = (P_{i,k+1}^2)^{-1} E_{ij,k}. \tag{3.21}$$

Proof From Lemma 3.1 and Theorem 3.1, (3.15) holds if and only if the following inequality holds

$$\begin{bmatrix} \bar{\Psi}_{i,k}^{11} & * & * & * & * \\ \Psi_{i,k}^{21} & \Psi_{i,k}^{22} & * & * & * \\ \Psi_{i,k}^{31} & \Psi_{i,k}^{32} & \mathscr{P}_{i,k+1}^{-1} & * & * \\ 0 & 0 & \mathscr{B}_{i,k}^T & \gamma^2 T_{i,k} & * \\ \mathscr{C}_{i,k} & 0 & 0 & 0 & \mathscr{P}_{i,k+1}^{-1} \end{bmatrix} > 0. \tag{3.22}$$

Considering a partition with the following form $\mathscr{P}_{i,k} = \mathrm{diag}\left\{P_{i,k}^1, P_{i,k}^2\right\}$, perform the congruence transformation to (3.22) by $\mathrm{diag}\left\{I, I, \mathscr{P}_{i,k+1}, I, \mathscr{P}_{i,k+1}\right\}$ and denote that $E_{ij,k} = P_{i,k+1}^2 K_{ij,k}$, $F_{i,k} = P_{i,k+1}^2 L_{i,k}$, $\bar{\Psi}_{i,k}^{31} = \mathscr{P}_{i,k+1}\Psi_{i,k}^{31}$, $\bar{\Psi}_{i,k}^{32} = \mathscr{P}_{i,k+1}\Psi_{i,k}^{32}$, $\Psi_{i,k}^{43} = \mathscr{B}_{i,k}^T\mathscr{P}_{i,k+1}$, and $\Psi_{i,k}^{51} = \mathscr{P}_{i,k+1}\mathscr{C}_{i,k}$, then one obtains (3.18). Similarly, (3.16) holds if and only if the LMI (3.19) holds. Therefore, from Theorem 3.1 and Definition 3.1, the proof is complete.

Remark 3.5 From Theorem 3.2, the filter gains $L_{i,k}$ and $K_{ij,k}$ are dependent on $P_{i,k+1}^2$, $F_{i,k}$ and $E_{ij,k}$, which are the solutions of inequalities (3.18)-(3.19) at time instant k. Moreover, it is worth noting that the decision variables involve only $\mathscr{P}_{i,k+1}$, $E_{ij,k}$, $F_{i,k}$, and $\lambda_{i,k}$ at time k. The other information are known at time instant k, for instance, $\mathscr{P}_{j,k}$ ($j \in \mathscr{N}_i$) has been calculated at time instant $k-1$. Therefore, the proposed algorithm can be recursively implemented on every node i ($i \in \mathscr{V}$).

Remark 3.6 For distributed filtering algorithm, a fundamental concern is the issue of time complexity. In this chapter, the proposed scheme is presented for every node in terms of a set of RLMIs. As is known, the standard LMI system has a polynomial time complexity bounded by $O(\mathscr{M}\mathscr{N}^3\log(\mathscr{V}/\varepsilon))$, where $\mathscr{M}$ is the total row size of the LMI system, $\mathscr{N}$ is the total number of scalar decision variables, $\mathscr{V}$ is a data-dependent scaling factor, and ε is the relative accuracy set for the algorithm. It can be calculated from (3.18) and (3.19) that $\mathscr{M} = (8+2p_i)n_x + n_w + n_\vartheta + 1$ and $\mathscr{N} = (p_i+2)n_x^2 + n_x + n_x n_y + 1$ for node i. Over the time horizon $[0, n-1]$, the time complexity of the algorithm for node i is represented as $O(np_i^4 n_x^7 + np_i^3(n_w + n_\vartheta)n_x^6)$. Obviously, it is unrelated to the network size N, which shows the scalability of the scheme proposed in this chapter [12]. Considering the sparsity of sensor nodes, it can be further described as $O\left(nn_x^7 + n(n_w+n_\vartheta)n_x^6\right)$. In comparison with the time complexity $O(N^3 n_x^7 + N^4(n_w+n_\vartheta)n_x^6)$ in [6] and $O\left(nN^8 n_x^7\right)$ in [8], our scheme has prominent advantages.

3.3 An Illustrative Example

In this section, a simulation example is presented to illustrate the effectiveness of the proposed distributed filtering scheme. Consider an SN with 10 nodes, whose topology is represented by a digraph $\mathscr{G} = (\mathscr{V}, \mathscr{E}, \mathcal{A})$ with the set of nodes $\mathscr{V} = \{1, 2, 3, 4, 5, 6, 7, 8, 9, 10\}$, the set of edges $\mathscr{E} = \{(2, 1), (3, 2), (4, 3), (5, 4), (6, 5), (7, 6), (8, 7), (9, 8), (10, 9), (1, 10)\}$. Obviously, $p_i = q_i = 1, i = 1, 2, \ldots, 10$.

The concerned plant is of the form

$$
\begin{aligned}
x_{k+1} &= \begin{bmatrix} -0.87 & 0.4 \\ 0.15 & 0.98\sin(6k) \end{bmatrix} x_k + f_k(x_k, v_k) + \begin{bmatrix} 0.1 \\ 0.1 \end{bmatrix} 0.1\cos(5k), \\
y_{i,k} &= \zeta_{i,k}\begin{bmatrix} 0.3 & 0.2\sin(ik) \end{bmatrix} x_k + 0.1\cos(5k)
\end{aligned}
$$

where $\zeta_{i,k}$ obeying the Bernoulli distribution are independent and identically distributed for different i and k.

The stochastic nonlinear function $f_k(x_k, v_k)$ is chosen as

$$
f_k(x_k, v_k) = (0.1x_k^{(1)}v_k^{(1)} + 0.2x_k^{(2)}\sin(v_k^{(2)}))[0.06 \quad 0.09]^T
$$

where $x_k^{(i)}$ ($i = 1, 2$) denotes the i-th element of the system state, and $v_k^{(1)}$ and $v_k^{(2)}$ are two zero-mean, uncorrelated Gaussian white noise sequences with unity variances. It is easy to verify that

$$\mathbb{E}\{f_k(x_k, v_k|x_k)\} = 0,$$

$$\mathbb{E}\{f_k(x_k, v_k)f_k^T(x_k, v_k)|x_k\} = \begin{bmatrix} 0.06 \\ 0.09 \end{bmatrix} \begin{bmatrix} 0.06 \\ 0.09 \end{bmatrix}^T x_k^T \begin{bmatrix} 0.01 & 0 \\ 0 & 0.0173 \end{bmatrix} x_k.$$

The time horizon $n - 1$, probability β_i, H_∞ attenuation level γ, scalar sequence $\alpha_{i,k}$, the initial condition $\mathscr{P}_{i,0}$, the initial weighted matrix U_i and disturbance weighted matrix $T_{i,k}$ are given as $n = 30$, $\beta_i = 0.8$, $\gamma = 0.1$, $\alpha_{i,k} \equiv 0.9$, $\mathscr{P}_{i,0} = \text{diag}\{I, 5.1I\}$, $U_i = 100\mathscr{P}_{i,0}$, and $T_{i,k} \equiv \text{diag}\{1, 1\}$.

By means of the YALMIP toolbox in MATLAB [13], all the desired filter gains are recursively derived from Theorem 3.2. In addition, the trajectories of the local disagreement functions $\sum_{j\in\mathcal{N}_i} \|e_{j,k} - e_{i,k}\|^2$ and the filtering errors $e_{i,k}^{(1)} = x_k^{(1)} - \hat{x}_{i,k}^{(1)}$ are, respectively, shown in Figs. 3.1, 3.2, 3.3, 3.4, 3.5 and 3.6 under three Cases, that is, Case I: $Q_{i,k} \equiv 0$, $R_{i,k} \equiv I$; Case II: $Q_{i,k} \equiv I$, $R_{i,k} \equiv 0$; and Case III: $Q_{i,k} \equiv I$, $R_{i,k} \equiv I$.

Recalling Definition 3.1, the evaluated performance index provides an opportunity to observe both the consensus speed (showing by the attenuation speed of local disagreement functions) and the filtering accuracy for the three mentioned Cases. According to Figs. 3.1, 3.2, 3.3, 3.4, 3.5 and 3.6, one concludes that: (1) the attenuation speed of local disagreement functions in Case II is the fastest, and the list of these three Cases is Case II>Case III>Case I; (2) when achieving consensus (about

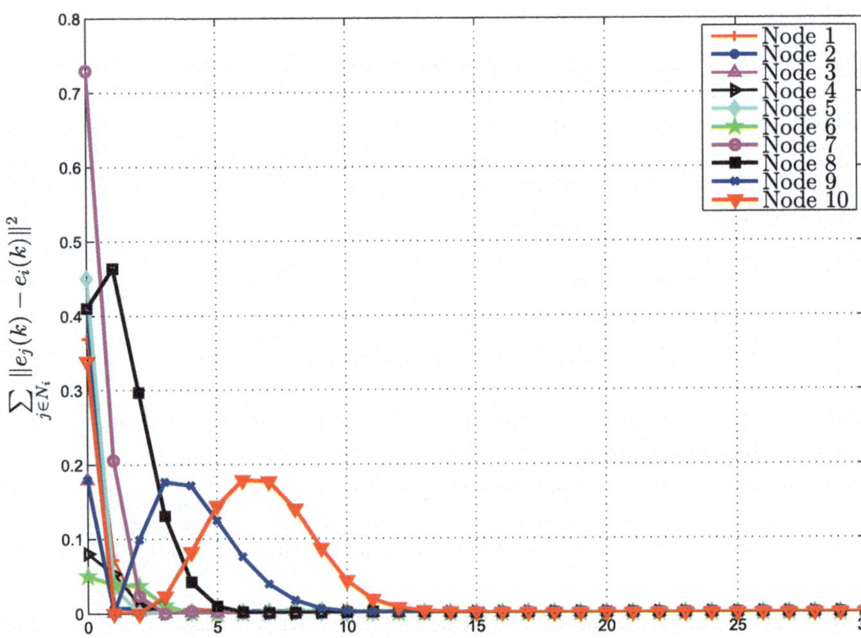

Fig. 3.1 The local disagreement functions (Case I)

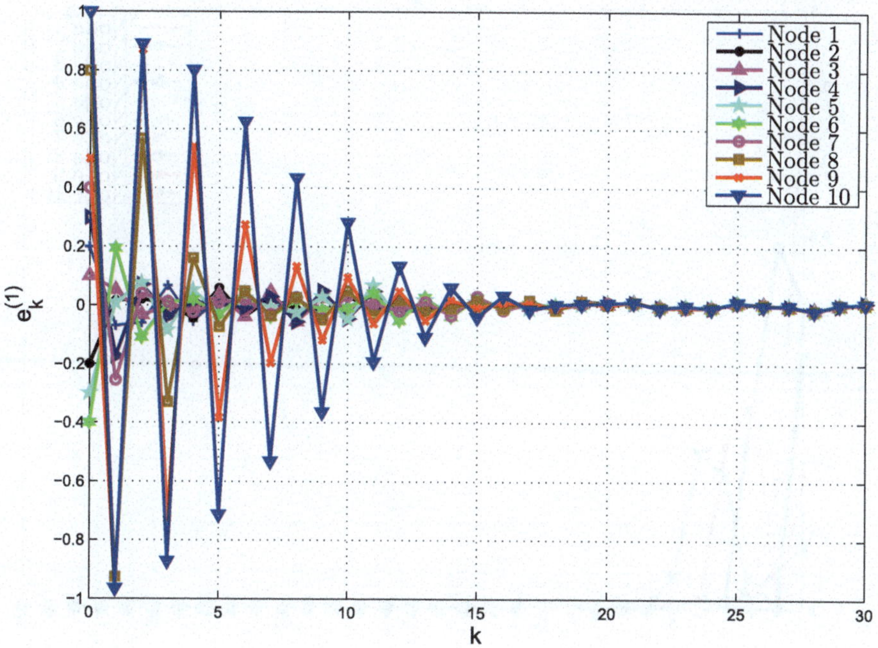

Fig. 3.2 The filtering errors $e_{i,k}^{(1)}$ (Case I)

after $k > 15$), the filtering errors of Case II is the biggest, at the same time their list is Case II>Case III>Case I. Obviously, it is impossible to obtain both the fastest consensus speed and the highest filtering accuracy, while Case III shows a compromise between them. In short, these simulation results reflect the important significance of taking two existing indices into a uniform framework. Also, this example verifies the effectiveness of developed distributed filtering approach.

3.4 Summary

In this chapter, we have investigated the distributed H_∞-consensus filtering problem for a class of discrete time-varying systems with stochastic nonlinearities and missing measurements observed by SN. The novel H_∞-consensus performance constraint has been proposed to simultaneously consider the requirements both the consensus and the filtering errors. By using the vector dissipativity theory, a set of sufficient conditions for the existence of the distributed filter has been established for each node such that the locally augmented dynamics satisfy the proposed H_∞-consensus performance constraint. Furthermore, the corresponding filter gains have been calculated on each node. As is well known, SN has only limited resources, for example,

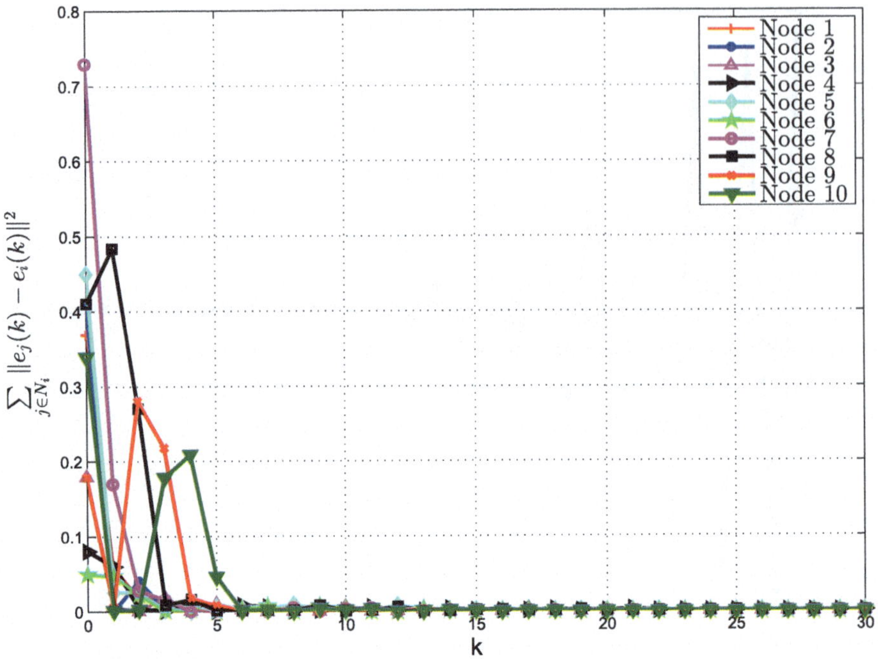

Fig. 3.3 The local disagreement functions (Case II)

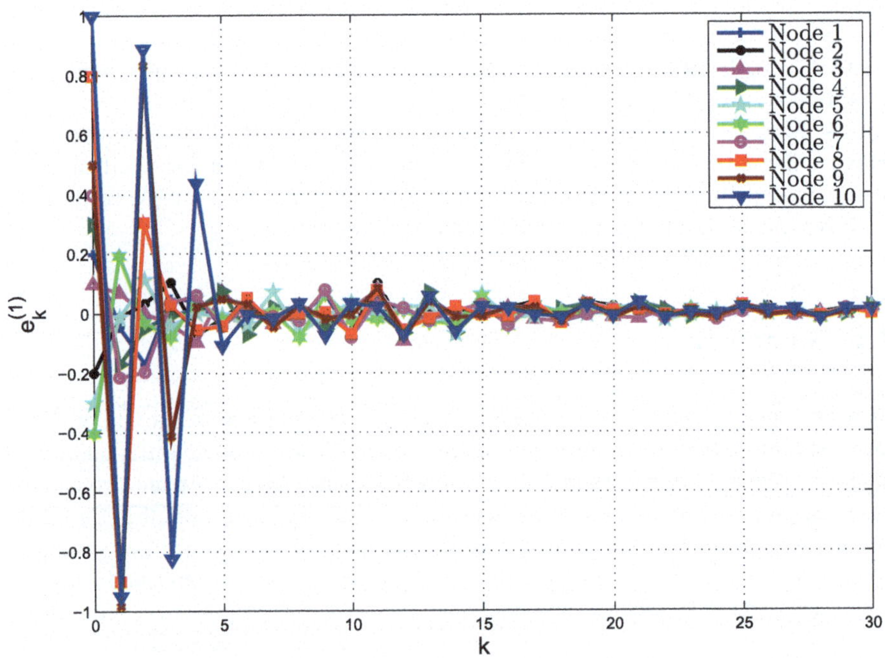

Fig. 3.4 The filtering errors $e_{i,k}^{(1)}$ (Case II)

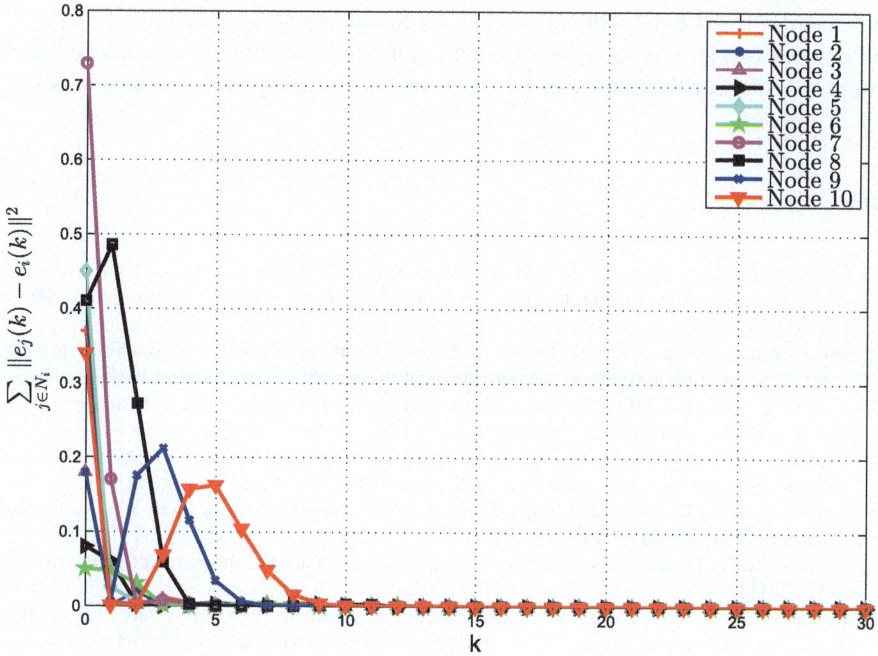

Fig. 3.5 The local disagreement functions (Case III)

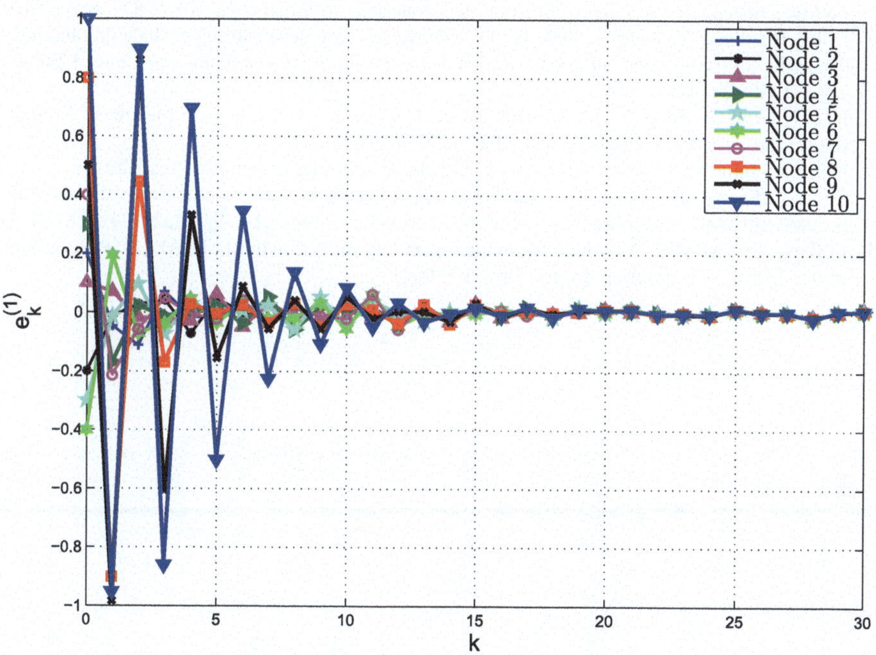

Fig. 3.6 The filtering errors $e_{i,k}^{(1)}$ (Case III)

the limited network bandwidth of wireless communication channel, limited energy of sensor node and etc. As such, how to reduce energy consumption and data collision would become a critical problem for the distributed filtering in the context of SN.

References

1. Ding, D., Wang, Z., Dong, H., Shu, H.: Distributed H_∞ state estimation with stochastic parameters and nonlinearities through sensor networks: the finite-horizon case. Automatica **48**(8), 1575–1585 (2012)
2. Hu, J., Wang, Z., Gao, H., Stergioulas, L.: Extended Kalman filtering with stochastic nonlinearities and multiple missing measurements. Automatica **48**(9), 2007–2015 (2012)
3. Yaz, Y., Yaz, E.: On LMI formulations of some problems arising in nonlinear stochastic system analysis. IEEE Trans. Autom. Control **44**(4), 813–816 (1999)
4. Jacobson, D.: A general result in stochastic optimal control of nonlinear discrete-time systems with quadratic performance criteria. J. Math. Anal. Appl. **47**(1), 153–161 (1974)
5. Ugrinovskii, V.: Distributed robust estimation over random switching networks using H_∞ consensus. Automatica **49**(1), 160–168 (2013)
6. Ugrinovskii, V.: Distributed robust filtering with H_∞ consensus of estimates. Automatica **47**(1), 1–13 (2011)
7. Ugrinovskii, V.: Gain-scheduled synchronization of parameter varying systems via relative H_∞ consensus with application to synchronization of uncertain bilinear systems. Automatica **50**(11), 2880–2887 (2014)
8. Shen, B., Wang, Z., Hung, Y.: Distributed H_∞-consensus filtering in sensor networks with multiple missing measurements: the finite-horizon case. Automatica **46**(10), 1682–1688 (2010)
9. Dong, H., Wang, Z., Lam, J., Gao, H.: Distributed filtering in sensor networks with randomly occurring saturations and successive packet dropouts. Int. J. Robust Nonlinear Control **24**(12), 1743–1759 (2014)
10. Han, F., Wei, G., Song, Y., Li, W.: Distributed H_∞-consensus filtering for piecewise discrete-time linear systems. J. Franklin Inst. **352**(5), 2029–2046 (2015)
11. Horn, R., Johnson, C.: Matrix Analysis, 2nd edn. Cambridge University Press (2013)
12. Yu, H., Zhuang, Y., Wang, W.: Distributed H_∞ filtering in sensor networks with randomly occurred missing measurements and communication link failures. Inf. Sci. **222**, 424–438 (2013)
13. Löfberg, J.: YALMIP: A toolbox for modeling and optimization in MATLAB. In: Proceedings of the CACSD Conference, Taipei, Taiwan (2004)

Chapter 4
Distributed Filtering for Random Parameter System with Event-Triggering Protocols

In the recent years, the event-triggering protocol has been proven to be an efficient strategy to save the precious resources i.e., network bandwidth and node energy. Within the framework of local performance analysis method, it is desirable to develop the distributed filtering with event-triggering protocol.

Chapter 3 has established the local performance analysis method and it is utilized to deal with the distributed filtering problem with stochastic nonlinearities and missing measurements. Along the same line, this chapter investigates the distributed H_∞-consensus filtering problem for a class of discrete time-varying linear systems with random parameters and event-triggering protocol. An event-triggering protocol for each node is employed to reduce the burden of the network communication. A novel matrix called information matrix is proposed to describe the complicated correlations among the elements of random matrix. By utilizing the information matrix, a weighted-covariance formula of random matrix is presented to analyze the system with random parameters. With the aid of the constructed dissipation matrix and a vector of supplied rate functions, a set of locally coupled conditions is obtained for each node to make the filtering error dynamics satisfy the stochastic vector dissipativity-like over the finite horizon. Meanwhile, the obtained sufficient conditions also solve the distributed H_∞-consensus filtering problem. Notably, the designed filtering algorithm can be implemented on each node to calculate the distributed filter gains. Finally, the effectiveness and applicability of the proposed algorithm are illustrated by a numerical simulation.

4.1 Problem Formulation

Consider a class of discrete time-varying linear systems defined on $k \in \mathscr{K} \triangleq \{0, 1, \ldots, n-1\}$:

F. Han et al., *Distributed Filtering, Control and Synchronization*,
Studies in Systems, Decision and Control 428,
https://doi.org/10.1007/978-3-030-97075-8_4

$$\begin{cases} x_{k+1} = A_k x_k + B_k w_k, \\ z_k = E_k x_k, \\ x(0) = x_0 \end{cases} \tag{4.1}$$

with N sensors modeled by

$$y_{i,k} = C_{i,k} x_k + D_{i,k} \vartheta_{i,k}, \quad i = 1, 2, \ldots, N \tag{4.2}$$

where $x_k \in \mathbb{R}^{n_x}$ is the state, $y_{i,k} \in \mathbb{R}^{n_y}$ is measurement from sensor node i, $w_k \in \mathbb{R}^{n_w}$ and $\vartheta_{i,k} \in \mathbb{R}^{n_{\vartheta_i}}$ are the external disturbance belonging to $l_2[0, n-1]$. $B_k \in \mathbb{R}^{n_x \times n_w}$, $D_{i,k} \in \mathbb{R}^{n_y \times n_{\vartheta_i}}$, and $E_k \in \mathbb{R}^{n_x \times n_w}$ are all known time-varying matrices.

Additionally, the time-varying parameter matrices A_k and $C_{i,k}$ are random and mutually independent. Without loss of generality, assume the random matrices A_k and $C_{i,k}$ have the following structures:

$$A_k = \bar{A}_k + \tilde{A}_k, \quad C_{i,k} = \bar{C}_{i,k} + \tilde{C}_{i,k}, \tag{4.3}$$

where $\mathbb{E}\{A_k\}=\bar{A}_k$, $\mathbb{E}\{C_{i,k}\}=\bar{C}_{i,k}$, $\mathbb{E}\{\tilde{A}_k\} = \mathbb{E}\{[\tilde{a}_k^{st}]\} = 0$, and $\mathbb{E}\{\tilde{C}_{i,k}\} = \mathbb{E}\{[\tilde{c}_{i,k}^{st}]\} = 0$. Furthermore, for the purpose of simplicity, denote the matrices composed of all covariance elements in $\tilde{A}_k$ and $\tilde{C}_{i,k}$ as $\mathcal{A}_k = [\mathbb{E}\{\tilde{a}_k^{st} \tilde{a}_k^{lm}\}] \triangleq [\hat{a}_k^{stlm}] \in \mathbb{R}^{n_x^2 \times n_x^2}$ and $\mathcal{C}_{i,k} = [\mathbb{E}\{\tilde{c}_{i,k}^{st} \tilde{c}_{i,k}^{lm}\}] \triangleq [\hat{c}_{i,k}^{stlm}] \in \mathbb{R}^{n_x n_y \times n_x n_y}$. These two new matrices are called the information matrices of $\tilde{A}_k$ and $\tilde{C}_{i,k}$, respectively. Also, according to the subscripts s and l, the information matrices $\mathcal{A}_k$ and $\mathcal{C}_{i,k}$ can be partitioned into matrix blocks, that is, $\hat{a}_k^{stlm}$ and $\hat{c}_{i,k}^{stlm}$, which stand for the tm-th component of the matrix blocks $\mathcal{A}_k^{sl}$ and $\mathcal{C}_{i,k}^{sl}$, respectively.

In this chapter, one constructs the following linear distributed filter:

$$\begin{cases} \hat{x}_{i,k+1} = \bar{A}_k \hat{x}_{i,k} + L_{i,k}(y_{i,k} - \bar{C}_{i,k} \hat{x}_{i,k}) + \sum\limits_{j \in \mathcal{N}_i} K_{ij,k}(\hat{x}_{j,k} - \hat{x}_{i,k}), \\ \hat{z}_{i,k} = E_k \hat{x}_{i,k} \end{cases} \tag{4.4}$$

where $\hat{x}_{i,k} \in \mathbb{R}^{n_x}$ is the estimated state on node i, $L_{i,k}$ and $K_{ij,k}$ are the filter gains to be calculated later.

From (4.4), the neighboring estimation $\hat{x}_{j,k}$ ($j \in \mathcal{N}_i$) is transmitted to the filter of node i. As is known to all, information transmission consumes a lot of energy and communication resource. Especially, these consumptions are non-negligible due to the large number of nodes in sensor network. Meanwhile, many transmissions are not needed in consideration of improving significantly the filtering performance. Thus, for the purpose of reducing unnecessary data transmissions among neighboring nodes, an event generator is constructed to determine whether the newly estimation will be sent to its neighboring nodes or not. In this chapter, such a relative event generator function $f(\cdot, \cdot)$ is defined for node $j \in \mathcal{N}_i$ as follows:

$$f(\sigma_{j,k},\delta_j) \triangleq \sigma_{j,k}^T \Omega_j \sigma_{j,k} - \delta_j \hat{x}_{j,k}^T \Omega_j \hat{x}_{j,k} \tag{4.5}$$

where Ω_j is a symmetric positive-definite weighted matrix, $\delta_j \in [0, 1)$ is a threshold parameter on node j, and $\sigma_{j,k} \triangleq \hat{x}_{j,s^j_{k_p}} - \hat{x}_{j,k}$ with $\hat{x}_{j,s^j_{k_p}}$ and $\hat{x}_{j,k}$ being the estimations of node j at latest event-triggered time $s^j_{k_p}$ and the current time k, respectively.

The transmission of the estimation $\hat{x}_{j,k}$ to node i is triggered as long as the following condition is satisfied:

$$f(\sigma_{j,k},\delta_j) > 0. \tag{4.6}$$

Based on this condition, the sequence of event-triggered instants of node j: $0 \leq s^j_{k_0} \leq s^j_{k_1} \leq \cdots \leq s^j_{k_p} \leq \cdots$ can be determined iteratively by

$$s^j_{k_{p+1}} \triangleq \inf\{s^j_k \in \mathbb{N} | s^j_k > s^j_{k_p}, f(\sigma_{j,k},\delta_j) > 0\}. \tag{4.7}$$

Remark 4.1 Different from the time-driven manner in the traditional distributed filtering problems, in this chapter, the event-triggering protocol is adopted to reduce the data communication frequency and network bandwidth occupations. With the event-triggering protocol applied here, some unnecessary transmissions could be avoided when the change rate of the estimations is relatively small. Obviously, the set of event-triggering instants is only a subset of the time sequences, i.e., $s^j_E \triangleq \{s^j_{k_0}, s^j_{k_1}, \ldots\} \subseteq s^j_S \triangleq \{0, 1, 2, \ldots\}$. Note that, when $\delta_j = 0$, all the estimations would be transmitted at every time instant, and the problem addressed here reduces to the traditional distributed filtering one.

To this end, the event-triggering-based estimator-type consensus filter (4.4) can be rewritten as follows:

$$\begin{cases} \hat{x}_{i,k+1} = \bar{A}_k \hat{x}_{i,k} + L_{i,k}(y_{i,k} - \bar{C}_{i,k}\hat{x}_{i,k}) + \sum\limits_{j\in\mathscr{N}_i} K_{ij,k}(\hat{x}_{j,s^j_{k_p}} - \hat{x}_{i,k}), \\ \hat{z}_i(k) = E_k \hat{x}_i(k). \end{cases} \tag{4.8}$$

For convenience of later analysis, letting $\eta_{i,k} \triangleq [x_k^T \quad \hat{x}_{i,k}^T]^T$, $\xi_{i,k} \triangleq [w_k^T \quad \vartheta_{i,k}^T]^T$, and $e_{i,k} \triangleq z_k - \hat{z}_{i,k}$, then one has

$$\begin{cases} \eta_{i,k+1} = \mathscr{A}_{i,k}\eta_{i,k} + \mathscr{B}_{i,k}\xi_{i,k} + \mathscr{L}_{i,k}\tilde{\mathscr{A}}_{i,k}\eta_{i,k} + \sum\limits_{j\in\mathscr{N}_i} \mathscr{K}_{ij,k}\eta_{j,k} + \sum\limits_{j\in\mathscr{N}_i} \tilde{\mathscr{K}}_{ij,k}\sigma_{j,k}, \\ e_{i,k} = \mathscr{E}_{i,k}\eta_{i,k} \end{cases} \tag{4.9}$$

where

$$\mathscr{A}_{i,k} = \begin{bmatrix} \bar{A}_k & 0 \\ L_{i,k}\bar{C}_{i,k} & \bar{A}_k - L_{i,k}\bar{C}_{i,k} - \sum_{j\in\mathscr{N}_i} K_{ij,k} \end{bmatrix},$$

$$\mathscr{E}_{i,k} = \begin{bmatrix} E_k^T \\ -E_k^T \end{bmatrix}^T, \quad \mathscr{L}_{i,k} = \begin{bmatrix} I & 0 \\ 0 & L_{i,k} \end{bmatrix}, \quad \mathscr{K}_{ij,k} = \begin{bmatrix} 0 & 0 \\ 0 & K_{ij,k} \end{bmatrix},$$

$$\mathscr{B}_{i,k} = \begin{bmatrix} B_k & 0 \\ 0 & L_{i,k}D_{i,k} \end{bmatrix}, \quad \tilde{\mathscr{K}}_{ij,k} = \begin{bmatrix} 0 \\ K_{ij,k} \end{bmatrix}, \quad \tilde{\mathscr{A}}_{i,k} = \begin{bmatrix} \tilde{A}_k & 0 \\ \tilde{C}_{i,k} & 0 \end{bmatrix}.$$

Accordingly, the distributed event-triggering communication condition (4.5) can be rewritten as

$$f(\sigma_{j,k}, \delta_j) = \sigma_{j,k}^T \Omega_j \sigma_{j,k} - \delta_j \eta_{j,k}^T \bar{\Omega}_j \eta_{j,k} \le 0 \tag{4.10}$$

where

$$\bar{\Omega}_j \triangleq \begin{bmatrix} 0 & 0 \\ 0 & \Omega_j \end{bmatrix}.$$

The distributed H_∞-consensus filtering problem under consideration in this chapter is to determine the filter gains $L_{i,k}$ and $K_{ij,k}$ such that the filtering error dynamics (4.9) subject to (4.10) satisfy the H_∞-consensus performance constraints over the finite horizon.

Definition 4.1 Given a disturbance attention level $\gamma > 0$, some families of initial weighted matrices $S_i \triangleq \text{diag}\{S_{1i}, S_{2i}\} > 0$, $T_{i,k} \triangleq \text{diag}\{T_{1i,k}, T_{2i,k}\} > 0$, $R_{i,k} > 0$ and $Q_{i,k} > 0$. The filtering error system (4.9) subject to (4.10) is said to satisfy the H_∞-consensus performance constraint over the finite horizon if the following inequality holds

$$\sum_{k=0}^{n-1} \mathbb{E}\{\Phi_{\mathscr{G}}(e_k)\} \le \frac{1}{N} \sum_{i=1}^{N} \left(\eta_{i,0}^T S_i \eta_{i,0} + \gamma^2 \sum_{k=0}^{n-1} \|\xi_{i,k}\|_{T_{i,k}}^2 \right) \tag{4.11}$$

where $\Phi_{\mathscr{G}}(e_k) \triangleq \frac{1}{N} \sum_{i=1}^{N} \left[\sum_{j\in\mathscr{N}_i} \|e_{i,k} - e_{j,k}\|_{Q_{i,k}}^2 + \|e_{i,k}\|_{R_{i,k}}^2 \right]$.

In this chapter, the aim is to design a distributed filter of the form (4.4) on each node i for the system (4.9) subject to (4.10). In other words, we are going to design the filter gains $L_{i,k}$ and $K_{ij,k}$ for each node such that the filtering error dynamics $e_{i,k}$ (4.9) subject to (4.10) satisfies the H_∞-consensus performance constraint (4.11) over the finite horizon $\mathscr{K}$.

4.2 Main Results

In this section, for given filter gains $L_{i,k}$ and $K_{ij,k}$, a set of locally coupled sufficient conditions is constructed for each node to guarantee the filtering error dynamics satisfies the H_∞-consensus performance constraints with a prescribed attenuation level γ. For this purpose, the basic idea comes from the local performance analysis method developed in Chap. 3. Namely, we devote oneself to seek some locally coupled conditions in the process of establishing the vector dissipativity and finally to solve the H_∞-consensus filtering problem. As a result, the first step is to present the conception of the vector dissipativity over the finite horizon for the filtering error dynamics (4.9) subject to (4.10).

Definition 4.2 The discrete time-varying dynamics (4.9) subject to (4.10) is stochastic vector dissipative-like over the finite horizon with respect to a vector of supply rate functions $\mathbf{S}(\eta_k, \xi_k) \triangleq [S_1^T(\eta_k, \xi_{1,k}) \quad \cdots \quad S_N^T(\eta_k, \xi_{N,k})]^T$, if there exists a vector of nonnegative definite storage functions $\mathbf{V}(\eta_k) \triangleq [V_1^T(\eta_{1,k}) \quad \cdots \quad V_N^T(\eta_{N,k})]^T$ with $V(0) = 0$, and a sequence of column substochastic nonsingular dissipation matrices $W_k \in \mathbb{R}^{N\times N}$ such that the following vector dissipation inequality is satisfied for any $k \in \mathscr{K}$:

$$\mathbb{E}\{\mathbf{V}(\eta_{k+1})\} \leq\leq W_k \mathbb{E}\{\mathbf{V}(\eta_k)\} + \mathbb{E}\{\mathbf{S}(\eta_k, \xi_k)\}. \tag{4.12}$$

In the forthcoming local performance analysis, the weighted-covariance matrices $\mathbb{E}\{\tilde{A}_{i,k}^T W \tilde{A}_{i,k}\}$ and $\mathbb{E}\{\tilde{C}_{i,k}^T R \tilde{C}_{i,k}\}$ need to be handled, where $W \in \mathbb{R}^{n_x\times n_x}$ and $R \in \mathbb{R}^{n_y\times n_y}$ are two deterministic matrices. For this problem, it is critical point to characterize the weighted-covariance matrices $\mathbb{E}\{\tilde{A}_{i,k}^T W \tilde{A}_{i,k}\}$ and $\mathbb{E}\{\tilde{C}_{i,k}^T R \tilde{C}_{i,k}\}$ by utilizing the known information matrices $\mathcal{A}_{i,k}$ and $\mathcal{C}_{i,k}$. In the following, via the Kronecker product and the Hadamard product, a formula is established to present the relationship between the information matrix and the weighted-covariance matrix.

Proposition 4.1 *For given two deterministic weighted matrices* $W \in \mathbb{R}^{n_x\times n_x} > 0$ *and* $R \in \mathbb{R}^{n_y\times n_y} > 0$, *one has*

$$\mathbb{E}\{\tilde{A}_{i,k}^T W \tilde{A}_{i,k}\} = \mathbf{I}_{n_x^2\times n_x}^T (W \otimes \mathbf{1}_{n_x\times n_x} \circ \mathcal{A}_{i,k}) \mathbf{I}_{n_x^2\times n_x}, \tag{4.13}$$

$$\mathbb{E}\{\tilde{C}_{i,k}^T R \tilde{C}_{i,k}\} = \mathbf{I}_{n_x n_y\times n_x}^T (R \otimes \mathbf{1}_{n_x\times n_x} \circ \mathcal{C}_{i,k}) \mathbf{I}_{n_x n_y\times n_x} \tag{4.14}$$

where $\mathbf{I}_{n_x^2\times n_x} \triangleq \mathbf{1}_{n_x\times 1} \otimes I_{n_x}$ *and* $\mathbf{I}_{n_x n_y\times n_x} \triangleq \mathbf{1}_{n_y\times 1} \otimes I_{n_x}$.

Proof First of all, denote

$$\begin{bmatrix} \tilde{a}_{i,k}^1 \\ \tilde{a}_{i,k}^2 \\ \vdots \\ \tilde{a}_{i,k}^{n_x} \end{bmatrix} \triangleq \begin{bmatrix} \tilde{a}_{i,k}^{11} & \tilde{a}_{i,k}^{12} & \cdots & \tilde{a}_{i,k}^{1n_x} \\ \tilde{a}_{i,k}^{21} & \tilde{a}_{i,k}^{22} & \cdots & \tilde{a}_{i,k}^{2n_x} \\ \vdots & \vdots & \ddots & \vdots \\ \tilde{a}_{i,k}^{n_x1} & \tilde{a}_{i,k}^{n_x2} & \cdots & \tilde{a}_{i,k}^{n_xn_x} \end{bmatrix} = \tilde{A}_{i,k}. \tag{4.15}$$

Next, calculate the weighted-covariance matrix from (4.15)

$$
\begin{aligned}
&\mathbb{E}\{\tilde{A}_{i,k}^T W \tilde{A}_{i,k}\}\\
&= \mathbb{E}\left\{ \begin{bmatrix} \tilde{a}_{i,k}^1 \\ \tilde{a}_{i,k}^2 \\ \vdots \\ \tilde{a}_{i,k}^{n_x} \end{bmatrix}^T \begin{bmatrix} w_{11} & w_{12} & \cdots & w_{1n_x} \\ w_{21} & w_{22} & \cdots & w_{2n_x} \\ \vdots & \vdots & \ddots & \vdots \\ w_{n_x1} & w_{n_x2} & \cdots & w_{n_xn_x} \end{bmatrix} \begin{bmatrix} \tilde{a}_{i,k}^1 \\ \tilde{a}_{i,k}^2 \\ \vdots \\ \tilde{a}_{i,k}^{n_x} \end{bmatrix} \right\}\\
&= \mathbb{E}\{w_{11}(\tilde{a}_{i,k}^1)^T \tilde{a}_{i,k}^1 + w_{12}(\tilde{a}_{i,k}^1)^T \tilde{a}_{i,k}^2 + \cdots + w_{n_xn_x}(\tilde{a}_{i,k}^{n_x})^T \tilde{a}_{i,k}^{n_x}\}\\
&= \sum_{s=1}^{n_x}\sum_{l=1}^{n_x} w_{sl}\mathbb{E}\{(\tilde{a}_{i,k}^s)^T \tilde{a}_{i,k}^l\}\\
&= \sum_{s=1}^{n_x}\sum_{l=1}^{n_x} w_{sl}\mathbb{E}\left\{ \begin{bmatrix} \tilde{a}_{i,k}^{s1}\tilde{a}_{i,k}^{l1} & \tilde{a}_{i,k}^{s1}\tilde{a}_{i,k}^{l2} & \cdots & \tilde{a}_{i,k}^{s1}\tilde{a}_{i,k}^{ln_x} \\ \tilde{a}_{i,k}^{s2}\tilde{a}_{i,k}^{l1} & \tilde{a}_{i,k}^{s2}\tilde{a}_{i,k}^{l2} & \cdots & \tilde{a}_{i,k}^{s2}\tilde{a}_{i,k}^{ln_x} \\ \vdots & \vdots & \ddots & \vdots \\ \tilde{a}_{i,k}^{sn_x}\tilde{a}_{i,k}^{l1} & \tilde{a}_{i,k}^{sn_x}\tilde{a}_{i,k}^{l2} & \cdots & \tilde{a}_{i,k}^{sn_x}\tilde{a}_{i,k}^{ln_x} \end{bmatrix} \right\}\\
&= \sum_{s=1}^{n_x}\sum_{l=1}^{n_x} w_{sl} \begin{bmatrix} \hat{a}_{i,k}^{s1l1} & \hat{a}_{i,k}^{s1l2} & \cdots & \hat{a}_{i,k}^{s1ln_x} \\ \hat{a}_{i,k}^{s2l1} & \hat{a}_{i,k}^{s2l2} & \cdots & \hat{a}_{i,k}^{s2ln_x} \\ \vdots & \vdots & \ddots & \vdots \\ \hat{a}_{i,k}^{sn_xl1} & \hat{a}_{i,k}^{sn_xl2} & \cdots & \hat{a}_{i,k}^{sn_xln_x} \end{bmatrix}\\
&= \sum_{s=1}^{n_x}\sum_{l=1}^{n_x} w_{sl}\mathcal{A}_{i,k}^{sl}\\
&= \mathbf{I}_{n_x^2\times n_x}^T (W \otimes \mathbf{1}_{n_x\times n_x} \circ \mathcal{A}_{i,k})\mathbf{I}_{n_x^2\times n_x}.
\end{aligned}
$$

The proof of (4.14) is similar and thus it is omitted. Therefore, the proof is complete.

Remark 4.2 Proposition 4.1 provides a method to calculate the weighted-covariance matrix of a random matrix. Actually, formulas (4.13) and (4.14) are quite general, which involve some special cases. For example, 1) if random matrix is diagonal, then formula (4.13) can be reduced to that of Lemma 1 in [1]; 2) if random matrix $\tilde{A}_{i,k} = f(\mu)\hat{A}_{i,k}$, where $f(\cdot)$ is a real valued function, $\mu \in \mathbb{R}$ is a random variable with $\mathbb{E}\{f^2(\mu)\} = \bar{\mu}$ and $\hat{A}_{i,k}$ is a deterministic matrix, then one has $\mathbb{E}\{\tilde{A}_{i,k}^T W \tilde{A}_{i,k}\} = \bar{\mu}\hat{A}_{i,k}^T W \hat{A}_{i,k}$; and 3) if random matrix $\tilde{A}_{i,k}$ is only independent in different rows, that is, $\mathbb{E}\{(\tilde{a}_{i,k}^s)^T \tilde{a}_{i,k}^s\} = \mathcal{A}_{i,k}^{ss} \neq 0$ and $\mathbb{E}\{(\tilde{a}_{i,k}^s)^T \tilde{a}_{i,k}^l\} = \mathcal{A}_{i,k}^{sl} = 0$ when $s \neq l$, then $\mathbb{E}\{\tilde{A}_{i,k}^T W \tilde{A}_{i,k}\} = \mathbf{I}_{n_x^2\times n_x}^T (\sum_{s=1}^{n_x} H_i^T W H_i \mathcal{A}_{i,k}^{ss})\mathbf{I}_{n_x^2\times n_x}$, where $H_i = [\underbrace{0 \ \dots \ 0}_{i-1} \ 1 \ \underbrace{0 \ \dots \ 0}_{n_x-i}]^T$.

Now, we are ready to establish the vector dissipativity for the locally augmented filtering dynamics (4.9). For the convenience of the following analysis, define an interval of out-degree q_i as

$$\mathbb{J}_{q_i} \triangleq \begin{cases} (0, \frac{1+q_i}{2q_i}), & \text{if } q_i \neq 0; \\ (0, 1], & \text{if } q_i = 0. \end{cases} \tag{4.16}$$

Lemma 4.1 *Given the prescribed disturbance attenuation level $\gamma > 0$, a scalar sequence $\alpha_{i,k} \in \mathbb{J}_{q_i}$, an initial weighted matrix $S_i > 0$, two sequences of weighted matrix $Q_{i,k}$ and $T_{i,k} > 0$, filter gain sequences $L_{i,k}$ and $K_{ij,k}$. The locally augmented dynamics (4.9) subject to (4.10) is stochastic vector dissipative-like over the finite horizon regarding the vector supply rate function $\mathbf{S}(\eta_k, \xi_k) \triangleq \text{col}(S_1(\eta_k, \xi_{1,k}), \ldots, S_N(\eta_k, \xi_{N,k}))$ with*

$$S_i(\eta_k, \xi_{i,k}) \triangleq \gamma^2 \xi_{i,k}^T T_{i,k} \xi_{i,k} - \sum_{j \in \mathcal{N}_i} \|\mathscr{E}_{i,k}\eta_{i,k} - \mathscr{E}_{j,k}\eta_{j,k}\|_{Q_{i,k}}^2 - \|\mathscr{E}_{i,k}\eta_{i,k}\|_{R_{i,k}}^2, \tag{4.17}$$

the vector storage function $\mathbf{V}(\eta_k) \triangleq \text{col}(V_1(\eta_{1,k}), \ldots, V_N(\eta_{N,k}))$ with $V_i(\eta_{i,k}) = \eta_{i,k}^T P_{i,k} \eta_{i,k}$, and the dissipation matrix sequence W_k with

$$W_k = \begin{bmatrix} \frac{1+q_1(1-\alpha_{1,k})}{1+q_1} & \frac{a_{12}\alpha_{2,k}}{1+q_2} & \cdots & \frac{a_{1N}\alpha_{N,k}}{1+q_N} \\ \frac{a_{21}\alpha_{1,k}}{1+q_1} & \frac{1+q_2(1-\alpha_{2,k})}{1+q_2} & \cdots & \frac{a_{2N}\alpha_{N,k}}{1+q_N} \\ \vdots & \vdots & \ddots & \vdots \\ \frac{a_{N1}\alpha_{N,k}}{1+q_1} & \frac{a_{N2}\alpha_{2,k}}{1+q_2} & \cdots & \frac{1+q_N(1-\alpha_{N,k})}{1+q_N} \end{bmatrix}, \tag{4.18}$$

if there exists a set of positive definite matrices $\{P_{i,k}\}_{k \in \mathscr{K} \cup \{n\}} > 0$ for any $i \in \mathscr{V}$, $k \in \mathscr{K}$ such that the following conditions hold

$$\Psi_{i,k} = \begin{bmatrix} \Psi_{i,k}^{11} & * & * & * \\ \Psi_{i,k}^{21} & \Psi_{i,k}^{22} & * & * \\ 0 & 0 & \Psi_{i,k}^{33} & * \\ \mathscr{A}_{i,k} & \Psi_{i,k}^{42} & \Psi_{i,k}^{43} & \Psi_{i,k}^{44} \end{bmatrix} < 0, \tag{4.19}$$

$$\mathscr{L}_{i,k}^T P_{i,k+1} \mathscr{L}_{i,k} \leq \mathscr{F}_{i,k} \tag{4.20}$$

where

$$\begin{aligned} \Psi_{i,k}^{11} &= -\frac{1+q_i(1-\alpha_{i,k})}{1+q_i} P_{i,k} + \mathbf{I}_{2n_x(n_x+n_y)\times 2n_x}^T (\mathscr{F}_{i,k} \otimes \mathbf{1}_{2n_x \times 2n_x} \circ \mathcal{A}_{i,k}) \mathbf{I}_{2n_x(n_x+n_y)\times 2n_x}, \\ &\quad + \mathscr{E}_{i,k}^T (p_i Q_{i,k} + R_{i,k}) \mathscr{E}_{i,k}, \quad \Psi_{i,k}^{21} = [-\mathscr{E}_{i,k}^T Q_{i,k} \mathscr{E}_{j_{i_1},k} \quad \cdots \quad -\mathscr{E}_{i,k}^T Q_{i,k} \mathscr{E}_{j_{p_i},k}]^T, \\ \Psi_{i,k}^{22} &= \text{diag}\{\delta_{j_{i_1}} \bar{\Omega}_{j_{i_1}} - \frac{\alpha_{j_{i_1},k}}{1+q_{j_{i_1}}} P_{j_{i_1},k} + \mathscr{E}_{j_{i_1},k}^T Q_{i,k} \mathscr{E}_{j_{i_1},k}, \ \ldots, \ \delta_{j_{i p_i}} \bar{\Omega}_{j_{i p_i}} - \frac{\alpha_{j_{i p_i},k}}{1+q_{j_{i p_i}}} P_{j_{i p_i},k} \\ &\quad + \mathscr{E}_{j_{i p_i},k}^T Q_{i,k} \mathscr{E}_{j_{i p_i},k}\}, \\ \Psi_{i,k}^{42} &= [\mathscr{K}_{i j_1,k} \quad \cdots \quad \mathscr{K}_{i j_{p_i},k}], \quad \Psi_{i,k}^{43} = [\tilde{\mathscr{K}}_{i j_1,k} \quad \cdots \quad \tilde{\mathscr{K}}_{i j_{p_i},k}], \\ \Psi_{i,k}^{33} &= \text{diag}\{-\Omega_{j_{i_1}}, \ldots, -\Omega_{j_{i p_i}}\}, \quad \Psi_{i,k}^{44} = -P_{i,k+1}^{-1} + \frac{1}{\gamma^2} \mathscr{B}_{i,k} T_{i,k}^{-1} \mathscr{B}_{i,k}^T. \end{aligned} \tag{4.21}$$

Proof According to the vector dissipativity theory, the key point of this proof is to verify the vector dissipation inequality (4.12). First of all, calculating the conditional expectation of $V_i(\eta_{i,k+1})$ along the trajectory of system (4.9) given $\eta_{i,k}$ and $\eta_{j,k}$, $j \in \mathscr{N}_i$, one has

$$\begin{aligned}
&\mathbb{E}\{V_i(\eta_{i,k+1})|\eta_{i,k}, \eta_{j,k}, j \in \mathscr{N}_i\} \\
&= \mathbb{E}\Bigg\{\Bigg[\mathscr{A}_{i,k}\eta_{i,k} + \mathscr{B}_{i,k}\xi_{i,k} + \mathscr{L}_{i,k}\tilde{\mathscr{A}}_{i,k}\eta_{i,k} + \sum_{j\in\mathscr{N}_i}\mathscr{K}_{ij,k}\eta_{j,k} + \sum_{j\in\mathscr{N}_i}\tilde{\mathscr{K}}_{ij,k}\sigma_{j,k}\Bigg]^T P_{i,k+1} \\
&\quad \times \Bigg[\mathscr{A}_{i,k}\eta_{i,k} + \mathscr{B}_{i,k}\xi_{i,k} + \mathscr{L}_{i,k}\tilde{\mathscr{A}}_{i,k}\eta_{i,k} + \sum_{j\in\mathscr{N}_i}\mathscr{K}_{ij,k}\eta_{j,k} + \sum_{j\in\mathscr{N}_i}\tilde{\mathscr{K}}_{ij,k}\sigma_{j,k}\Bigg] |\eta_{i,k}, \eta_{j,k}, j \in \mathscr{N}_i\Bigg\} \\
&\quad - \gamma^2\xi_{i,k}^T T_{i,k}\xi_{i,k} + \gamma^2\xi_{i,k}^T T_{i,k}\xi_{i,k} \\
&= \eta_{i,k}^T\mathscr{A}_{i,k}^T P_{i,k+1}\mathscr{A}_{i,k}\eta_{i,k} + \mathbb{E}\{\eta_{i,k}^T\tilde{\mathscr{A}}_{i,k}^T\mathscr{L}_{i,k}^T P_{i,k+1}\mathscr{L}_{i,k}\tilde{\mathscr{A}}_{i,k}\eta_{i,k}\} \\
&\quad + 2\eta_{i,k}^T\mathscr{A}_{i,k}^T P_{i,k+1}\sum_{j\in\mathscr{N}_i}\mathscr{K}_{ij,k}\eta_{j,k} + 2\eta_{i,k}^T\mathscr{A}_{i,k}^T P_{i,k+1}\sum_{j\in\mathscr{N}_i}\tilde{\mathscr{K}}_{ij,k}\sigma_{j,k} \\
&\quad + \sum_{j\in\mathscr{N}_i}\eta_{j,k}^T\mathscr{K}_{ij,k}^T P_{i,k+1}\sum_{j\in\mathscr{N}_i}\mathscr{K}_{ij,k}\eta_{j,k} + \sum_{j\in\mathscr{N}_i}\sigma_{j,k}^T\tilde{\mathscr{K}}_{ij,k}^T P_{i,k+1}\sum_{j\in\mathscr{N}_i}\tilde{\mathscr{K}}_{ij,k}\sigma_{j,k} \\
&\quad + \xi_{i,k}^T\left(\mathscr{B}_{i,k}^T P_{i,k+1}\mathscr{B}_{i,k} - \gamma^2 T_{i,k}\right)\xi_{i,k} + 2\xi_{i,k}^T\mathscr{B}_{i,k}^T P_{i,k+1}\sum_{j\in\mathscr{N}_i}\mathscr{K}_{ij,k}\eta_{j,k} \\
&\quad + 2\xi_{i,k}^T\mathscr{B}_{i,k}^T P_{i,k+1}\sum_{j\in\mathscr{N}_i}\tilde{K}_{ij,k}\sigma_{j,k} + 2\xi_{i,k}^T\mathscr{B}_{i,k}P_{i,k+1}\mathscr{A}_{i,k}\eta_{i,k} \\
&\quad + 2\sum_{j\in\mathscr{N}_i}\eta_{j,k}^T\mathscr{K}_{ij,k}^T P_{i,k+1}\sum_{j\in\mathscr{N}_i}\tilde{\mathscr{K}}_{ij,k}\sigma_{j,k} + \gamma^2\xi_{i,k}^T T_{i,k}\xi_{i,k}.
\end{aligned}$$

From Proposition 4.1 and (4.20), compute the weighted-covariance matrix of $\tilde{\mathscr{A}}_{i,k}$ as follows:

$$\begin{aligned}
&\mathbb{E}\left\{\eta_{i,k}^T\tilde{\mathscr{A}}_{i,k}^T\mathscr{L}_{i,k}^T P_{i,k+1}\mathscr{L}_{i,k}\tilde{\mathscr{A}}_{i,k}\eta_{i,k}\right\} \\
&= \eta_{i,k}^T\mathbf{I}_{2n_x(n_x+n_y)\times 2n_x}^T(\mathscr{L}_{i,k}^T P_{i,k+1}\mathscr{L}_{i,k}\otimes\mathbf{1}_{2n_x\times 2n_x}\circ A_{i,k})\mathbf{I}_{2n_x(n_x+n_y)\times 2n_x}\eta_{i,k} \\
&\le \eta_{i,k}^T\mathbf{I}_{2n_x(n_x+n_y)\times 2n_x}^T(\mathscr{F}_{i,k}\otimes\mathbf{1}_{2n_x\times 2n_x}\circ A_{i,k})\mathbf{I}_{2n_x(n_x+n_y)\times 2n_x}\eta_{i,k}.
\end{aligned} \tag{4.22}$$

It follows from (4.19) that $\Psi_{i,k}^{44} = -P_{i,k+1}^{-1} + \frac{1}{\gamma^2}\mathscr{B}_{i,k}T_{i,k}^{-1}\mathscr{B}_{i,k}^T < 0$. In term of the Schur Complement Lemma, one has $\Phi_{i,k} \triangleq \gamma^2 T_{i,k} - \mathscr{B}_{i,k}^T P_{i,k+1}\mathscr{B}_{i,k} > 0$. Thus, via the completing-square technique, the disturbance under the worst case is given as

$$\begin{aligned}
\xi_{i,k}^* &\triangleq \Phi_{i,k}^{-1}\mathscr{B}_{i,k}^T P_{i,k+1}\sum_{j\in\mathscr{N}_i}\mathscr{K}_{ij,k}\eta_{j,k} + \Phi_{i,k}^{-1}\mathscr{B}_{i,k}^T P_{i,k+1}\sum_{j\in\mathscr{N}_i}\tilde{\mathscr{K}}_{ij,k}\sigma_{j,k} \\
&\quad + \Phi_{i,k}^{-1}\mathscr{B}_{i,k}^T P_{i,k+1}\mathscr{A}_{i,k}\eta_{i,k}.
\end{aligned}$$

Then, it obtains

$$\begin{aligned}
&\xi_{i,k}^T\left(\mathscr{B}_{i,k}^T P_{i,k+1}\mathscr{B}_{i,k}-\gamma^2 T_{i,k}\right)\xi_{i,k}+2\xi_{i,k}^T\mathscr{B}_{i,k}^T P_{i,k+1}\sum_{j\in\mathscr{N}_i}\mathscr{K}_{ij,k}\eta_{j,k}\\
&+2\xi_{i,k}^T\mathscr{B}_{i,k}^T P_{i,k+1}\sum_{j\in\mathscr{N}_i}\tilde{\mathscr{K}}_{ij,k}\sigma_{j,k}+2\xi_{i,k}^T\mathscr{B}_{i,k}^T P_{i,k+1}\mathscr{A}_{i,k}\eta_{i,k}\\
\le&\sum_{j\in\mathscr{N}_i}\eta_{j,k}^T\mathscr{K}_{ij,k}^T P_{i,k+1}\mathscr{B}_{i,k}\Phi_{i,k}^{-1}\mathscr{B}_{i,k}^T P_{i,k+1}\sum_{j\in\mathscr{N}_i}\mathscr{K}_{ij,k}\eta_{j,k}\\
&+\eta_{i,k}^T\mathscr{A}_{i,k}^T P_{i,k+1}\mathscr{B}_{i,k}\Phi_{i,k}^{-1}\mathscr{B}_{i,k}^T P_{i,k+1}\mathscr{A}_{i,k}\eta_{i,k}\\
&+\sum_{j\in\mathscr{N}_i}\sigma_{j,k}^T\tilde{\mathscr{K}}_{ij,k}^T P_{i,k+1}\mathscr{B}_{i,k}\Phi_{i,k}^{-1}\mathscr{B}_{i,k}^T P_{i,k+1}\sum_{j\in\mathscr{N}_i}\tilde{\mathscr{K}}_{ij,k}\sigma_{j,k}\\
&+2\sum_{j\in\mathscr{N}_i}\eta_{j,k}^T\mathscr{K}_{ij,k}^T P_{i,k+1}\mathscr{B}_{i,k}\Phi_{i,k}^{-1}\mathscr{B}_{i,k}^T P_{i,k+1}\sum_{j\in\mathscr{N}_i}\tilde{\mathscr{K}}_{ij,k}\sigma_{j,k}\\
&+2\eta_{i,k}^T\mathscr{A}_{i,k}^T P_{i,k+1}\mathscr{B}_{i,k}\Phi_{i,k}^{-1}\mathscr{B}_{i,k}^T P_{i,k+1}\sum_{j\in\mathscr{N}_i}\mathscr{K}_{ij,k}\eta_{j,k}\\
&+2\eta_{i,k}^T\mathscr{A}_{i,k}^T P_{i,k+1}\mathscr{B}_{i,k}\Phi_{i,k}^{-1}\mathscr{B}_{i,k}^T P_{i,k+1}\sum_{j\in\mathscr{N}_i}\tilde{\mathscr{K}}_{ij,k}\sigma_{j,k}.
\end{aligned}\tag{4.23}$$

By means of the matrix inverse formula:

$$P_{i,k+1}\mathscr{B}_{i,k}\Phi_{i,k}^{-1}\mathscr{B}_{i,k}^T P_{i,k+1}+P_{i,k+1}=(P_{i,k+1}^{-1}-\frac{1}{\gamma^2}\mathscr{B}_{i,k}T_{i,k}^{-1}\mathscr{B}_{i,k}^T)^{-1},$$

one obtains

$$\begin{aligned}
&\eta_{i,k}^T\mathscr{A}_{i,k}^T P_{i,k+1}\mathscr{B}_{i,k}\Phi_{i,k}^{-1}\mathscr{B}_{i,k}^T P_{i,k+1}\mathscr{A}_{i,k}\eta_{i,k}+\eta_{i,k}^T\mathscr{A}_{i,k}^T P_{i,k+1}\mathscr{A}_{i,k}\eta_{i,k}\\
&+2\eta_{i,k}^T\mathscr{A}_{i,k}^T P_{i,k+1}\mathscr{B}_{i,k}\Phi_{i,k}^{-1}\mathscr{B}_{i,k}^T P_{i,k+1}\sum_{j\in\mathscr{N}_i}\mathscr{K}_{ij,k}\eta_{j,k}\\
&+2\eta_{i,k}^T\mathscr{A}_{i,k}^T P_{i,k+1}\sum_{j\in\mathscr{N}_i}\mathscr{K}_{ij,k}\eta_{j,k}\\
&+2\eta_{i,k}^T\mathscr{A}_{i,k}^T P_{i,k+1}\mathscr{B}_{i,k}\Phi_{i,k}^{-1}\mathscr{B}_{i,k}^T P_{i,k+1}\sum_{j\in\mathscr{N}_i}\tilde{\mathscr{K}}_{ij,k}\sigma_{j,k}\\
&+2\eta_{i,k}^T\mathscr{A}_{i,k}^T P_{i,k+1}\sum_{j\in\mathscr{N}_i}\tilde{\mathscr{K}}_{ij,k}\sigma_{j,k}\\
&+\sum_{j\in\mathscr{N}_i}\sigma_{j,k}^T\tilde{\mathscr{K}}_{ij,k}^T P_{i,k+1}\mathscr{B}_{i,k}\Phi_{i,k}^{-1}\mathscr{B}_{i,k}^T P_{i,k+1}\sum_{j\in\mathscr{N}_i}\tilde{\mathscr{K}}_{ij,k}\sigma_{j,k}\\
&+\sum_{j\in\mathscr{N}_i}\sigma_{j,k}^T\tilde{\mathscr{K}}_{ij,k}^T P_{i,k+1}\sum_{j\in\mathscr{N}_i}\tilde{\mathscr{K}}_{ij,k}\sigma_{j,k}
\end{aligned}$$

$$
\begin{aligned}
&+2\sum_{j\in\mathcal{N}_i}\eta_{j,k}^T\mathcal{K}_{ij,k}^T P_{i,k+1}\mathcal{B}_{i,k}\Phi_{i,k}^{-1}\mathcal{B}_{i,k}^T P_{i,k+1}\sum_{j\in\mathcal{N}_i}\tilde{\mathcal{K}}_{ij,k}\sigma_{j,k}\\
&+2\sum_{j\in\mathcal{N}_i}\eta_{j,k}^T\mathcal{K}_{ij,k}^T P_{i,k+1}\sum_{j\in\mathcal{N}_i}\tilde{\mathcal{K}}_{ij,k}\sigma_{j,k}\\
&+\sum_{j\in\mathcal{N}_i}\eta_{j,k}^T\mathcal{K}_{ij,k}^T P_{i,k+1}\mathcal{B}_{i,k}\Phi_{i,k}^{-1}\mathcal{B}_{i,k}^T P_{i,k+1}\sum_{j\in\mathcal{N}_i}\mathcal{K}_{ij,k}\eta_{j,k}\\
&+\sum_{j\in\mathcal{N}_i}\eta_{j,k}^T\mathcal{K}_{ij,k}^T P_{i,k+1}\sum_{j\in\mathcal{N}_i}\mathcal{K}_{ij,k}\eta_{j,k}\\
=&\,\eta_{i,k}^T\mathcal{A}_{i,k}^T(P_{i,k+1}^{-1}-\frac{1}{\gamma^2}\mathcal{B}_{i,k}T_{i,k}^{-1}\mathcal{B}_{i,k}^T)^{-1}\mathcal{A}_{i,k}\eta_{i,k}\\
&+\sum_{j\in\mathcal{N}_i}\eta_{j,k}^T\mathcal{K}_{ij,k}^T(P_{i,k+1}^{-1}-\frac{1}{\gamma^2}\mathcal{B}_{i,k}T_{i,k}^{-1}\mathcal{B}_{i,k}^T)^{-1}\sum_{j\in\mathcal{N}_i}\mathcal{K}_{ij,k}\eta_{j,k}\\
&+\sum_{j\in\mathcal{N}_i}\sigma_{j,k}^T\tilde{\mathcal{K}}_{ij,k}^T(P_{i,k+1}^{-1}-\frac{1}{\gamma^2}\mathcal{B}_{i,k}T_{i,k}^{-1}\mathcal{B}_{i,k}^T)^{-1}\sum_{j\in\mathcal{N}_i}\tilde{\mathcal{K}}_{ij,k}\sigma_{j,k}\\
&+2\sum_{j\in\mathcal{N}_i}\eta_{j,k}^T\mathcal{K}_{ij,k}^T(P_{i,k+1}^{-1}-\frac{1}{\gamma^2}\mathcal{B}_{i,k}T_{i,k}^{-1}\mathcal{B}_{i,k}^T)^{-1}\sum_{j\in\mathcal{N}_i}\tilde{\mathcal{K}}_{ij,k}\sigma_{j,k}\\
&+2\sum_{j\in\mathcal{N}_i}\eta_{j,k}^T\mathcal{K}_{ij,k}^T(P_{i,k+1}^{-1}-\frac{1}{\gamma^2}\mathcal{B}_{i,k}T_{i,k}^{-1}\mathcal{B}_{i,k}^T)^{-1}\mathcal{A}_{i,k}\eta_{j,k}\\
&+2\sum_{j\in\mathcal{N}_i}\sigma_{j,k}^T\mathcal{K}_{ij,k}^T(P_{i,k+1}^{-1}-\frac{1}{\gamma^2}\mathcal{B}_{i,k}T_{i,k}^{-1}\mathcal{B}_{i,k}^T)^{-1}\mathcal{A}_{i,k}\eta_{j,k}\\
\triangleq&\,\Upsilon_{i,k}.
\end{aligned}
\tag{4.24}
$$

Furthermore, from (4.6), it follows that $\delta_j\eta_{j,k}^T\bar{\Omega}_j\eta_{j,k}-\sigma_{j,k}^T\Omega_j\sigma_{j,k}\geq 0,\ j\in\mathcal{N}_i$. Thus, it obtains

$$
\begin{aligned}
&\mathbb{E}\{V_i(\eta_{i,k+1})|\eta_{i,k},\eta_{j,k},\ j\in\mathcal{N}_i\}\\
\leq&\Upsilon_{i,k}+\sum_{j\in\mathcal{N}_i}(\delta_j\eta_{j,k}^T\bar{\Omega}_j\eta_{j,k}-\sigma_{j,k}^T\Omega_j\sigma_{j,k})\\
&+S_i(\eta_k,\xi_{i,k})+\sum_{j\in\mathcal{N}_i}\|\mathcal{E}_{i,k}\eta_{i,k}-\mathcal{E}_{j,k}\eta_{j,k}\|_{Q_{i,k}}^2+\|\mathcal{E}_{i,k}\eta_{i,k}\|_{R_{i,k}}^2\\
&+\eta_{i,k}^T\mathbf{I}_{2n_x(n_x+n_y)\times 2n_x}^T(\mathcal{F}_{i,k}\otimes\mathbf{1}_{2n_x\times 2n_x}\circ\mathcal{A}_{i,k})\mathbf{I}_{2n_x(n_x+n_y)\times 2n_x}\eta_{i,k}.
\end{aligned}
\tag{4.25}
$$

Denoting $\xi_{i,k}\triangleq[\eta_{i,k}^T\ \ \eta_{\mathcal{N}_{i,k}}^T\ \ \sigma_{\mathcal{N}_{i,k}}^T]^T$, $\eta_{\mathcal{N}_{i,k}}\triangleq[\eta_{j_{i_1},k}^T\ \ \cdots\ \ \eta_{j_{ip_i},k}^T]^T$, $\sigma_{\mathcal{N}_{i,k}}\triangleq[\sigma_{j_{i_1},k}^T\ \ \cdots\ \ \sigma_{j_{ip_i},k}^T]^T$, it then follows from (4.19) that

$$\xi_{i,k}^T \left\{ \begin{bmatrix} \Psi_{i,k}^{11} & * & * \\ \Psi_{i,k}^{21} & \Psi_{i,k}^{22} & * \\ 0 & 0 & \Psi_{i,k}^{33} \end{bmatrix} - \begin{bmatrix} \mathscr{A}_{i,k}^T \\ (\Psi_{i,k}^{42})^T \\ (\Psi_{i,k}^{43})^T \end{bmatrix} (\Psi_{i,k}^{44})^{-1} \begin{bmatrix} \mathscr{A}_{i,k}^T \\ (\Psi_{i,k}^{42})^T \\ (\Psi_{i,k}^{43})^T \end{bmatrix}^T \right\} \xi_{i,k} < 0. \quad (4.26)$$

Or equivalently, one has

$$\begin{aligned} & \Upsilon_{i,k} + \sum_{j \in \mathscr{N}_i} (\delta_j \eta_{j,k}^T \bar{\Omega}_j \eta_{j,k} - \sigma_{j,k}^T \Omega_j \sigma_{j,k}) \\ & + \sum_{j \in \mathscr{N}_i} \|\mathscr{E}_{i,k}\eta_{i,k} - \mathscr{E}_{j,k}\eta_{j,k}\|_{Q_{i,k}}^2 + \|\mathscr{E}_{i,k}\eta_{i,k}\|_{R_{i,k}}^2 \\ & + \eta_{i,k}^T \mathbf{I}_{2n_x(n_x+n_y)\times 2n_x}^T (\mathscr{F}_{i,k} \otimes \mathbf{1}_{2n_x\times 2n_x} \circ \mathcal{A}_{i,k}) \mathbf{I}_{2n_x(n_x+n_y)\times 2n_x} \eta_{i,k} \\ < & \sum_{j \in \mathscr{N}_i} \frac{\alpha_{j,k}}{1+q_j} \eta_{j,k}^T P_{j,k} \eta_{j,k} + \frac{1+q_i(1-\alpha_{i,k})}{1+q_i} \eta_{i,k}^T P_{i,k} \eta_{i,k} \end{aligned} \quad (4.27)$$

where $\frac{\alpha_{j,k}}{1+q_j} > 0 (j \in \mathscr{N}_i)$ indicates that node i always has close relationships with its neighboring nodes.

By means of (4.27), one next gets

$$\begin{aligned} & \mathbb{E}\{V_i(\eta_{i,k+1}) | \eta_{i,k}, \eta_{j,k}, j \in \mathscr{N}_i\} \\ \leq & S_i(\eta_k, \xi_{i,k}) + \sum_{j \in \mathscr{N}_i} \frac{\alpha_{j,k}}{1+q_j} \eta_{j,k}^T P_{j,k} \eta_{j,k} + \frac{1+q_i(1-\alpha_{i,k})}{1+q_i} \eta_{i,k}^T P_{i,k} \eta_{i,k} \\ = & S_i(\eta_k, \xi_{i,k}) + \begin{bmatrix} \frac{a_{i1}\alpha_{1,k}}{1+q_1} & \cdots & \frac{1+q_i(1-\alpha_{i,k})}{1+q_i} & \cdots & \frac{a_{iN}\alpha_{N,k}}{1+q_N} \end{bmatrix} V(\eta_k). \end{aligned} \quad (4.28)$$

From (4.28), matrix W_k of the form (4.18) can be constructed. Obviously, W_k is strictly diagonal dominant due to $\alpha_{i,k} \in \mathfrak{I}_{q_i}$, and W_k is also a substochastic matrix. Hence, W_k is exactly the dissipation matrix.

As a result, via the dissipation matrix W_k, (4.28) becomes

$$\mathbb{E}\{V_i(\eta_{i,k+1}) | \eta_{i,k}, \eta_{j,k}, j \in \mathscr{N}_i\} \leq [W_k \mathbf{V}(\eta_k)]_i + S_i(\eta_k, \xi_{i,k}) \quad (4.29)$$

where $[W_k \mathbf{V}(\eta_k)]_i$ is the i-th row of the vector $W_k \mathbf{V}(\eta_k)$. Furthermore, by means of the properties of the conditional expectation, one obtains

$$\mathbb{E}\{V_i(\eta_{i,k+1})\} \leq \mathbb{E}\{[W_k \mathbf{V}(\eta_k)]_i\} + \mathbb{E}\{S_i(\eta_k, \xi_{i,k})\}. \quad (4.30)$$

Since (4.30) holds for arbitrary node $i \in \mathscr{V}$, it can be rewritten as a vector form by augmenting the inequalities of all nodes:

$$\mathbb{E}\{\mathbf{V}(\eta_{k+1})\} \leq\leq \mathbb{E}\{W_k \mathbf{V}(\eta_k)\} + \mathbb{E}\{\mathbf{S}(\eta_k, \xi_k)\}. \quad (4.31)$$

From Definition 4.2, the proof is therefore complete.

Theorem 4.1 *Given the prescribed disturbance attenuation level* $\gamma > 0$*, the scalar sequence* $\alpha_{i,k} \in \mathfrak{I}_{q_i}$*, the initial weighted matrix* $S_i > 0$*, two weighted matrix sequences* $Q_{i,k}$ *and* $T_{i,k} > 0$ *and filter gains* $L_{i,k}$ *and* $K_{ij,k}$*, the local augmented dynamics* (4.9) *subject to* (4.10) *satisfy the* H_∞*-consensus performance constraint* (4.11)*, if there exists a family of positive definite matrices* $\{P_{i,k}\}_{k\in\mathscr{K}\cup\{n\}} > 0$ *with the initial condition* $P_{i,0} \le \gamma^2 S_i$ *such that both* (4.19) *and* (4.20) *hold.*

Proof First and foremost, it is pre-multiplied by $\mathbf{1}^T$ on both sides of the vector dissipation inequality (4.31), then one has a scalar form

$$\mathbf{1}^T\mathbb{E}\{\mathbf{V}(\eta_{k+1})\} \le \mathbf{1}^T\mathbb{E}\{W_k\mathbf{V}(\eta_k)\} + \mathbf{1}^T\mathbb{E}\{\mathbf{S}(\eta_k, \xi_k)\}.$$

Denoting $v(\eta_k) \triangleq \mathbf{1}^T\mathbb{E}\{\mathbf{V}(\eta_k)\}$, it then follows from (4.17) and $\mathbf{1}^T W_k = \mathbf{1}^T$ that

$$v(\eta_{k+1}) - v(\eta_k) + N\mathbb{E}\{\Phi_{\mathscr{G}}(\eta_k)\} \le \sum_{i=1}^{N} \gamma^2 \xi_{i,k}^T T_{i,k} \xi_{i,k}, \tag{4.32}$$

which implies

$$v(\eta_n) - v(\eta_0) + \sum_{k=0}^{n-1} N\mathbb{E}\{\Phi_{\mathscr{G}}(\eta_k)\} \le \sum_{k=0}^{n-1}\sum_{i=1}^{N} \gamma^2 \xi_{i,k}^T T_{i,k} \xi_{i,k}. \tag{4.33}$$

That is,

$$\sum_{k=0}^{n-1} \mathbb{E}\{\Phi_{\mathscr{G}}(\eta_k)\} \le \frac{1}{N}\sum_{i=1}^{N}(\eta_{i,0}^T P_{i,0}\eta_{i,0} + \gamma^2 \|\xi_i\|_{T_i}^2).$$

Finally, considering the initial condition $P_{i,0} \le \gamma^2 S_i$, the H_∞-consensus performance constraint (4.11) is satisfied. The proof is thus complete.

Now, the performance analysis of the filtering error dynamics has been completed. A set of locally coupled conditions has been found to guarantee that the filtering error dynamics satisfies the proposed H_∞-consensus performance constraint. Next, we begin to provide a method to design the filter gains $L_{i,k}$ and $K_{ij,k}$, which is to be presented in the following theorem.

Theorem 4.2 *Given the prescribed disturbance attenuation level* $\gamma > 0$*, two weighted matrix sequences* $Q_{i,k}$ *and* $T_{i,k} > 0$*, two initial weighted matrices* $S_{1i} > 0$ *and* $S_{2i} > 0$*, the scalar sequence* $\alpha_{i,k} \in \mathfrak{I}_{q_i}$*. The finite-horizon distributed* H_∞*-consensus filtering problem is solvable if there exist two matrix sequences* $E_{ij,k}$ *and* $F_{i,k}$*, positive definite matrix sequence* $\mathscr{F}_{i,k}^2 > 0$ *and two positive definite matrix sequences* $\{P_{i,k}^1\}_{k\in\mathscr{K}\cup\{n\}}$ *and* $\{P_{i,k}^2\}_{k\in\mathscr{K}\cup\{n\}}$ *with the initial condition* $P_{i,0}^1 \le \gamma^2 S_{1i}$ *and* $P_{i,0}^2 \le \gamma^2 S_{2i}$*, such that the following conditions hold*

$$\begin{bmatrix} \Psi_{i,k}^{11} & * & * & * & * \\ \Psi_{i,k}^{21} & \Psi_{i,k}^{22} & * & * & * \\ 0 & 0 & \Psi_{i,k}^{33} & * & * \\ \Psi_{i,k}^{41} & \bar{\Psi}_{i,k}^{42} & \bar{\Psi}_{i,k}^{43} & -P_{i,k+1} & * \\ 0 & 0 & 0 & \Psi_{i,k}^{54} & -\gamma^2 T_{i,k} \end{bmatrix} < 0, \tag{4.34}$$

$$\begin{bmatrix} -\mathscr{F}_{i,k}^2 & * \\ F_{i,k} & -P_{i,k+1}^2 \end{bmatrix} \le 0 \tag{4.35}$$

where

$$P_{i,k} = \begin{bmatrix} P_{i,k}^1 & 0 \\ 0 & P_{i,k}^2 \end{bmatrix}, \quad \Psi_{i,k}^{41} = \begin{bmatrix} P_{i,k+1}^1 \bar{A}_k & 0 \\ F_{i,k}\bar{C}_{i,k} & P_{i,k+1}^2 \bar{A}_k - F_{i,k}\bar{C}_{i,k} - \sum_{j\in\mathcal{N}_i} E_{ij,k} \end{bmatrix},$$
$$\bar{\Psi}_{i,k}^{42} = \begin{bmatrix} 0 & 0 & \cdots 0 & 0 \\ 0 & E_{ij_1,k} & \cdots 0 & E_{ij_{p_i},k} \end{bmatrix}, \quad \Psi_{i,k}^{43} = \begin{bmatrix} 0 & \cdots & 0 \\ E_{ij_1,k} & \cdots & E_{ij_{p_i},k} \end{bmatrix},$$
$$\Psi_{i,k}^{54} = \begin{bmatrix} B_k^T P_{i,k+1}^1 & 0 \\ 0 & D_{i,k}^T F_{i,k}^T \end{bmatrix}, \quad P_{i,k+1}\mathscr{L}_{i,k} = \begin{bmatrix} P_{i,k+1}^1 & 0 \\ 0 & F_{i,k} \end{bmatrix}. \tag{4.36}$$

Moreover, the filter gains can be computed as follows:

$$L_{i,k} = (P_{i,k+1}^2)^{-1} F_{i,k}, \quad K_{ij,k} = (P_{i,k+1}^2)^{-1} E_{ij,k}.$$

Proof From the Schur Complement Lemma, (4.19) holds if and only if the following inequality holds

$$\begin{bmatrix} \Psi_{i,k}^{11} & * & * & * & * \\ \Psi_{i,k}^{21} & \Psi_{i,k}^{22} & * & * & * \\ 0 & 0 & \Psi_{i,k}^{33} & * & * \\ \mathscr{A}_{i,k} & \Psi_{i,k}^{42} & \Psi_{i,k}^{43} & -P_{i,k+1}^{-1} & * \\ 0 & 0 & 0 & \mathscr{B}_{i,k}^T & -\gamma^2 T_{i,k} \end{bmatrix} < 0, \tag{4.37}$$

and then pre- and post-multiply both sides of (4.37) by diag$\{I, I, I, P_{i,k+1}, I\}$. Denoting $\Psi_{i,k}^{41} = P_{i,k+1}\mathscr{A}_{i,k}$, $\bar{\Psi}_{i,k}^{42} = P_{i,k+1}\Psi_{i,k}^{42}$, $\bar{\Psi}_{43,k} = P_{i,k+1}\Psi_{i,k}^{43}$, $\Psi_{i,k}^{54} = \mathscr{B}_{i,k}^T P_{i,k+1}$, and $F_{i,k} = P_{i,k+1}^2 L_{i,k}$, $E_{ij,k} = P_{i,k+1}^2 K_{ij,k}$, we can obtain (4.34).

Next, from (4.20), one has

$$\mathscr{L}_{i,k}^T P_{i,k+1}\mathscr{L}_{i,k} \le \mathscr{F}_{i,k} \Leftrightarrow \begin{bmatrix} P_{i,k+1}^1 & * \\ 0 & L_{i,k}^T P_{i,k+1}^2 L_{i,k} \end{bmatrix} \le \mathscr{F}_{i,k}. \tag{4.38}$$

Then, $\mathscr{F}_{i,k}$ takes the form as

$$\mathscr{F}_{i,k} = \begin{bmatrix} P_{i,k+1}^1 & * \\ 0 & \mathscr{F}_{i,k}^2 \end{bmatrix}. \tag{4.39}$$

As a result, one obtains (4.35) in terms of the famous Schur complement lemma. In the end, in view of Lemma 4.1 and Theorem 4.1, the proof of this theorem is complete.

Remark 4.3 Generally speaking, distributed filtering algorithm should be implemented in a distributed manner, which is conducive to taking full advantages of distributed algorithm. Our scheme is not only distributed in the filtering form but also in the algorithm implementation. From (4.34) to (4.35), the information from neighboring nodes is known at time instant k, such as $\alpha_{i,k}$, δ_j, Ω_j, and $P_{j,k}$ $(j \in \mathscr{N}_i)$. As a matter of fact, the decision variables include $P^1_{i,k+1}$, $P^2_{i,k+1}$, $\mathscr{F}^2_{i,k}$, $F_{i,k}$, and $E_{ij,k}$. Therefore, node i is able to independently calculate its own filter gains on the premise of obtaining the information from its neighboring nodes, which is an indeed distributed manner.

Remark 4.4 This chapter considers the distributed H_∞-consensus filtering problem with both random parameters and event-triggering protocol. We firstly propose the information matrix, the element of which describes the covariance of any two elements of a random matrix. Via the information matrix, the weighted-covariance matrix formula has been proposed and applied to the process of the following local performance analysis. Inspired by the vector dissipativity theory, the framework of local performance analysis has been established. Then, the locally augmented way is employed to handle event-triggering protocol. Based on the performance analysis of single node and its neighboring nodes, the locally coupled design conditions have been established via the vector dissipativity theory. The advantages of this distributed filtering algorithm involve the following aspects: (1) scalability, that is, its computational complexity is unrelated to the network size; (2) a distributedly implemented manner, namely, this algorithm can be implemented on sensor node without centralized computation; and (3) a locally augmented way, which is utilized to deal with the complicated system dynamics and network-induced phenomena, such as random parameters and event-triggering protocol.

4.3 An Illustrative Example

In this section, a simulation example is presented to illustrate the effectiveness of the distributed filtering design method developed in this chapter.

Consider an SN which has six nodes, whose topology is represented by a digraph $\mathscr{G} = (\mathscr{V}, \mathscr{E}, \mathcal{A})$ with the set of nodes $\mathscr{V} = \{1, 2, 3, 4, 5, 6\}$, the set of edges $\mathscr{E} = \{(2, 1), (3, 2), (4, 3), (5, 4), (6, 5), (1, 6)\}$.

The time-varying dynamics of the system is described by (4.1) with parameters as

$$A_k(\mu_k) = \bar{A}_k + \tilde{A}_k(\mu_k), \quad E_k = [0.1\ 0.1],$$
$$\bar{A}_k = \begin{bmatrix} 0.87 & -0.4 \\ 0.15 & 0.98 + 0.1 * \sin(6k) \end{bmatrix}, \quad \tilde{A}_k(\mu_k) = 0.1 \begin{bmatrix} \mu_k^1\ \mu_k^2 \\ \mu_k^3\ \mu_k^4 \end{bmatrix}.$$

The dynamics of nodes is formulated by (4.2) with parameters as

$$\begin{aligned} C_{i,k}(\varepsilon_{i,k}) =& \bar{C}_{i,k} + \tilde{C}_{i,k}(\varepsilon_{i,k}) = \bar{C}_{i,k} + 0.1 * [\varepsilon_{i,k}^1\ \varepsilon_{i,k}^2], \\ \bar{C}_{1,k} =& [0.5\ 0.5], \quad D_{1,k} = 0.1, \\ \bar{C}_{2,k} =& [0.5\ 0.5], \quad D_{2,k} = 0.1, \\ \bar{C}_{3,k} =& [0.5\ 0.5], \quad D_{3,k} = 0.1, \\ \bar{C}_{4,k} =& [0.8\ 0.7], \quad D_{4,k} = 0.1, \\ \bar{C}_{5,k} =& [0.8\ 0.7], \quad D_{5,k} = 0.1, \\ \bar{C}_{6,k} =& [0.8\ 0.7], \quad D_{6,k} = 0.1 \end{aligned}$$

where μ_k^i $(i = 1, 2, 3, 4)$, $\varepsilon_{i,k}^1$, and $\varepsilon_{i,k}^2$ are mutually independent zero-mean Gaussian noise sequences with unity covariances. Also, it is verified that

$$\mathcal{A}_{i,k} = \text{diag}_4\{0.01\}, \quad \mathcal{C}_{i,k} = \text{diag}_2\{0.01\}.$$

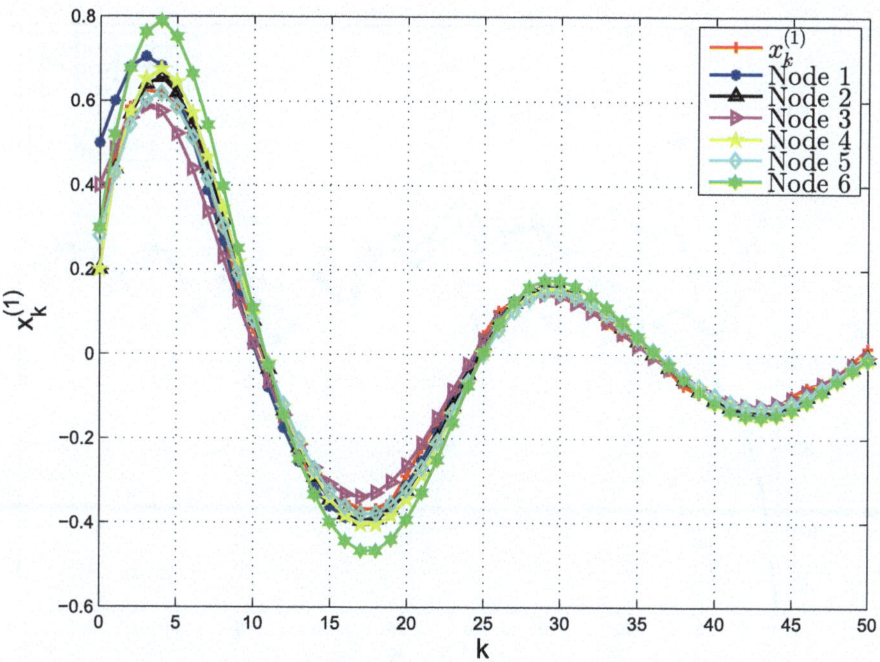

Fig. 4.1 The trajectories for the 1st component of the system

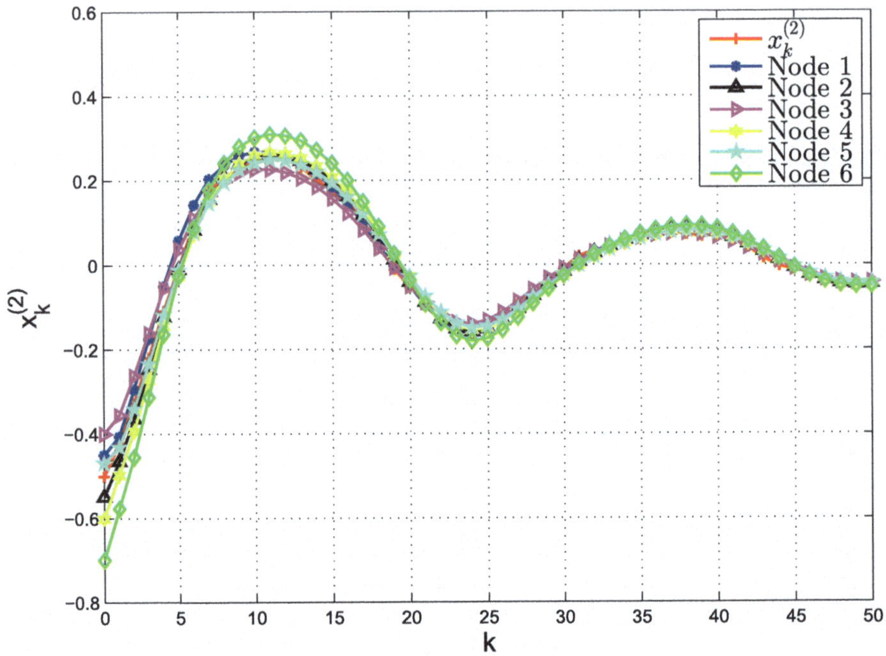

Fig. 4.2 The trajectories for the 2nd component of the system

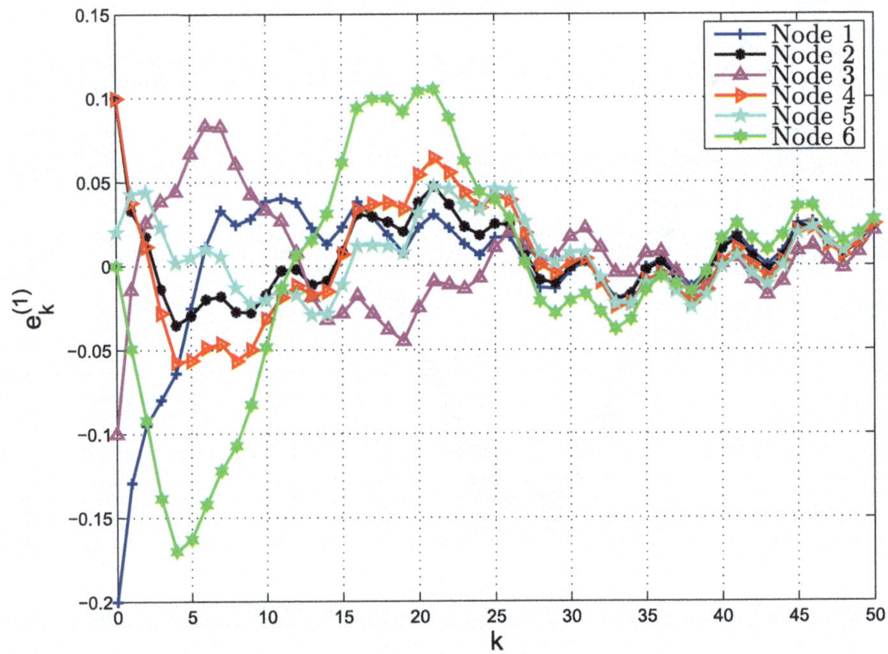

Fig. 4.3 The filtering errors for the 1st component of the system

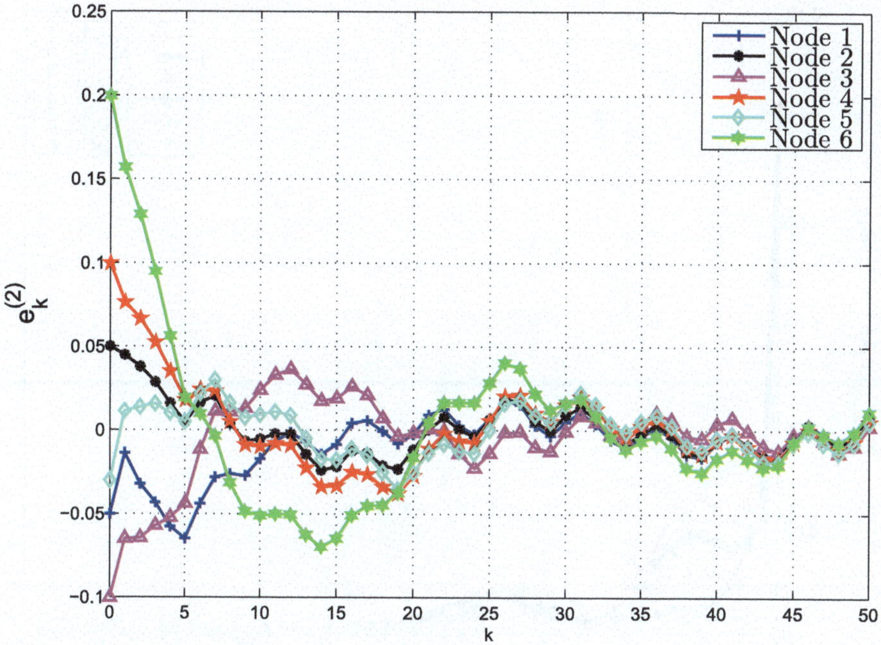

Fig. 4.4 The filtering errors for the 2nd component of the system

The disturbance attenuation level γ^2 and time horizon n are taken as 0.001 and 50, respectively. In addition, the performance weighted matrices $Q_{i,k}$ and $T_{i,k}$, the initial weighted matrices S_i and $P_{i,0}$ are given as $Q_{i,k} \equiv 1$, $T_{i,k} \equiv I$, $S_i = P_{i,0} = 5I$ $(i = 1, 2, 3, 4, 5, 6)$, $\Omega_i = I$, $\delta_1 = \delta_2 = \delta_5 = 0.2$, $\delta_3 = \delta_4 = \delta_6 = 0.3$. By using the YALMIP toolbox, one derives the filter gains of all nodes from Theorem 4.2.

In this simulation example, the initial conditions are set as $x_0 = [0.3 \;\; -0.5]^T$, $\hat{x}_{1,0} = [0.5 \;\; -0.45]^T$, $\hat{x}_{2,0} = [0.2 \;\; -0.55]^T$, $\hat{x}_{3,0} = [0.4 \;\; -0.4]^T$, $\hat{x}_{4,0} = [0.2 \;\; -0.6]^T$, $\hat{x}_{5,0} = [0.28 \;\; -0.47]^T$, and $\hat{x}_{6,0} = [0.3 \;\; -0.7]^T$. Simulation results are shown in Figs. 4.1, 4.2, 4.3, 4.4, 4.5 and 4.6. Figures 4.1 and 4.2 depict the trajectories of the states and filters, respectively. The filtering errors between the states and filters of each node are plotted in Figs. 4.3 and 4.4, respectively. Figures 4.5 and 4.6 describe the disagreement function and the triggering instants for each node, respectively. From these figures, the filtering accuracy and the consensus simultaneously have a satisfactory presentation. Meanwhile, the transmission number of data among neighboring nodes is significantly reduced. As a result, it has been confirmed that the proposed distributed filtering scheme in this chapter is indeed effective.

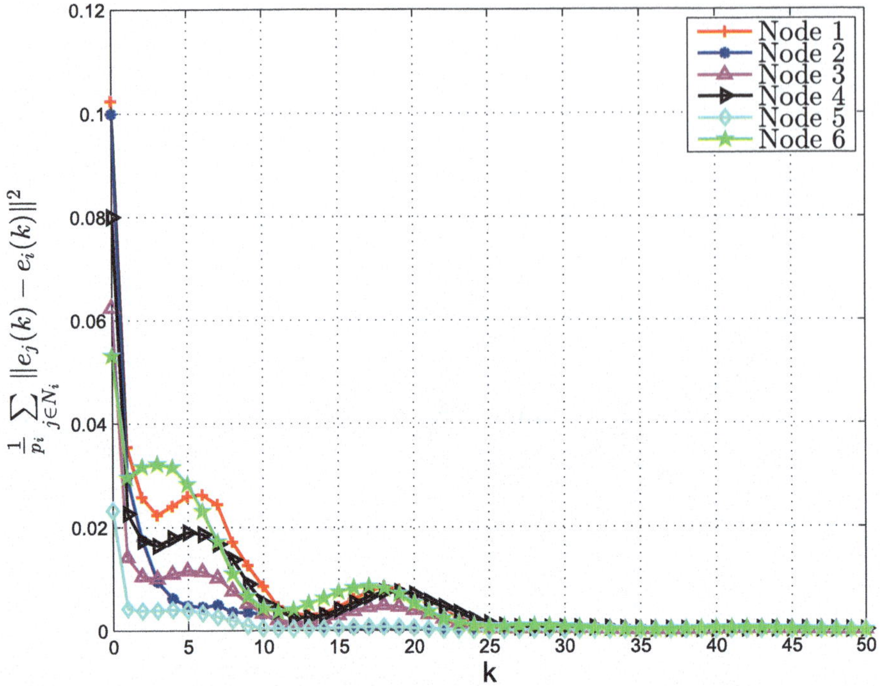

Fig. 4.5 The local disagreement function for every node

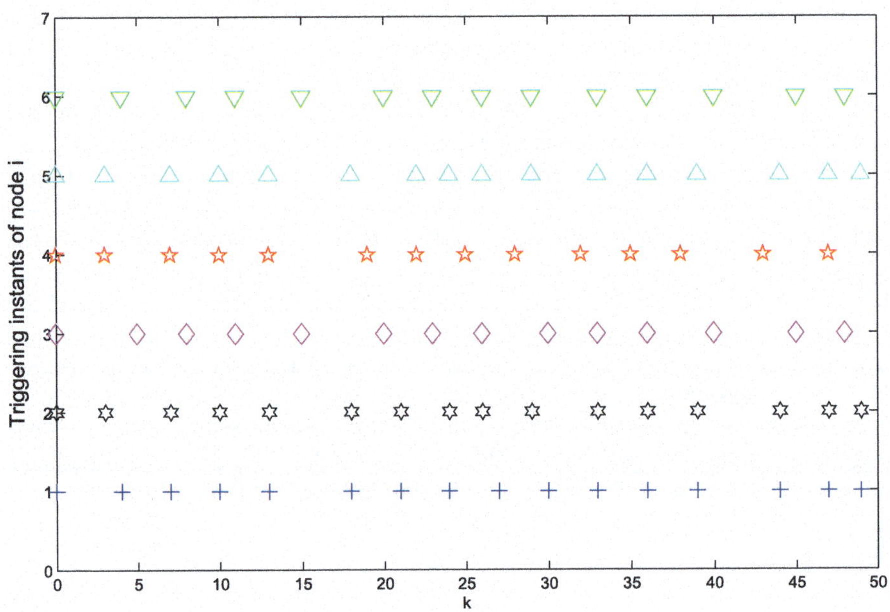

Fig. 4.6 The triggering-time instants for every node

4.4 Summary

In this chapter, the local performance analysis method based on the vector dissipation theory has been utilized to deal with the distributed filtering problem for a class of discrete time-varying random parameter systems with the event-triggering protocol. The statistical property of random matrix has been first formulated as the information matrix, which is employed to calculate the weighted-covariance matrix by means of Kroneck product and Hadrmard product. A novel event-triggering protocol has been introduced for neighboring nodes to reduce the energy consumption during the information transmission in order to prolong the life of sensor node. A set of locally coupled design sufficient conditions has been derived for each node such that the filtering error dynamics satisfy the vector dissipativity over the finite horizon, as well as the proposed H_∞-consensus constraint. In addition, the explicit expressions of the desired filter gains have been obtained. Finally, a simulation example has been employed to demonstrate the effectiveness of the proposed filtering algorithm. In the future, one of possible research directions would be to investigate the distributed filtering problem with various communication protocols, which are widely applied in the networked control systems.

Reference

1. Hu, J., Wang, Z., Shen, B., Gao, H.: Gain-constrained recursive filtering with stochastic nonlinearities and probabilistic sensor delays. IEEE Trans. Signal Process. **61**(5), 1230–1238 (2013)

Chapter 5
Scalable Consensus Filtering for Uncertain Systems with Round-Robin Protocol

Other than the event-triggering protocol, there are various communication protocols, which are also employed to schedule the data transmission over the network. As discussed in Chap. 1, Round-Robin protocol (RRP) has become the most widely used protocol since it is very simple and easy to carry out. As such, a desirable goal is to investigate the influence of the Round-Robin protocol on the design of the distributed filtering.

This chapter is concerned with the scalable distributed H_∞-consensus filtering problem for a class of discrete time-varying systems over sensor networks with the Round-Robin protocol. The challenge comes from the fact that the time-varying parameters of the network are subject to randomly occurring norm-bounded uncertainties and the measurement outputs of the sensor nodes are saturated due to the sector nonlinearities. For preventing data collisions and saving energy, the Round-Robin protocol determines which neighboring node can access the shared network for information transmission at each time step. An H_∞ performance index is proposed to characterize the disturbance attenuation level of the resulting filtering error dynamics. By stochastic analysis in combination with the recursive matrix inequality approach, a distributed filtering algorithm is developed for each individual sensor node to ensure the pre-specified estimation performance. Finally, an illustrative simulation example is shown to verify the effectiveness and applicability of the theoretical results.

5.1 Problem Formulation

Consider the following discrete-time stochastic system with randomly occurring norm-bounded uncertainties defined on $k \in \mathscr{H} \triangleq \{0, 1, \ldots, n-1\}$:

F. Han et al., *Distributed Filtering, Control and Synchronization*,
Studies in Systems, Decision and Control 428,
https://doi.org/10.1007/978-3-030-97075-8_5

$$x_{k+1} = (A_k + \beta_k \Delta A_k)x_k + (B_k + \beta_k \Delta B_k)w_k, \tag{5.1}$$

with N saturated sensor measurements modeled by

$$y_{i,k} = \sigma_i(C_{i,k}x_k) + D_{i,k}\vartheta_{i,k}, \quad i \in \mathscr{V} \tag{5.2}$$

where $x_k \in \mathbb{R}^{n_x}$ is the state, $y_{i,k} \in \mathbb{R}^{n_y}$ is the measurement of node i, $w_k \in \mathbb{R}^{n_w}$ and $\vartheta_{i,k} \in \mathbb{R}^{n_\vartheta}$ are the external disturbances belonging to $l_2[0, n-1]$. A_k, B_k, $C_{i,k}$ and $D_{i,k}$ are known time-varying matrices with compatible dimensions.

The time-varying parameter uncertainties ΔA_k and ΔB_k satisfy

$$[\Delta A_k \quad \Delta B_k] = M_k F_k [N_{1k} \quad N_{2k}] \tag{5.3}$$

where M_k, N_{1k}, and N_{2k} are known time-varying matrices with appropriate dimensions, and F_k is an unknown time-varying matrix satisfying the constraint

$$F_k F_k^T \leq I, \ k \in \mathscr{H}. \tag{5.4}$$

The random variable β_k is a Bernoulli distributed sequence taking values on 0 or 1 with the following probabilities:

$$\text{Prob}\{\beta_k = 0\} = 1 - \bar{\beta}_k, \quad \text{Prob}\{\beta_k = 1\} = \bar{\beta}_k \tag{5.5}$$

where $\bar{\beta}_k$ is a known constant belonging to $[0, 1]$.

The saturation function $\sigma_i(\cdot) : \mathbb{R}^{n_y} \to \mathbb{R}^{n_y}$ is defined as

$$\sigma_i(z) = \left[\sigma_{i1}(z_1)\ \sigma_{i2}(z_2)\ \cdots\ \sigma_{in_y}(z_{n_y})\right]^T, \tag{5.6}$$

with $\sigma_{ij}(z_j) = \text{sign}(z_j)\min\{z_{j,\max}, |z_j|\}$, where z_j is the j-th element of the vector z and $z_{j,\max}$ is the saturation level of z_j.

Assuming that there exist diagonal matrices H_{i1} and H_{i2} such that $0 \leq H_{i1} < I \leq H_{i2}$, the saturation function $\sigma_i(C_{i,k}x_k)$ in (5.2) can be decomposed into a linear part and a nonlinear part as

$$\sigma_i(C_{i,k}x_k) = \bar{H}_i C_{i,k}x_k + \Phi_i(C_{i,k}x_k) \tag{5.7}$$

where $\Phi_i(C_{i,k}x_k)$ is a nonlinear vector-valued function satisfying

$$\Phi_i^T(C_{i,k}x_k)\Phi_i(C_{i,k}x_k) \leq x_k^T C_{i,k}^T H_i^T H_i C_{i,k}x_k \tag{5.8}$$

with $\bar{H}_i = \frac{1}{2}(H_{i2} + H_{i1})$ and $H_i = \frac{1}{2}(H_{i2} - H_{i1})$.

Considering the limited capacity of the communication channel, the RRP is adopted to schedule the data transmission and make full utilization of the shared network resources. With the neighbor set of node i, denoted as $\mathscr{N}_i \triangleq \{j_{i1}, j_{i2}, \ldots, j_{ip_i}\}$, let the first transmission node be j_{i1}. In addition, denote by $\hbar_{i,k} \triangleq \mod(k-1, p_i)+1$ the subscript of the selected neighbor of node i having access to the shared communication network. Obviously, $j_{\hbar_{i,k}} \in \mathcal{N}_i$.

In this chapter, the following distributed filter is designed for node i:

$$\hat{x}_{i,k+1} = A_k \hat{x}_{i,k} + L_{i,k}(y_{i,k} - \bar{H}_i C_{i,k} \hat{x}_{i,k}) + K_{ij_{\hbar_{i,k}},k}(\hat{x}_{j_{\hbar_{i,k}},k} - \hat{x}_{i,k}) \tag{5.9}$$

where $\hat{x}_{i,k} \in \mathbb{R}^{n_x}$ and $\hat{x}_{j_{\hbar_{i,k}},k} \in \mathbb{R}^{n_x}$ are, respectively, the estimates of x_k on node i and on its neighboring node $j_{\hbar_{i,k}} \in \mathcal{N}_i$. Here, the filter gains $L_{i,k}$ and $K_{ij_{\hbar_{i,k}},k}$ are to be determined later.

Assumption 5.1 If a node has no neighboring node, then it is uniformly observable. If a node has neighboring nodes, then it is uniformly distributed observable under the RRP.

To further simplify the expressions, denote $K_{ij_\hbar,k} \triangleq K_{ij_{\hbar_{i,k}},k}$ and $\hat{x}_{j_{\hbar_i},k} \triangleq \hat{x}_{j_{\hbar_{i,k}},k}$. Let $e_{i,k} \triangleq x_k - \hat{x}_{i,k}$ be the filtering error of node i. Then, combing (5.1), (5.2), (5.7) and (5.9), one has the following filtering error dynamics:

$$\begin{aligned} e_{i,k+1} = {} & (A_k - L_{i,k}\bar{H}_i C_{i,k} - K_{ij_{\hbar_i},k})e_{i,k} + B_k w_k - L_{i,k}\Phi_i(C_{i,k}x_k) \\ & + K_{ij_{\hbar_i},k} e_{j_{\hbar_i},k} - L_{i,k} D_{i,k} v_{i,k} + \beta_k \Delta B_k w_k + \beta_k \Delta A_k x_k. \end{aligned} \tag{5.10}$$

Denoting $\eta_{i,k} \triangleq [x_k^T \;\; e_{i,k}^T]^T$, $\xi_{i,k} \triangleq [w_k^T \;\; \vartheta_{i,k}^T]^T$ and $\tilde{\beta}_k \triangleq \beta_k - \bar{\beta}_k$, one has

$$\begin{aligned} \eta_{i,k+1} = {} & (\mathscr{A}_{i,k} + \bar{\beta}_k \Delta \mathscr{A}_k)\eta_{i,k} + \mathscr{K}_{ij_\hbar,k}\eta_{j_{\hbar_i},k} + \tilde{\beta}_k \Delta \mathscr{A}_k \eta_{i,k} \\ & + \mathscr{L}_{i,k}\bar{\Phi}_i(\mathscr{C}_{i,k}\eta_{i,k}) + (\mathscr{B}_{i,k} + \bar{\beta}_k \Delta \mathscr{B}_k)\xi_{i,k} + \tilde{\beta}_k \Delta \mathscr{B}_k \xi_{i,k} \end{aligned} \tag{5.11}$$

where

$$\begin{aligned} \mathscr{A}_{i,k} &= \mathrm{diag}\{A_k, A_k - L_{i,k}\bar{H}_i C_{i,k} - K_{ij_{\hbar},k}\}, \\ \mathscr{C}_{i,k} &= \mathrm{diag}\{C_{i,k}, 0\}, \quad \mathscr{K}_{ij_\hbar,k} = \mathrm{diag}\{0, K_{ij_\hbar,k}\}, \\ \mathscr{B}_{i,k} &= \begin{bmatrix} B_k & 0 \\ B_k & -L_{i,k}D_{i,k} \end{bmatrix}, \quad \Delta\mathscr{A}_k = \begin{bmatrix} \Delta A_k & 0 \\ \Delta A_k & 0 \end{bmatrix}, \\ \Delta\mathscr{B}_k &= \begin{bmatrix} \Delta B_k & 0 \\ \Delta B_k & 0 \end{bmatrix}, \quad \mathscr{L}_{i,k} = \begin{bmatrix} 0 & 0 \\ -L_{i,k} & 0 \end{bmatrix}, \\ \bar{\Phi}_i(\mathscr{C}_{i,k}\eta_{i,k}) &= \bar{\Phi}_i\left(\begin{bmatrix} C_{i,k}x_k \\ 0 \end{bmatrix}\right) = \begin{bmatrix} \Phi_i(C_{i,k}x_k) \\ 0 \end{bmatrix}. \end{aligned}$$

Also, (5.8) is rewritten as

$$\bar{\Phi}_i^T(\mathscr{C}_{i,k}\eta_{i,k})\bar{\Phi}_i(\mathscr{C}_{i,k}\eta_{i,k}) \le \eta_{i,k}^T \mathscr{C}_{i,k}^T \mathscr{H}_i^T \mathscr{H}_i \mathscr{C}_{i,k}\eta_{i,k} \tag{5.12}$$

where $\mathscr{H}_i \triangleq \mathrm{diag}\{H_i, 0\}$.

By defining

$$z_{i,k} = [0\ I]\eta_{i,k} \triangleq E\eta_{i,k}, \tag{5.13}$$

one obtains the following dynamics associated with node i:

$$\begin{cases} \eta_{i,k+1} = (\mathscr{A}_{i,k} + \bar{\beta}_k \Delta\mathscr{A}_k)\eta_{i,k} + \mathscr{K}_{ij_{\hbar},k}\eta_{j_{\hbar_i},k} + \mathscr{L}_{i,k}\bar{\Phi}_i(\mathscr{C}_{i,k}\eta_{i,k}) \\ \qquad\quad + \tilde{\beta}_k \Delta\mathscr{A}_k \eta_{i,k} + (\mathscr{B}_{i,k} + \bar{\beta}_k \Delta\mathscr{B}_k)\xi_{i,k} + \tilde{\beta}_k \Delta\mathscr{B}_k \xi_{i,k}, \\ z_{i,k} = E\eta_{i,k}. \end{cases} \tag{5.14}$$

Before proceeding further, introduce the following performance index to characterize the disturbance attenuation level of the filtering error dynamics (5.14) against the external disturbances.

Definition 5.1 Let the disturbance attenuation level $\gamma > 0$ and the weighted matrices $\mathscr{U}_i$, $R_{i,k}$, $Q_{i,k}$, and $\mathscr{T}_{i,k}$ be given. The filtering error system (5.14) is said to satisfy the H_∞-consensus performance constraint over the finite horizon $\mathscr{H}$ with the RRP if the following inequality holds:

$$\mathbb{E}\left\{\sum_{k=0}^{n-1}\sum_{i=1}^{N}[\|z_{j_{\hbar_i},k} - z_{i,k}\|^2_{R_{i,k}} + \|z_{i,k}\|^2_{Q_{i,k}}]\right\} \leq \gamma^2 \sum_{i=1}^{N}\left(\|\eta_{i,0}\|^2_{\mathscr{U}_i} + \sum_{k=0}^{n-1}\|\xi_{i,k}\|^2_{\mathscr{T}_{i,k}}\right) \tag{5.15}$$

where $\mathscr{U}_i \triangleq \text{diag}\{U_{1i}, U_{2i}\}$, $\mathscr{T}_{i,k} \triangleq \text{diag}\{T_{1i,k}, T_{2i,k}\}$, and U_{1i}, U_{2i}, $R_{i,k}$, $Q_{i,k}$, $T_{1i,k}$, and $T_{2i,k}$ are all known positive definite matrices.

The main objective in this chapter is to determine the filter gains $L_{i,k}$ and $K_{ij_{\hbar},k}$ such that the filtering error dynamics (5.14) satisfies the proposed H_∞-consensus performance constraint over the finite horizon.

Remark 5.1 The consensus-based performance index defined in (5.15) distinguishes itself from those existing in the literature through evaluating the effect from the RRP, which can be confirmed from the subscript $j_{\hbar_i}$ of the term $\sum_{k=0}^{n-1}\sum_{i=1}^{N}$ $\|z_{j_{\hbar_i},k} - z_{i,k}\|^2_{R_{i,k}}$. Accordingly, in case that the RRP is not considered, the proposed performance index is degenerated to the one in [1], which means that the performance measure here is more general. Furthermore, from the engineering point of view, the proposed performance cost is essentially different from the existing ones for the following *four* reasons: (1) it accounts for the dynamical change of neighboring sensors; (2) the introduction of the weighted matrices $Q_{i,k}$ and $R_{i,k}$ provides a desirable tradeoff between the consensus and the filtering accuracy; (3) the performance cost enables one to only investigate the impacts on the filtering accuracy by setting $Q_{i,k} = I$ and $R_{i,k} = 0$ or the consensus with $Q_{i,k} = 0$ and $R_{i,k} = I$; and (4) with the new cost function, the selected vector-type storage functions and vector-type supply rate functions, to be discussed later, depend on a time-varying subset of the neighbors of sensor i.

5.2 Main Results

In this section, a set of sufficient conditions is obtained for the filtering error dynamics (5.14) to satisfy the H_∞-consensus performance criterion (5.15).

For convenience, denote

$$S_i(z_{i,k}, z_{j_{\hbar_i},k}, \xi_{i,k}) \triangleq \gamma^2\|\xi_{i,k}\|^2_{\mathcal{T}_{i,k}} - \|z_{i,k}\|^2_{Q_{i,k}} - \|z_{j_{\hbar_i},k} - z_{i,k}\|^2_{R_{i,k}}. \tag{5.16}$$

Definition 5.2 The filtering error system (5.14) is said to be stochastically vector-dissipative over the finite horizon $\mathcal{H}$ with respect to the following vector of supply rate functions:

$$\mathbf{S}(\eta_k, \xi_k) \triangleq [S_1(z_{1,k}, z_{j_{\hbar_1},k}, \xi_{1,k}) \quad \cdots \quad S_N(z_{N,k}, z_{j_{\hbar_N},k}, \xi_{N,k})]^T,$$

if there exists a vector of nonnegative definite storage functions $\mathbf{V}(\eta_k) \triangleq [V_1(\eta_{1,k}) \ \cdots \ V_N(\eta_{N,k})]^T$ (with $\mathbf{V}(0) = 0$) and a sequence of nonsingular column substochastic dissipation matrices $W_k \in \mathbb{R}^{N\times N}$ such that the following vector dissipation inequality is satisfied for any $k \in \mathcal{H}$:

$$\mathbb{E}\{\mathbf{V}(\eta_{k+1})\} \leq\leq W_k\mathbb{E}\{\mathbf{V}(\eta_k)\} + \mathbb{E}\{\mathbf{S}(z_k, \xi_k)\}. \tag{5.17}$$

Next, a set of local conditions is established for each node such that the filtering error dynamics (5.14) is stochastically vector-dissipative over the finite horizon $\mathcal{H}$. Define an interval-valued function $\mathfrak{I}_{q_i}$ with respect to out-degree q_i as follows:

$$\mathfrak{I}_{q_i} = \begin{cases} \left(0, \frac{1+q_i}{2q_i}\right), & \text{if } q_i \neq 0; \\ (0, 1], & \text{if } q_i = 0. \end{cases} \tag{5.18}$$

Theorem 5.1 *Let the disturbance attenuation level $\gamma > 0$, the scalar sequence $\alpha_{i,k} \in \mathfrak{I}_{q_i}$, the weighted matrices $R_{i,k}$, $Q_{i,k}$, $\mathcal{U}_i$, and $\mathcal{T}_{i,k}$, and the filter gain sequences $L_{i,k}$ and $K_{ij,k}$ be given. The system (5.14) is stochastic vector-dissipative over the finite horizon $\mathcal{H}$ with respect to the vector supply rate function $\mathbf{S}(z_k, \xi_k)$ and also satisfies the H_∞-consensus performance criterion (5.15), if there exists a sequence of positive scalars $\lambda_{i,k}$ and a vector of storage functions $\mathbf{V}(\eta_k)$ with the form $V_i(\eta_{i,k}) = \eta_{i,k}^T\mathcal{P}_{i,k}\eta_{i,k}$, where $\{\mathcal{P}_{i,k}\}_{k\in\mathcal{H}\cup\{n\}}$ is a sequence of positive definite matrices with initial conditions $\mathcal{P}_{i,0} \leq \gamma^2\mathcal{U}_i$, such that the following conditions hold for all $k \in \mathcal{H}$, $i \in \mathcal{V}$:*

$$\Xi_{ij_\hbar,k} = \begin{bmatrix} \Xi^{11}_{ij_\hbar,k} & * & * & * \\ \Xi^{21}_{ij_\hbar,k} & \Xi^{22}_{ij_\hbar,k} & * & * \\ \Xi^{31}_{ij_\hbar,k} & \Xi^{32}_{ij_\hbar,k} & \Xi^{33}_{ij_\hbar,k} & * \\ \Xi^{41}_{ij_\hbar,k} & \Xi^{42}_{ij_\hbar,k} & \Xi^{43}_{ij_\hbar,k} & \Xi^{44}_{ij_\hbar,k} \end{bmatrix} \leq 0 \tag{5.19}$$

where

$$
\begin{aligned}
\Xi^{11}_{ij\hbar,k} &= (\mathscr{A}_{i,k} + \bar{\beta}_k \Delta\mathscr{A}_k)^T \mathscr{P}_{i,k+1}(\mathscr{A}_{i,k} + \bar{\beta}_k \Delta\mathscr{A}_k) - \theta_{i,k}\mathscr{P}_{i,k} + E^T R_{i,k} E + E^T Q_{i,k} E \\
&\quad + \varphi_k \Delta\mathscr{A}_k^T \mathscr{P}_{i,k+1} \Delta\mathscr{A}_k + \lambda_{i,k}\mathscr{C}_{i,k}^T \mathscr{H}_i^T \mathscr{H}_i \mathscr{C}_{i,k}, \quad \theta_{i,k} = \frac{1 + q_i(1 - \alpha_{i,k})}{1 + q_i}, \\
\Xi^{21}_{ij\hbar,k} &= \mathscr{K}^T_{ij\hbar,k}\mathscr{P}_{i,k+1}(\mathscr{A}_{i,k} + \bar{\beta}_k \Delta\mathscr{A}_k) - E^T R_{i,k} E, \\
\Xi^{22}_{ij\hbar,k} &= \mathscr{K}^T_{ij\hbar,k}\mathscr{P}_{i,k+1}\mathscr{K}_{ij\hbar,k} + E^T R_{i,k} E - \frac{\alpha_{j\hbar_i,k}}{1 + q_{j\hbar}} \mathscr{P}_{j\hbar_i,k}, \\
\Xi^{31}_{ij\hbar,k} &= \mathscr{L}^T_{i,k}\mathscr{P}_{i,k+1}(\mathscr{A}_{i,k} + \bar{\beta}_k \Delta\mathscr{A}_k) + \frac{1}{2}\lambda_{i,k}\mathscr{H}_i\mathscr{C}_{i,k}, \quad \varphi_k = \bar{\beta}_k - \bar{\beta}_k^2, \\
\Xi^{32}_{ij\hbar,k} &= \mathscr{L}^T_{i,k}\mathscr{P}_{i,k+1}\mathscr{K}_{ij\hbar,k}, \quad \Xi^{33}_{ij\hbar,k} = \mathscr{L}^T_{i,k}\mathscr{P}_{i,k+1}\mathscr{L}_{i,k} - \lambda_{i,k} I, \\
\Xi^{41}_{ij\hbar,k} &= (\mathscr{B}_{i,k} + \bar{\beta}_k \Delta\mathscr{B}_k)^T \mathscr{P}_{i,k+1}(\mathscr{A}_{i,k} + \bar{\beta}_k \Delta\mathscr{A}_k) + \varphi_k \Delta\mathscr{B}_k^T \mathscr{P}_{i,k+1}\Delta\mathscr{A}_k, \\
\Xi^{42}_{ij\hbar,k} &= (\mathscr{B}_{i,k} + \bar{\beta}_k \Delta\mathscr{B}_k)^T \mathscr{P}_{i,k+1}\mathscr{K}_{ij\hbar,k}, \quad \Xi^{43}_{ij\hbar,k} = (\mathscr{B}_{i,k} + \bar{\beta}_k \Delta\mathscr{B}_k)^T \mathscr{P}_{i,k+1}\mathscr{L}_{i,k}, \\
\Xi^{44}_{ij\hbar,k} &= (\mathscr{B}_{i,k} + \bar{\beta}_k \Delta\mathscr{B}_k)^T \mathscr{P}_{i,k+1}(\mathscr{B}_{i,k} + \bar{\beta}_k \Delta\mathscr{B}_k) - \gamma^2 \mathscr{T}_{i,k} + \varphi_k \Delta\mathscr{B}_k^T \mathscr{P}_{i,k+1}\Delta\mathscr{B}_k.
\end{aligned}
$$

Proof This proof is divided into two steps.

Step (1) Proof of the stochastic vector-dissipativity over the finite horizon $\mathscr{H}$.

First, it is easy to verify that

$$
\begin{aligned}
\mathbb{E}\{\tilde{\beta}_k\} &= \mathbb{E}\{\beta_k - \bar{\beta}_k\} = 0, \\
\mathbb{E}\{\tilde{\beta}_k^2\} &= \mathbb{E}\{\beta_k - \bar{\beta}_k\}^2 = \varphi_k.
\end{aligned}
$$

Next, calculate the storage function concerning i-th node along the trajectory of system (5.14) as follows:

$$
\begin{aligned}
&\mathbb{E}\{V_{i,k+1} | \eta_{i,k}, \eta_{j\hbar_i,k}\} \\
&= \mathbb{E}\{((\mathscr{A}_{i,k} + \bar{\beta}_k \Delta\mathscr{A}_k)\eta_{i,k} + \mathscr{K}_{ij\hbar,k}\eta_{j\hbar_i,k} + (\mathscr{B}_{i,k} + \bar{\beta}_k \Delta\mathscr{B}_k)\xi_{i,k} \\
&\quad + \tilde{\beta}_k \Delta\mathscr{A}_k \eta_{i,k} + \tilde{\beta}_k \Delta\mathscr{B}_k \xi_{i,k} + \mathscr{L}_{i,k}\bar{\Phi}_i(\mathscr{C}_{i,k}\eta_{i,k}))^T \mathscr{P}_{i,k+1} \\
&\quad \times ((\mathscr{A}_{i,k} + \bar{\beta}_k \Delta\mathscr{A}_k)\eta_{i,k} + \mathscr{K}_{ij\hbar,k}\eta_{j\hbar_i,k} + (\mathscr{B}_{i,k} + \bar{\beta}_k \Delta\mathscr{B}_k)\xi_{i,k} \\
&\quad + \tilde{\beta}_k \Delta\mathscr{A}_k \eta_{i,k} + \tilde{\beta}_k \Delta\mathscr{B}_k \xi_{i,k} + \mathscr{L}_{i,k}\bar{\Phi}_i(\mathscr{C}_{i,k}\eta_{i,k})) | \eta_{i,k}, \eta_{j\hbar_i,k}\} \\
&= \eta_{i,k}^T (\mathscr{A}_{i,k} + \bar{\beta}_k \Delta\mathscr{A}_k)^T \mathscr{P}_{i,k+1}(\mathscr{A}_{i,k} + \bar{\beta}_k \Delta\mathscr{A}_k)\eta_{i,k} \\
&\quad + \eta^T_{j\hbar_i,k}\mathscr{K}^T_{ij\hbar,k}\mathscr{P}_{i,k+1}\mathscr{K}_{ij\hbar,k}\eta_{j\hbar_i,k} \\
&\quad + \varphi_k \eta_{i,k}^T \Delta\mathscr{A}_k^T \mathscr{P}_{i,k+1}\Delta\mathscr{A}_k \eta_{i,k} \\
&\quad + 2\varphi_k \xi_{i,k}^T \Delta\mathscr{B}_k^T \mathscr{P}_{i,k+1}\Delta\mathscr{A}_k \eta_{i,k} \\
&\quad + \bar{\Phi}_i^T(\mathscr{C}_{i,k}\eta_{i,k})\mathscr{L}^T_{i,k}\mathscr{P}_{i,k+1}\mathscr{L}_{i,k}\bar{\Phi}_i(\mathscr{C}_{i,k}\eta_{i,k}) \\
&\quad + 2\eta_{i,k}^T(\mathscr{A}_{i,k} + \bar{\beta}_k \Delta\mathscr{A}_k)^T \mathscr{P}_{i,k+1}\mathscr{K}_{ij\hbar,k}\eta_{j\hbar_i,k} \\
&\quad + 2\xi_{i,k}^T(\mathscr{B}_{i,k} + \bar{\beta}_k \Delta\mathscr{B}_k)^T \mathscr{P}_{i,k+1}(\mathscr{A}_{i,k} + \bar{\beta}_k \Delta\mathscr{A}_k)\eta_{i,k} \\
&\quad + 2\eta_{i,k}^T(\mathscr{A}_{i,k} + \bar{\beta}_k \Delta\mathscr{A}_k)^T \mathscr{P}_{i,k+1}\mathscr{L}_{i,k}\bar{\Phi}_i(\mathscr{C}_{i,k}\eta_{i,k})
\end{aligned}
$$

$$
\begin{aligned}
& + 2\xi_{i,k}^T(\mathscr{B}_{i,k} + \bar{\beta}_k \Delta\mathscr{B}_k)^T \mathscr{P}_{i,k+1}\mathscr{K}_{ij\hbar,k}\eta_{j_{\hbar_i},k} \\
& + 2\eta_{j_{\hbar_i},k}^T \mathscr{K}_{ij\hbar,k}^T \mathscr{P}_{i,k+1}\mathscr{L}_{i,k}\bar{\Phi}_i(\mathscr{C}_{i,k}\eta_{i,k}) \\
& + 2\xi_{i,k}^T(\mathscr{B}_{i,k} + \bar{\beta}_k \Delta\mathscr{B}_k)^T \mathscr{P}_{i,k+1}\mathscr{L}_{i,k}\bar{\Phi}_i(\mathscr{C}_{i,k}\eta_{i,k}) \\
& + \xi_{i,k}^T((\mathscr{B}_{i,k} + \bar{\beta}_k \Delta\mathscr{B}_k)^T \mathscr{P}_{i,k+1}(\mathscr{B}_{i,k} + \bar{\beta}_k \Delta\mathscr{B}_k) \\
& + \varphi_k \Delta\mathscr{B}_k^T \mathscr{P}_{i,k+1}\Delta\mathscr{B}_k - \gamma^2 \mathscr{T}_{i,k})\xi_{i,k} + \gamma^2 \xi_{i,k}^T \mathscr{T}_{i,k}\xi_{i,k} \\
\triangleq\ & \zeta_{ij\hbar,k}^T \Omega_{ij\hbar,k}\zeta_{ij\hbar,k} + \gamma^2 \xi_{i,k}^T \mathscr{T}_{i,k}\xi_{i,k}
\end{aligned} \tag{5.20}
$$

where

$$
\Omega_{ij\hbar,k} = \begin{bmatrix} \Omega_{ij\hbar,k}^{11} & * & * & * \\ \Omega_{ij\hbar,k}^{21} & \Omega_{ij\hbar,k}^{22} & * & * \\ \Omega_{ij\hbar,k}^{31} & \Omega_{ij\hbar,k}^{32} & \Omega_{ij\hbar,k}^{33} & * \\ \Omega_{ij\hbar,k}^{41} & \Omega_{ij\hbar,k}^{42} & \Omega_{ij\hbar,k}^{43} & \Omega_{ij\hbar,k}^{44} \end{bmatrix},
$$

$$
\begin{aligned}
\zeta_{ij\hbar,k} &= \begin{bmatrix} \eta_{i,k}^T & \eta_{j_{\hbar_i},k}^T & \bar{\Phi}_i^T(\mathscr{C}_{i,k}\eta_{i,k}) & \xi_{i,k}^T \end{bmatrix}^T, \quad \Omega_{ij\hbar,k}^{33} = \mathscr{L}_{i,k}^T \mathscr{P}_{i,k+1}\mathscr{L}_{i,k}, \\
\Omega_{ij\hbar,k}^{11} &= (\mathscr{A}_{i,k} + \bar{\beta}_k \Delta\mathscr{A}_k)^T \mathscr{P}_{i,k+1}(\mathscr{A}_{i,k} + \bar{\beta}_k \Delta\mathscr{A}_k) + \varphi_k \Delta\mathscr{A}_k^T \mathscr{P}_{i,k+1}\Delta\mathscr{A}_k, \\
\Omega_{ij\hbar,k}^{21} &= \mathscr{K}_{ij\hbar,k}^T \mathscr{P}_{i,k+1}(\mathscr{A}_{i,k} + \bar{\beta}_k \Delta\mathscr{A}_k), \quad \Omega_{ij\hbar,k}^{22} = \mathscr{K}_{ij\hbar,k}^T \mathscr{P}_{i,k+1}\mathscr{K}_{ij\hbar,k}, \\
\Omega_{ij\hbar,k}^{31} &= \mathscr{L}_{i,k}^T \mathscr{P}_{i,k+1}(\mathscr{A}_{i,k} + \bar{\beta}_k \Delta\mathscr{A}_k), \quad \Omega_{ij\hbar,k}^{32} = \mathscr{L}_{i,k}^T \mathscr{P}_{i,k+1}\mathscr{K}_{ij\hbar,k}, \\
\Omega_{ij\hbar,k}^{41} &= (\mathscr{B}_{i,k} + \bar{\beta}_k \Delta\mathscr{B}_k)^T \mathscr{P}_{i,k+1}(\mathscr{A}_{i,k} + \bar{\beta}_k \Delta\mathscr{A}_k) + \varphi_k \Delta\mathscr{B}_k^T \mathscr{P}_{i,k+1}\Delta\mathscr{A}_k, \\
\Omega_{ij\hbar,k}^{42} &= (\mathscr{B}_{i,k} + \bar{\beta}_k \Delta\mathscr{B}_k)^T \mathscr{P}_{i,k+1}\mathscr{K}_{ij\hbar,k}, \quad \Omega_{ij\hbar,k}^{43} = (\mathscr{B}_{i,k} + \bar{\beta}_k \Delta\mathscr{B}_k)^T \mathscr{P}_{i,k+1}\mathscr{L}_{i,k}, \\
\Omega_{ij\hbar,k}^{44} &= (\mathscr{B}_{i,k} + \bar{\beta}_k \Delta\mathscr{B}_k)^T \mathscr{P}_{i,k+1}(\mathscr{B}_{i,k} + \bar{\beta}_k \Delta\mathscr{B}_k) + \varphi_k \Delta\mathscr{B}_k^T \mathscr{P}_{i,k+1}\Delta\mathscr{B}_k - \gamma^2 \mathscr{T}_{i,k}.
\end{aligned}
$$

Taking into account (5.19), one further has

$$
\begin{aligned}
& \zeta_{ij\hbar,k}^T \Omega_{ij\hbar,k}\zeta_{ij\hbar,k} + \|z_{i,k}\|_{Q_{i,k}}^2 + \|z_{j_{\hbar_i},k} - z_{i,k}\|_{R_{i,k}}^2 - \frac{\alpha_{j_{\hbar_i},k}}{1 + q_{j_{\hbar_i}}}\eta_{j_{\hbar_i},k}^T \mathscr{P}_{j_{\hbar_i},k}\eta_{j_{\hbar_i},k} \\
& - \theta_{i,k}\eta_{i,k}^T \mathscr{P}_{i,k}\eta_{i,k} - \lambda_{i,k}(\bar{\Phi}_i^T(\mathscr{C}_{i,k}\eta_{i,k})\bar{\Phi}_i(\mathscr{C}_{i,k}\eta_{i,k}) - \eta_{i,k}^T \mathscr{C}_{i,k}^T \mathscr{H}_i^T \mathscr{H}_i \mathscr{C}_{i,k}\eta_{i,k}) \\
=\ & \zeta_{ij\hbar,k}^T \Xi_{ij\hbar,k}\zeta_{ij\hbar,k} \le 0.
\end{aligned}
$$

On the other hand, it follows from (5.19) that

$$
\begin{aligned}
& \mathbb{E}\{V_{i,k+1}|\eta_{i,k}, \eta_{j_{\hbar_i},k}\} \\
\le\ & \zeta_{ij\hbar,k}^T \Omega_{ij\hbar,k}\zeta_{ij\hbar,k} + \gamma^2 \xi_{i,k}^T \mathscr{T}_{i,k}\xi_{i,k} - \lambda_{i,k}(\bar{\Phi}_i^T(\mathscr{C}_{i,k}\eta_{i,k})\bar{\Phi}_i(\mathscr{C}_{i,k}\eta_{i,k}) - \eta_{i,k}^T \mathscr{C}_{i,k}^T \mathscr{H}_i^T \mathscr{H}_i \mathscr{C}_{i,k}\eta_{i,k}) \\
\le\ & \frac{\alpha_{j_{\hbar_i},k}}{1 + q_{j_{\hbar_i}}}\eta_{j_{\hbar_i},k}^T \mathscr{P}_{j_{\hbar_i},k}\eta_{j_{\hbar_i},k} + \theta_{i,k}\eta_{i,k}^T \mathscr{P}_{i,k}\eta_{i,k} + S_i(z_{i,k}, z_{j_{\hbar_i},k}, \xi_{i,k}) \\
=\ & \begin{bmatrix} 0 & \cdots & \theta_{i,k} & \cdots & \dfrac{\alpha_{j_{\hbar_i},k}}{1 + q_{j_{\hbar_i}}} & \cdots & 0 \end{bmatrix} \mathbf{V}(\eta_k) + S_i(z_{i,k}, z_{j_{\hbar_i},k}, \xi_{i,k}).
\end{aligned}
$$

Then, defining a new matrix W_k with the i-th row being

$$\begin{bmatrix} 0 & \cdots & \theta_{i,k} & \cdots & \dfrac{\alpha_{j_{\hbar_i},k}}{1+q_{j_{\hbar_i}}} & \cdots & 0 \end{bmatrix},$$

it follows from [2, 3] that W_k is the desired dissipation matrix. Consequently,

$$\mathbb{E}\{V_{i,k+1}|\eta_{i,k},\eta_{j_{\hbar_i},k}\} \leq [W_k \mathbf{V}(\eta_k)]_i + S_i(z_{i,k}, z_{j_{\hbar_i},k}, \xi_{i,k})$$

where $[W_k \mathbf{V}(\eta_k)]_i$ is the i-th element of vector $W_k \mathbf{V}(\eta_k)$. By using properties of the conditional expectation, the following inequality can be obtained:

$$\mathbb{E}\{V_{i,k+1}\} \leq \mathbb{E}\{[W_k \mathbf{V}(\eta_k)]_i\} + \mathbb{E}\{S_i(z_{i,k}, z_{j_{\hbar_i},k}, \xi_{i,k})\},$$

which indicates

$$\mathbb{E}\{\mathbf{V}(\eta_{k+1})\} \leq\leq W_k \mathbb{E}\{\mathbf{V}(\eta_k)\} + \mathbb{E}\{\mathbf{S}(z_k, \xi_k)\}. \tag{5.21}$$

Therefore, from Definition 5.2, the filtering error dynamics (5.14) is stochastic vector-dissipative.

Step (2) Proof of the guaranteed H_∞-consensus performance index

Left-multiplying $\mathbf{1}^T$ to both sides of (5.21) yields

$$\mathbf{1}^T \mathbb{E}\{\mathbf{V}(\eta_{k+1})\} \leq \mathbf{1}^T \mathbb{E}\{\mathbf{S}(z_k, \xi_k)\} + \mathbf{1}^T W_k \mathbb{E}\{\mathbf{V}(\eta_k)\}.$$

Denote $v(\eta_k) \triangleq \mathbf{1}^T \mathbb{E}\{\mathbf{V}(\eta_k)\}$. From $S_i(z_{i,k}, z_{j_{\hbar_i},k}, \xi_{i,k})$, it follows that

$$v(\eta_{k+1}) - v(\eta_k) + \sum_{i=1}^{N} \mathbb{E}\left\{\|z_{i,k}\|^2_{Q_{i,k}} + \|z_{j_{\hbar_i},k} - z_{i,k}\|^2_{R_{i,k}}\right\} \leq \gamma^2 \sum_{i=1}^{N} \xi_{i,k}^T \mathscr{T}_{i,k} \xi_{i,k},$$

which further implies that

$$v(\eta_n) - v(\eta_0) + \sum_{i=1}^{N}\sum_{k=0}^{n-1} \mathbb{E}\left\{\|z_{i,k}\|^2_{Q_{i,k}} + \|z_{j_{\hbar_i},k} - z_{i,k}\|^2_{R_{i,k}}\right\} \leq \gamma^2 \sum_{i=1}^{N}\sum_{k=0}^{n-1} \|\xi_{i,k}\|^2_{\mathscr{T}_{i,k}}.$$

Therefore,

$$\sum_{i=1}^{N}\sum_{k=0}^{n-1} \mathbb{E}\left\{\|z_{i,k}\|^2_{Q_{i,k}} + \|z_{j_{\hbar_i},k} - z_{i,k}\|^2_{R_{i,k}}\right\} \leq \sum_{i=1}^{N}\left(\eta_{i,0}^T \mathscr{P}_{i,0} \eta_{i,0} + \gamma^2 \sum_{k=0}^{n-1} \|\xi_{i,k}\|^2_{\mathscr{T}_{i,k}}\right).$$

Taking the initial condition $\mathscr{P}_{i,0} \leq \gamma^2 \mathscr{U}_i$ into consideration, the H_∞-consensus performance constraint (5.15) is satisfied, which completes the proof.

Remark 5.2 In the vector dissipativity theory proposed in [4], the dissipation matrix has been used to formulate the coupling relationship among the neighboring nodes in an SN. In fact, the construction of the dissipation matrix plays a critical role in achieving the scalable design of the distributed filtering. The novelty of constructing the dissipation matrix in this chapter is clarified as follows. Note that, in a representative publication [1], the dissipation matrix W has been constructed as a negative diagonally dominant matrix with the i-th row being

$$\begin{bmatrix} \frac{2\alpha_1 a_{i1}}{1+q_1} - \hat{\delta} & \cdots & -\frac{2q_i\alpha_i}{1+q_i} - \hat{\delta} & \cdots & \frac{2\alpha_N a_{iN}}{1+q_N} - \hat{\delta} \end{bmatrix}$$

where $\hat{\delta}$ guarantees that W is negative diagonally dominant and $\mathbf{1}^T W \ll -\epsilon\mathbf{1}$ ($\epsilon > 0$). In this chapter, a substochastic matrix W_k is constructed, with the i-th row being

$$\begin{bmatrix} 0 & \cdots & \frac{1+q_i(1-\alpha_{i,k})}{1+q_i} & \cdots & \frac{\alpha_{j_{\hbar_i},k}}{1+q_{j_{\hbar_i}}} & \cdots & 0 \end{bmatrix}.$$

It can be seen that the new structure reflects the distinctive scheduling of the RRP, i.e., every row of this matrix has only two nonzero elements: one is the diagonal element $\frac{1+q_i(1-\alpha_{i,k})}{1+q_i}$, and the other is $\frac{\alpha_{j_{\hbar_i},k}}{1+q_{j_{\hbar_i}}}$, where $j_{\hbar_i}$ is the chosen neighboring node of i according to the RRP. Moreover, it can be easily found that $\mathbf{1}^T W_k \ll \mathbf{1}$ and W_k is positive diagonally dominant, i.e.,

$$\frac{1+q_i(1-\alpha_{i,k})}{1+q_i} > \sum_{j=1}^{N} \frac{a_{ji}}{1+q_i} = \frac{q_i}{1+q_i} \geq \sum_{j=1}^{N} W_{ji,k},$$

since $0 < \alpha_{i,k} < \frac{1+q_i}{2q_i}$, where $W_{ji,k}$ is the (j,i)-th element of W_k. From the above comparisons, one can conclude that the dissipation matrix constructed in this chapter is essentially different from the existing ones and are easier to use.

In the following, the distributed filter gains $L_{i,k}$ and $K_{ij,k}$ are designed based on Theorem 5.1. Before giving the design algorithm, the following lemma is established, which is useful in subsequent derivation.

Lemma 5.1 (S-procedure *[5]*) *Let $\mathscr{U}$, $\mathscr{V}_k$, $\mathscr{W}$, and $\mathscr{Q}$ be real matrices of appropriate dimensions with $\mathscr{Q}$ satisfying $\mathscr{Q} = \mathscr{Q}^T$. Then, for all $\mathscr{V}_k\mathscr{V}_k^T \leq I$,*

$$\mathscr{Q} + \mathscr{U}\mathscr{V}_k\mathscr{W} + \mathscr{W}^T\mathscr{V}_k^T\mathscr{U}^T < 0$$

holds if and only if there exists a positive scalar μ such that

$$\mathscr{Q} + \mu^{-1}\mathscr{U}\mathscr{U}^T + \mu\mathscr{W}^T\mathscr{W} < 0.$$

Theorem 5.2 *Given scalars $\alpha_{i,k} \in \mathfrak{I}_{q_i}$ and matrices U_{1i}, U_{2i}, $R_{i,k}$, $Q_{i,k}$, and $\mathscr{T}_{i,k}$, the filtering error system* (5.14) *satisfies the H_∞-consensus performance constraint* (5.15) *with the pre-specified disturbance attenuation level $\gamma > 0$, if there exist matrices $E_{ij_\hbar,k}$ and $F_{i,k}$, positive scalars $\lambda_{i,k}$ and $\mu_{i,k}$, and sequences of positive definite matrices $\{P^1_{i,k}\}_{k\in\mathscr{H}\cup\{n\}}$ and $\{P^2_{i,k}\}_{k\in\mathscr{H}\cup\{n\}}$ with initial conditions $P^1_{i,0} \leq \gamma^2 U_{1i}$ and $P^2_{i,0} \leq \gamma^2 U_{2i}$, such that the following conditions hold for all $k \in \mathscr{H}$, $i \in \mathscr{V}$:*

$$\begin{bmatrix} \Upsilon^a_{ij_\hbar,k} & * \\ \bar{\Upsilon}^b_{ij_\hbar,k} & \bar{\Upsilon}^c_{ij_\hbar,k} \end{bmatrix} < 0 \tag{5.22}$$

where

$$\Upsilon^a_{ij_\hbar,k} = \begin{bmatrix} \Upsilon^{11}_{ij_\hbar,k} & * & * & * \\ \Upsilon^{21}_{ij_\hbar,k} & \Upsilon^{22}_{ij_\hbar,k} & * & * \\ 0 & 0 & -\lambda_{i,k} I & * \\ 0 & 0 & 0 & \Upsilon^{44}_{ij_\hbar,k} \end{bmatrix},$$

$$\bar{\Upsilon}^b_{ij_\hbar,k} = \begin{bmatrix} \bar{\mathscr{A}}_{i,k} & \bar{\mathscr{K}}_{ij_\hbar,k} & \bar{\mathscr{L}}_{i,k} & \bar{\mathscr{B}}_{i,k} \\ 0 & 0 & 0 & 0 \\ 0 & 0 & 0 & 0 \end{bmatrix},$$

$$\bar{\Upsilon}^c_{ij_\hbar,k} = \begin{bmatrix} -\mathscr{P}_{i,k+1} & * & * \\ 0 & -\mathscr{P}_{i,k+1} & * \\ \bar{\beta}_k \bar{\mathscr{M}}^T_k & \sqrt{\varphi_k}\bar{\mathscr{M}}^T_k & -\frac{\mu_{i,k}}{2} I \end{bmatrix},$$

$$\Upsilon^{11}_{ij_\hbar,k} = E^T Q_{i,k} E + E^T R_{i,k} E - \theta_{i,k}\mathscr{P}_{i,k} + \mu_{i,k}\mathscr{N}^T_{1k}\mathscr{N}_{1k} + \lambda_{i,k}\mathscr{C}^T_{i,k}\mathscr{H}^T_i \mathscr{H}_i\mathscr{C}_{i,k},$$

$$\Upsilon^{21}_{ij_\hbar,k} = -E^T R_{i,k} E, \quad \Upsilon^{22}_{ij_\hbar,k} = E^T R_{i,k} E - \frac{\alpha_{j_{\hbar_i},k}}{1+q_{j_{\hbar_i}}}\mathscr{P}_{j_{\hbar_i},k},$$

$$\Upsilon^{44}_{ij_\hbar,k} = -\gamma^2\mathscr{T}_{i,k} + \mu_{i,k}\mathscr{N}^T_{2k}\mathscr{N}_{2k},$$

$$\bar{\mathscr{A}}_{i,k} = \mathrm{diag}\{P^1_{i,k+1}A_k,\ P^2_{i,k+1}A_k - F_{i,k}\bar{H}_i C_{i,k} - E_{ij_\hbar,k}\},$$

$$\bar{\mathscr{K}}_{ij_\hbar,k} = \mathrm{diag}\{0,\ E_{ij_\hbar,k}\}, \quad \mathscr{N}_{1k} = [N_{1k}\ N_{1k}],$$

$$\mathscr{N}_{2k} = [N_{2k}\ \ N_{2k}], \quad \mathscr{P}_{i,k} = \mathrm{diag}\{P^1_{i,k+1},\ P^2_{i,k+1}\},$$

$$\bar{\mathscr{B}}_{i,k} = \begin{bmatrix} P^1_{i,k+1}B_k & 0 \\ P^2_{i,k+1}B_k & -F_{i,k}D_{i,k} \end{bmatrix}, \quad \mathscr{M}_k = \begin{bmatrix} M_k \\ M_k \end{bmatrix},$$

$$\bar{\mathscr{L}}_{i,k} = \begin{bmatrix} 0 & 0 \\ -F_{i,k} & 0 \end{bmatrix}, \quad \bar{\mathscr{M}}_k = \begin{bmatrix} P^1_{i,k+1}M_k \\ P^2_{i,k+1}M_k \end{bmatrix}.$$

Moreover, the filter gains $L_{i,k}$ and $K_{ij,k}$ are given as follows:

$$L_{i,k} = (P^2_{i,k+1})^{-1}F_{i,k}, \quad K_{ij_\hbar,k} = (P^2_{i,k+1})^{-1}E_{ij_\hbar,k}. \tag{5.23}$$

Proof First and foremost, according to (5.3), one has

$$[\Delta\mathscr{A}_k \ \ \Delta\mathscr{B}_k] = \mathscr{M}_k\mathscr{F}_k[\mathscr{N}_{1k} \ \ \mathscr{N}_{2k}] \tag{5.24}$$

where $\mathscr{F}_k \triangleq \mathrm{diag}\{F_k, F_k\}$. To cope with the uncertainties, (5.19) is rewritten as

$$\Xi_{ij\hbar,k} = \Pi_{ij\hbar,k} + \tilde{\mathscr{M}}_k \tilde{\mathscr{F}}_k \tilde{\mathscr{N}}_k + \tilde{\mathscr{N}}_k^T \tilde{\mathscr{F}}_k \tilde{\mathscr{M}}_k^T \tag{5.25}$$

where

$$\begin{aligned}
\Pi_{ij\hbar,k} &= \begin{bmatrix} \Pi^a_{ij\hbar,k} & * \\ \Pi^b_{ij\hbar,k} & \Pi^c_{ij\hbar,k} \end{bmatrix}, \\
\Pi^a_{ij\hbar,k} &= \begin{bmatrix} \Pi^{11}_{ij\hbar,k} & * & * & * \\ \Pi^{21}_{ij\hbar,k} & \Pi^{22}_{ij\hbar,k} & * & * \\ \Pi^{31}_{ij\hbar,k} & 0 & -\lambda_{i,k} I & * \\ 0 & 0 & 0 & -\gamma^2 \mathscr{T}_{i,k} \end{bmatrix}, \\
\Pi^b_{ij\hbar,k} &= \begin{bmatrix} \mathscr{A}_{i,k} & \mathscr{K}_{ij\hbar,k} & \mathscr{L}_{i,k} & \mathscr{B}_{i,k} \\ 0 & 0 & 0 & 0 \end{bmatrix}, \\
\Pi^c_{ij\hbar,k} &= \begin{bmatrix} -\mathscr{P}^{-1}_{i,k+1} & * \\ 0 & -\mathscr{P}^{-1}_{i,k+1} \end{bmatrix}, \\
\tilde{\mathscr{M}}_k &= \begin{bmatrix} 0 & * * & * & * * \\ 0 & 0 * & * & * * \\ 0 & 0 0 & * & * * \\ 0 & 0 0 & 0 & * * \\ \bar{\beta}_k \mathscr{M}_k & 0 0 & \bar{\beta}_k \mathscr{M}_k & 0 * \\ \sqrt{\varphi_k} \mathscr{M}_k & 0 0 & \sqrt{\varphi_k} \mathscr{M}_k & 0 0 \end{bmatrix}, \\
\tilde{\mathscr{F}}_k &= \mathrm{diag}\{\mathscr{F}_k, 0, 0, \mathscr{F}_k, 0, 0\}, \\
\tilde{\mathscr{N}}_k &= \mathrm{diag}\{\mathscr{N}_{1k}, 0, 0, \mathscr{N}_{2k}, 0, 0\}, \\
\Pi^{11}_{ij\hbar,k} &= E^T Q_{i,k} E + E^T R_{i,k} E - \theta_{i,k} \mathscr{P}_{i,k} + \lambda_{i,k} \mathscr{C}^T_{i,k} \mathscr{H}_i^T \mathscr{H}_i \mathscr{C}_{i,k}, \\
\Pi^{22}_{ij\hbar,k} &= E^T R_{i,k} E - \frac{\alpha_{j\hbar_i,k}}{1 + q_{j\hbar_i}} \mathscr{P}_{j\hbar_i,k}, \\
\Pi^{21}_{ij\hbar,k} &= -\, E^T R_{i,k} E, \quad \Pi^{31}_{ij\hbar,k} = \frac{1}{2} \lambda_{i,k} H_i \mathscr{C}_{i,k}.
\end{aligned}$$

Since $\mathscr{F}_k \mathscr{F}_k^T \leq I$, it is inferred from Lemma 5.1 that inequality (5.19) holds if and only if there exists a positive constant $\mu_{i,k}$ such that the following inequality holds:

$$\Pi_{ij\hbar,k} + \mu^{-1}_{i,k} \tilde{\mathscr{M}}_k \tilde{\mathscr{M}}_k^T + \mu_{i,k} \tilde{\mathscr{N}}_k^T \tilde{\mathscr{N}}_k < 0, \tag{5.26}$$

which, by the Schur Complement Lemma, (5.26) holds if and only if the following inequality holds

$$\begin{bmatrix} \Upsilon^a_{ij\hbar,k} & * \\ \Upsilon^b_{ij\hbar,k} & \Upsilon^c_{ij\hbar,k} \end{bmatrix} < 0 \tag{5.27}$$

where

$$\Upsilon^{a}_{ij_{\hbar},k} = \begin{bmatrix} \Upsilon^{11}_{ij_{\hbar},k} & * & * & * \\ \Upsilon^{21}_{ij_{\hbar},k} & \Upsilon^{22}_{ij_{\hbar},k} & * & * \\ \Upsilon^{31}_{ij_{\hbar},k} & 0 & -\lambda_{i,k} I & * \\ 0 & 0 & 0 & \Upsilon^{44}_{ij_{\hbar},k} \end{bmatrix},$$

$$\Upsilon^{b}_{ij_{\hbar},k} = \begin{bmatrix} \mathscr{A}_{i,k} & \mathscr{K}_{ij_{\hbar},k} & \mathscr{L}_{i,k} & \mathscr{B}_{i,k} \\ 0 & 0 & 0 & 0 \\ 0 & 0 & 0 & 0 \end{bmatrix},$$

$$\Upsilon^{c}_{ij_{\hbar},k} = \begin{bmatrix} -\mathscr{P}^{-1}_{i,k+1} & * & * \\ 0 & -\mathscr{P}^{-1}_{i,k+1} & * \\ \bar{\beta}_k \mathscr{M}^T_k & \sqrt{\varphi_k}\mathscr{M}^T_k & -\frac{\mu_{i,k}}{2} I \end{bmatrix}.$$

Then, one can arrive at (5.22) by performing the congruent transformation $\text{diag}\{I, I, I, I, \mathscr{P}_{i,k+1}, \mathscr{P}_{i,k+1}, I\}$ to (5.27). To this end, according to Theorem 5.1, the filtering error system achieves the required H_∞-consensus performance criterion and the proof is thus complete.

Now, consider the following two cases to show the design flexibility of the proposed scheme.

Case 1: For node i with $p_i = 0$ and $q_i > 0$, the corresponding filter structure is simplified to

$$\hat{x}_{i,k+1} = A_k \hat{x}_{i,k} + L_{i,k}(y_{i,k} - H_{i1} C_{i,k} \hat{x}_{i,k}).$$

In this case, the corresponding design algorithm for filter parameter $L_{i,k}$ is presented in the following corollary with proof omitted for conciseness.

Corollary 5.1 *Let the scalar sequence $\alpha_{i,k} \in \mathbb{J}_{q_i}$ and matrices U_{1i}, U_{2i}, $Q_{i,k}$, and $\mathscr{T}_{i,k}$ be given. The filtering error system (5.14) satisfies the H_∞-consensus performance constraint (5.15) with the pre-specified disturbance attenuation level $\gamma > 0$, if there exists matrix $F_{i,k}$, positive scalars $\lambda_{i,k}$ and $\mu_{i,k}$, and sequences of positive definite matrices $\{P^1_{i,k}\}_{k \in \mathscr{H} \cup \{n\}}$ and $\{P^2_{i,k}\}_{k \in \mathscr{H} \cup \{n\}}$ with the initial conditions $P^1_{i,0} \le \gamma^2 U_{1i}$ and $P^2_{i,0} \le \gamma^2 U_{2i}$, such that the following conditions hold for all $k \in \mathscr{H}$, $i \in \mathscr{V}$:*

$$\begin{bmatrix} \widehat{\Upsilon}^a_{i,k} & \\ \widehat{\Upsilon}^b_{i,k} & \Upsilon^c_{i,k} \end{bmatrix} < 0 \tag{5.28}$$

where

$$\widehat{\Upsilon}^a_{i,k} = \begin{bmatrix} \bar{\Upsilon}^{11}_{i,k} & * & * \\ \Upsilon^{31}_{i,k} & -\lambda_{i,k} I & * \\ 0 & 0 & \Upsilon^{44}_{i,k} \end{bmatrix}, \quad \widehat{\Upsilon}^b_{i,k} = \begin{bmatrix} \bar{\mathscr{A}}_{i,k} & \bar{\mathscr{L}}_{i,k} & \bar{\mathscr{B}}_{i,k} \\ 0 & 0 & 0 \\ 0 & 0 & 0 \end{bmatrix},$$

$$\bar{\Upsilon}^{11}_{i,k} = E^T Q_{i,k} E - \theta_{i,k} \mathscr{P}_{i,k} + \mu_{i,k} \mathscr{N}^T_{1k} \mathscr{N}_{1k}.$$

Moreover, the filter gains $L_{i,k}$ is given by

$$L_{i,k} = (P^2_{i,k+1})^{-1} F_{i,k}. \tag{5.29}$$

<u>**Case 2:**</u> When the RRP is not implemented, the filter structure of node i is reformulated as

$$\hat{x}_{i,k+1} = A_k \hat{x}_{i,k} + L_{i,k}(y_{i,k} - H_{i1} C_{i,k} \hat{x}_{i,k}) + \sum_{j \in \mathcal{N}_i} \mathcal{K}_{ij,k}(\hat{x}_{i,k} - \hat{x}_{j,k}).$$

The corresponding filtering scheme is summarized in the following corollary with proof omitted.

Corollary 5.2 *Let the scalar sequence $\alpha_{i,k} \in \mathfrak{I}_{q_i}$ and matrices U_{1i}, U_{2i}, $R_{i,k}$, $Q_{i,k}$, and $\mathcal{T}_{i,k}$ be given. The filtering error system* (5.14) *satisfies the H_∞-consensus performance constraint* (5.15) *with the pre-specified disturbance attenuation level $\gamma > 0$, if there exist matrices $F_{i,k}$, $E_{ij_1,k}$, ..., and $E_{ij_{p_i},k}$, positive scalars $\lambda_{i,k}$ and $\mu_{i,k}$, and sequences of positive definite matrices $\{P^1_{i,k}\}_{k \in \mathcal{H} \cup \{n\}}$ and $\{P^2_{i,k}\}_{k \in \mathcal{H} \cup \{n\}}$ with the initial conditions $P^1_{i,0} \leq \gamma^2 U_{1i}$ and $P^2_{i,0} \leq \gamma^2 U_{2i}$, such that the following conditions hold for all $k \in \mathcal{H}$, $i \in \mathcal{V}$:*

$$\begin{bmatrix} \tilde{\Upsilon}^a_{i,k} & * \\ \tilde{\Upsilon}^b_{i,k} & \bar{\Upsilon}^c_{i,k} \end{bmatrix} < 0 \tag{5.30}$$

where

$$\tilde{\Upsilon}^a_{i,k} = \begin{bmatrix} \Upsilon^{11}_{ij,k} & * & * & * \\ \tilde{\Upsilon}^{21}_{ij,k} & \tilde{\Upsilon}^{22}_{ij,k} & * & * \\ \Upsilon^{31}_{ij,k} & 0 & -\lambda_{i,k} I & * \\ 0 & 0 & 0 & \Upsilon^{44}_{ij,k} \end{bmatrix}, \quad \tilde{\Upsilon}^b_{i,k} = \begin{bmatrix} \bar{\mathscr{A}}_{i,k} & \tilde{\mathcal{K}}_{ij,k} & \bar{\mathscr{L}}_{i,k} & \bar{\mathscr{B}}_{i,k} \\ 0 & 0 & 0 & 0 \\ 0 & 0 & 0 & 0 \end{bmatrix},$$

$$\tilde{\Upsilon}^{22}_{ij,k} = \mathrm{diag}\left\{ E^T R_{i,k} E - \frac{\alpha_{j_{i_1},k}}{1 + q_{j_{i_1}}} \mathscr{P}_{j_{i_1},k}, \ldots, E^T R_{i,k} E - \frac{\alpha_{j_{ip_i},k}}{1 + q_{j_{ip_i}}} \mathscr{P}_{j_{ip_i},k} \right\},$$

$$\tilde{\mathcal{K}}_{ij,k} = \begin{bmatrix} 0 & 0 & \cdots & 0 & 0 \\ 0 & E_{ij_1,k} & \cdots & 0 & E_{ij_{p_i},k} \end{bmatrix}, \quad \tilde{\Upsilon}^{21}_{ij,k} = -[\underbrace{E^T R_{i,k} E \quad \cdots \quad E^T R_{i,k} E}_{p_i}]^T.$$

Moreover, the filter gains $L_{i,k}$ and $K_{ij,k}$ are given as follows:

$$L_{i,k} = (P^2_{i,k+1})^{-1} F_{i,k}, \quad K_{ij,k} = (P^2_{i,k+1})^{-1} E_{ij,k}, \quad j \in \mathcal{N}_i.$$

Remark 5.3 It is observed from Theorem 5.2 and Corollaries 5.1 and 5.2 that a distinctive feature of the above filter design algorithm is its flexible structure. It is worth noting that only the second row and the second column of (5.22) include the information from neighbors of node i. That is, the second row and the second column of (5.22) can be locally adjusted to adapt to the changes of the neighboring nodes. Meanwhile, the above main results are independent of the network size (i.e., the number of sensor nodes in a sensor network). Therefore, it follows from [6, 7] that the distributed algorithm proposed in this chapter is scalable.

Remark 5.4 So far, the design problem of scalable distributed H_∞-consensus filters has been studied for a class of time-varying systems having randomly occurring uncertainties with the RRP. The norm-bounded uncertainties considered in the network occur in a random manner. For the state estimation problem, each node receives the information sent from one neighboring node at each time step, which is determined by the RRP. Then, by resorting to the stochastic analysis method, the filtering algorithm is developed to be executed independently on each node with a pre-specified performance constraint in the H_∞ measure.

Remark 5.5 Compared with the existing results, the distinctive features of the proposed algorithm include: (1) at each time step, only one neighboring node propagates its estimate for consensus thanks to the implementation of the RRP, and this can effectively avoid data collision and reduce communication burden; (2) a new performance index is proposed to characterize the noise attenuation level of filtering errors against external disturbances; and (3) the designed algorithm achieves the desired scalability and possesses high flexibility to the dynamical changes of the network topology. This investigation represents the first of the few attempts to develop local design methods for the distributed H_∞-consensus filtering problem with the RRP.

5.3 An Illustrative Example

In this section, a simulation example is presented to illustrate the effectiveness of the proposed design scheme.

Consider an SN with topology represented by a digraph $\mathscr{G} = (\mathscr{V}, \mathscr{E}, \mathcal{A})$ having the set of nodes $\mathscr{V} = \{1, 2, 3, 4, 5\}$, and the set of edges $\mathscr{E} = \{(2, 1), (3, 2), (4, 3), (5, 4), (5, 1), (2, 4), (5, 3)\}$, as shown in Fig. 5.1. Obviously, $p_1 = 0$, $p_2 = 2$, $p_3 = p_4 = 1$, $p_5 = 3$, and $q_1 = q_3 = q_4 = 2$, $q_2 = 1$, $q_5 = 0$. Clearly, node 2 can receive information from nodes 1 and 4 in turn; node 5 can receive information from nodes 1, 3 and 4 in turn. With the RRP, for a certain time step k, if $\mathrm{mod}(k-1, 2) = 0$, then node 2 can receive information from node 1, otherwise from node 4; if $\mathrm{mod}(k-1, 3) = 0$, then node 5 can receive information from node 1, else if $\mathrm{mod}(k-1, 3) = 1$, then node 5 can receive information from node 3, otherwise from node 4.

As described in [1], the parameter matrices for the continuous-time system are

$$A_c = \begin{bmatrix} -3.2 & 10 & 0 \\ 1 & -1 & 1 \\ 0 & -14.87 & 0 \end{bmatrix}, \quad B_c = \begin{bmatrix} -0.1246 \\ -0.4461 \\ 0.3350 \end{bmatrix}.$$

The nominal part of the above system represents the well-known Chua electronic circuit. By setting the sampling period $\Delta = 0.2$, the parameter matrices of the corresponding discrete system are obtained as:

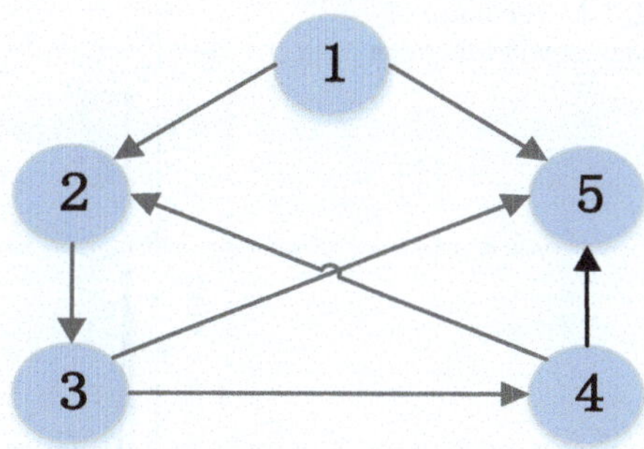

Fig. 5.1 Topology of a Sensor Network

$$A = \begin{bmatrix} 0.6472 & 1.2598 & 0.1492 \\ 0.1260 & 0.7025 & 0.1737 \\ -0.2219 & -2.5834 & 0.7270 \end{bmatrix}, \quad B = \begin{bmatrix} -0.0825 \\ -0.0732 \\ 0.1845 \end{bmatrix}.$$

To reflect the frequently occurred parameter fluctuations, we allow certain time-varying variations and modify the parameter matrices as follows:

$$A_k = \begin{bmatrix} 0.6472 + 0.1\sin(k) & 1.2598 & 0.1492 \\ 0.1260 & 0.7025 & 0.1737 \\ -0.2219 & -2.5834 & 0.7270 \end{bmatrix}, \quad B_k = \begin{bmatrix} -0.0825 + 0.01\sin(k) \\ -0.0732 \\ 0.1845 \end{bmatrix}.$$

The other parameters are set as follows: $C_{1,k} = [0.0032 \ -0.0047 \ 0.001]$, $C_{2,k} = [0.01 \ -0.1 \ -0.013]$, $C_{3,k} = [0.02 \ -0.02 \ 0.03]$, $C_{4,k} = [0.03 \ -0.03 \ 0.03]$, $C_{5,k} = [0.1 \ 0 \ 0]$, $D_{i,k} = 0.025$, $\Delta A_k = \text{diag}\{0.04\sin(k), 0, 0\}$, $\Delta B_k = [0.4\sin(k) \ 0 \ 0]^T$, $F_k = \sin(k)$, $M_k = [0.2 \ 0]^T$, $N_{1k} = [0.2 \ 0]$, $N_{2k} = 0.2$, $z_{i,\max} = 0.3$, $H_{11} = 0.95$, $H_{21} = 0.8$, $H_{31} = 0.75$, $H_{41} = 0.65$, $H_{51} = 0.7$, $H_{12} = 1.65$, $H_{22} = 1.5$, $H_{32} = 1.45$, $H_{42} = 1.35$, $H_{52} = 1.4$, $n = 51$, $\bar{\beta}_k \equiv 0.9$, $\gamma = 0.3$, $\alpha_{i,k} \equiv 0.9$, $\mathscr{P}_{i,0} = \text{diag}\{I, 5I\}$, $R_{i,k} \equiv 0.5I$, $Q_{i,k} \equiv 0.5I$, $U_i = 100\mathscr{P}_{i,0}$, and $\mathscr{T}_{i,k} \equiv \text{diag}\{1, 1\}$. The disturbances are set as $w_k = 0.1\cos(5k)$ and $\xi_{i,k} = 0.1\cos(5k)$. By resorting to the YALMIP toolbox in MATLAB, all the desired filter gains are recursively computed based on Theorem 5.2.

The simulation results are shown in Figs. 5.2, 5.3, 5.4, 5.5 and 5.6, where Fig. 5.2 plots the norms of the disagreement functions, Fig. 5.3, depicts the norms of the filtering errors, and Figs. 5.4, 5.5 and 5.6 show the elements of x_k and their estimations. The j-th element of x_k and its estimation from node i are denoted as $x_k^{(j)}$ and $\hat{x}_{i,k}^{(j)}$ ($j = 1, 2, 3$; $i = 1, \ldots, 5$), respectively. The norms of the filtering errors in the above mentioned two special cases. From the simulation results, it can be observed that the proposed distributed filtering algorithm is indeed effective.

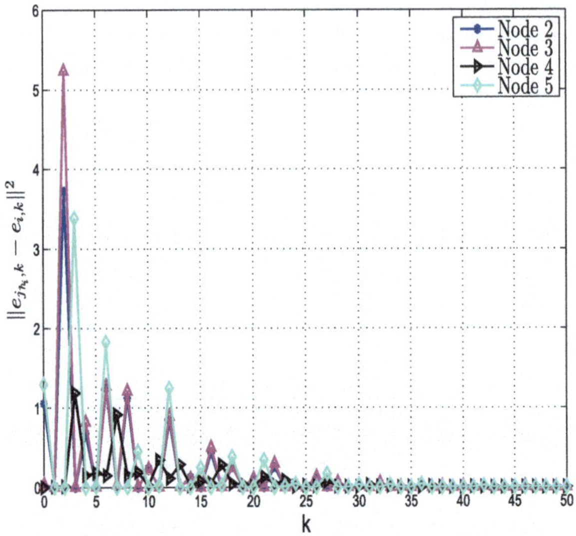

Fig. 5.2 The norm of disagreement functions

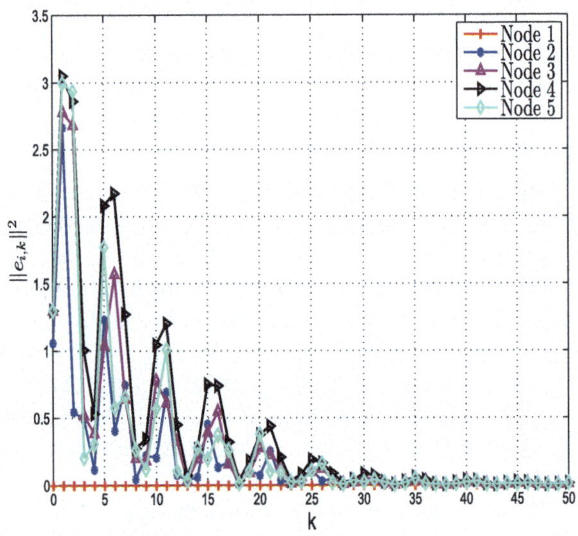

Fig. 5.3 The norm of filtering errors

5.4 Summary

In this chapter, the distributed H_∞-consensus filtering problem has been investigated for a class of discrete time-varying systems with randomly occurring norm-bounded uncertainties and sensor saturations with the RRP over sensor networks. Typical norm-bounded uncertainties have been considered to enter into the network in a random manner. Sensor saturations have been involved, reflecting the limited measurement capacity of nodes in SN. In order to reduce the usage of network resources and

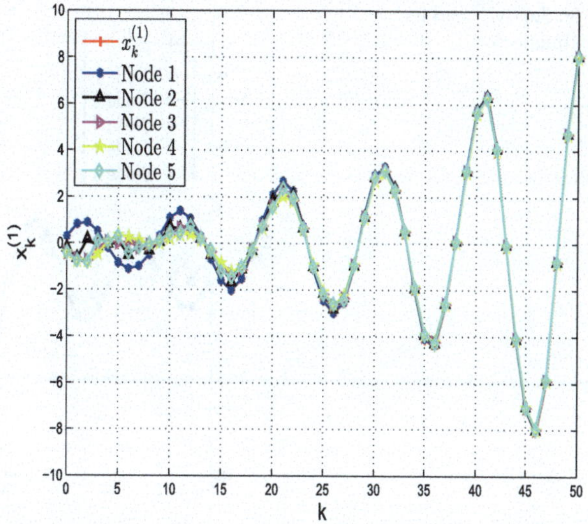

Fig. 5.4 $x_k^{(1)}$ and its estimations

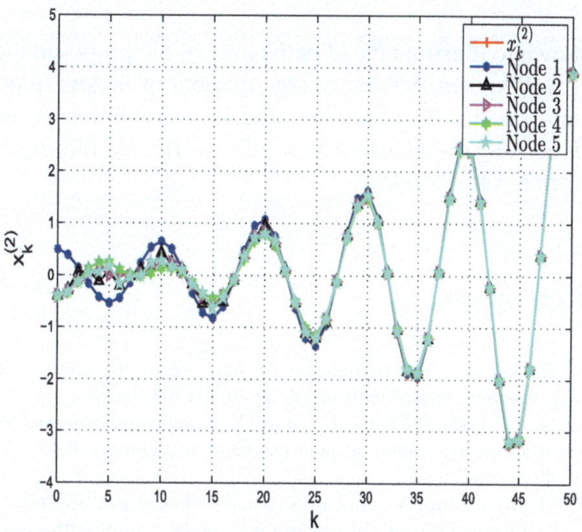

Fig. 5.5 $x_k^{(2)}$ and its estimations

sensor energy, the RRP has been employed to allow every node to receive information only from one neighboring node at each time step. Then, to establish the vector dissipativity of the filtering error dynamics, the supply rate functions have been chosen according to the desirable performance index. Next, the vector dissipation inequality has been transformed to the desirable H_∞-consensus performance index using properties of the dissipation matrix. Sufficient conditions have been established for the filtering error dynamics to achieve certain pre-specified disturbance attenuation criterion in the H_∞ measure. Two special cases have been examined to show the flexibility and scalability of the designed scheme. Finally, a simulation example has been

Fig. 5.6 $x_k^{(3)}$ and its estimations

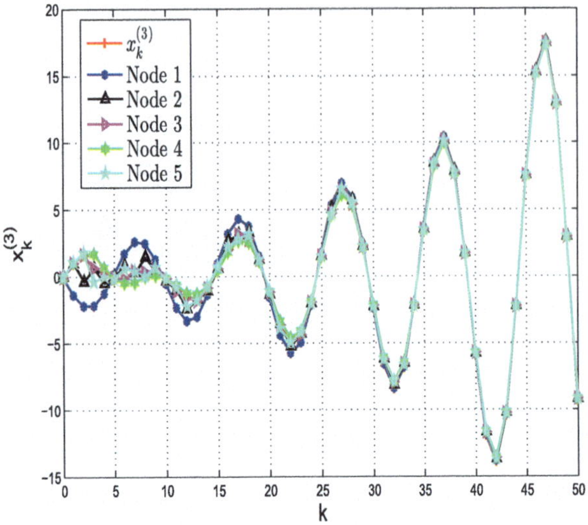

shown to illustrate the effectiveness of the proposed distributed filtering scheme. As is well known, SN has a large number of nodes. In many practical cases, a part of sensor nodes could not measure the information of the plant. As such, it is also an interesting problem to develop the distributed filtering based on the information only from a part of nodes.

References

1. Ugrinovskii, V., Fridman, E.: A round-robin type protocol for distributed estimation with H_∞ consensus. Syst. Control Lett. **69**, 103–110 (2014)
2. Han, F., Wei, G., Ding, D., Song, Y.: Local condition-based consensus filtering with stochastic nonlinearities and multiple missing measurements. IEEE Trans. Autom. Control **62**(9), 4784–4790 (2017)
3. Han, F., Song, Y., Zhang, S., Li, W.: Local condition-based finite-horizon distributed H_∞-consensus filtering for random parameter system with event-triggering protocols. Neurocomputing **219**, 221–231 (2017)
4. Haddad, W., Chellabolna, V., Nersesov, S.: Vector dissipativity theory and stability of feedback interconnections for large-scale non-linear dynamical systems. Int. J. Control **77**(10), 907–919 (2004)
5. Boyd, S., Ghaoui, L., Feron, E., Balakrishnan, V.: Linear Matrix Inequalities in System and Control Theory. SIAM, Philadelphia, PA (1994)
6. Olfati-Saber, R.: Distributed Kalman filter with embedded consensus filters. In: Proceedings of the 44th IEEE Conference on Decision and Control, and the European Control Conference, pp. 8179–8184 (2005)
7. Olfati-Saber, R.: Kalman-consensus filter: optimality, stability and performance. In: Proceedings of the 48th IEEE Conference on Decision and Control, and 28th Chinese Control Conference, pp. 7036–7042 (2009)

Chapter 6
Partial-Nodes-Based Scalable H_∞-Consensus Filtering with Censored Measurements

The local performance analysis method is conducted for each node based on its available information. As such, such design idea reflects the differences among nodes in an SN. When only a partial of nodes accesses the measurement information, the distributed filtering scheme via the local performance analysis method should embody the differences with or without the measurement information except as the structure of distributed filter.

This chapter deals with the scalable distributed H_∞-consensus filtering problem for a class of discrete time-varying systems subject to multiplicative noises and censored measurements over SNs. For the underlying SN, it is assumed that only the measurement outputs from partial sensor nodes are available. Also, the phenomenon of censored measurements is taken into account to reflect the limited capability in measuring. A novel H_∞-consensus performance index is put forward to evaluate the disturbance rejection level of the filters against the simultaneous presence of external disturbances, initial conditions as well as censoring effects. By utilizing the vector dissipativity theory and the recursive matrix inequality technique, sufficient conditions are established under which the prescribed H_∞-consensus performance index is achieved. The parameters of the desired distributed filters are calculated via solving certain matrix inequalities, where such a calculation is conducted in a local sense so as to preserve the scalability of the filter design. Finally, an illustrative simulation example is provided to demonstrate the effectiveness and applicability of the proposed filtering strategy.

F. Han et al., *Distributed Filtering, Control and Synchronization*,
Studies in Systems, Decision and Control 428,
https://doi.org/10.1007/978-3-030-97075-8_6

6.1 Problem Formulation

First of all, consider a class of discrete time-varying systems defined on $k \in \mathscr{H} \triangleq \{0, 1, \ldots, n-1\}$:

$$x_{k+1} = \left(A_{0,k} + \sum_{t=1}^{r} \phi_{t,k} A_{t,k}\right) x_k + B_k w_k \tag{6.1}$$

where $x_k \in \mathbb{R}^{n_x}$ is the system state, $w_k \in \mathbb{R}^{n_w}$ is the external disturbance belonging to $l_2[0, n-1]$ and $\phi_{t,k}$ $(t = 1, 2, \ldots, r) \in \mathbb{R}$ are mutually independent multiplicative noises with zero means and unity variances. $A_{0,k}$, $A_{t,k}$, and B_k are known time-varying matrices with compatible dimensions.

Without loss of generality, for the SN under consideration, it is assumed that the measurement outputs of the first l_0 $(l_0 < N)$ nodes are available with the following form:

$$y^*_{i,k} = C_{i,k} x_k + D_{i,k} \xi_{i,k}, \quad \forall i = 1, 2, \ldots, l_0 \tag{6.2}$$

where $y^*_{i,k} \triangleq [y^*_{i,k}(1)\ y^*_{i,k}(2)\ \cdots\ y^*_{i,k}(n_y)\,]^T \in \mathbb{R}^{n_y}$ is the measurement output of node i and $\xi_{i,k} \in \mathbb{R}^{n_\vartheta}$ is the external disturbance belonging to $l_2[0, n-1]$. $C_{i,k}$ and $D_{i,k}$ are known time-varying matrices with compatible dimensions.

As stated in [1–5], in practical engineering, sensors often suffer from the measurement censoring, which can be formulated by the following well-known Tobit type I model:

$$y_{i,k}(t) = \begin{cases} y^*_{i,k}(t), & \text{if } y^*_{i,k}(t) > \Gamma_i(t); \\ \Gamma_i(t), & \text{otherwise} \end{cases} \tag{6.3}$$

where $\Gamma_i(t) \in \mathbb{R}$ is a known threshold and $t = 1, 2, \ldots, n_y$.

For convenience, introduce the following indicative variable $p_{i,k}(t)$:

$$p_{i,k}(t) \triangleq \begin{cases} 1, & \text{if } y^*_{i,k}(t) > \Gamma_i(t); \\ 0, & \text{otherwise.} \end{cases} \tag{6.4}$$

Then, according to (6.3)-(6.4), the actual measurement output $y_{i,k}(t)$ is described as follows:

$$y_{i,k}(t) = p_{i,k}(t) y^*_{i,k}(t) + (1 - p_{i,k}(t))\Gamma_i(t). \tag{6.5}$$

Denoting

$$\begin{aligned} \Gamma_i &\triangleq [\Gamma_i(1)\ \Gamma_i(2)\ \cdots\ \Gamma_i(n_y)]^T, \\ y_{i,k} &\triangleq [y_{i,k}(1)\ y_{i,k}(2)\ \cdots\ y_{i,k}(n_y)]^T, \\ p_{i,k} &\triangleq \text{diag}\{p_{i,k}(1), p_{i,k}(2), \ldots, p_{i,k}(n_y)\}, \end{aligned}$$

one has

$$y_{i,k} = p_{i,k} y_{i,k}^* + (I - p_{i,k})\Gamma_i. \tag{6.6}$$

Noticing that the measurement outputs are only available from partial nodes, the following PNB-based distributed filters are constructed:

$$\begin{cases} \hat{x}_{i,k+1} = A_{0,k}\hat{x}_{i,k} + L_{i,k}(y_{i,k} - C_{i,k}\hat{x}_{i,k}) + \sum\limits_{j\in\mathcal{N}_i} K_{ij,k}(\hat{x}_{j,k} - \hat{x}_{i,k}), \\ \qquad i = 1, 2, \ldots, l_0, \\ \hat{x}_{i,k+1} = A_{0,k}\hat{x}_{i,k} + \sum\limits_{j\in\mathcal{N}_i} K_{ij,k}(\hat{x}_{j,k} - \hat{x}_{i,k}), \\ \qquad i = l_0 + 1, l_0 + 2, \ldots, N \end{cases} \tag{6.7}$$

where $\hat{x}_{i,k} \in \mathbb{R}^{n_x}$ is the estimate of x_k by node i, and $L_{i,k}$ and $K_{ij,k}$ are the filter gains to be determined later.

Setting $e_{i,k} \triangleq x_k - \hat{x}_{i,k}$, the filtering error dynamics of node i is described from (6.1)-(6.7) as follows:

$$\begin{cases} e_{i,k+1} = \left(A_{0,k} - L_{i,k}C_{i,k} - \sum\limits_{j\in\mathcal{N}_i} K_{ij,k}\right)e_{i,k} + \sum\limits_{t=1}^{r}\phi_{t,k}A_{t,k}x_k - L_{i,k}p_{i,k}D_{i,k}\xi_{i,k} \\ \qquad + B_k w_k - L_{i,k}(I - p_{i,k})(\Gamma_i - C_{i,k}x_k) + \sum\limits_{j\in\mathcal{N}_i} K_{ij,k}e_{j,k}, \\ \qquad i = 1, 2, \ldots, l_0, \\ e_{i,k+1} = \left(A_{0,k} - \sum\limits_{j\in\mathcal{N}_i} K_{ij,k}\right)e_{i,k} + \sum\limits_{t=1}^{r}\phi_{t,k}A_{t,k}x_k + B_k w_k + \sum\limits_{j\in\mathcal{N}_i} K_{ij,k}e_{j,k}, \\ \qquad i = l_0 + 1, l_0 + 2, \ldots, N. \end{cases} \tag{6.8}$$

Denoting $\eta_{i,k} \triangleq [x_k^T\ e_{i,k}^T]^T$, $v_{i,k} \triangleq [w_k^T\ \xi_{i,k}^T]^T$, $\varpi_i \triangleq [0\ \Gamma_i^T]^T$, $z_{i,k} \triangleq \mathcal{F}\eta_{i,k}$, and $\mathcal{F} \triangleq [0\ I]$, the filtering error system is obtained with the following compact form:

$$\begin{cases} \eta_{i,k+1} = \left(\mathscr{A}_{0,k} + \mathscr{C}_{i,k} - \sum_{j\in\mathscr{N}_i} \mathscr{K}_{ij,k}\right)\eta_{i,k} + \sum_{j\in\mathscr{N}_i} \mathscr{K}_{ij,k}\eta_{j,k} \\ \qquad + (\mathscr{B}_k + \mathscr{D}_{i,k})v_{i,k} - \mathscr{E}_{i,k}\varpi_i + \sum_{t=1}^{r} \phi_{t,k}\mathscr{A}_{t,k}\eta_{i,k}, \\ \qquad i = 1, 2, \ldots, l_0, \\ \eta_{i,k+1} = \left(\mathscr{A}_{0,k} - \sum_{j\in\mathscr{N}_i} \mathscr{K}_{ij,k}\right)\eta_{i,k} + \mathscr{B}_k v_{i,k} + \sum_{t=1}^{r} \phi_{t,k}\mathscr{A}_{t,k}\eta_{i,k} \\ \qquad + \sum_{j\in\mathscr{N}_i} \mathscr{K}_{ij,k}\eta_{j,k}, \\ \qquad i = l_0 + 1, l_0 + 2, \ldots, N, \\ z_{i,k} = \mathscr{F}\eta_{i,k}, \quad i = 1, 2, \ldots, N \end{cases} \tag{6.9}$$

where

$$\begin{aligned} \mathscr{A}_{0,k} &= \mathrm{diag}\{A_{0,k}, A_{0,k}\}, \quad \mathscr{D}_{i,k} = \mathrm{diag}\{0, -L_{i,k}p_{i,k}D_{i,k}\}, \\ \mathscr{E}_{i,k} &= \mathrm{diag}\{0, L_{i,k}(1 - p_{i,k})\}, \quad \mathscr{K}_{ij,k} = \mathrm{diag}\{0, K_{ij,k}\}, \\ \mathscr{A}_{t,k} &= \begin{bmatrix} A_{t,k} & 0 \\ A_{t,k} & 0 \end{bmatrix}, \quad \mathscr{B}_k = \begin{bmatrix} B_k & 0 \\ B_k & 0 \end{bmatrix}, \\ \mathscr{C}_{i,k} &= \begin{bmatrix} 0 & 0 \\ L_{i,k}(1 - p_{i,k})C_{i,k} & -L_{i,k}C_{i,k} \end{bmatrix}. \end{aligned}$$

In what follows, we are to examine the impact from the censored measurements on the system performance. Before proceeding, the following notations are introduced:

$$\begin{aligned} v_k &\triangleq [v_{1,k}^T \cdots v_{N,k}^T]^T, \\ z_k &\triangleq [z_{1,k}^T \cdots z_{N,k}^T]^T, \\ \bar{\mathbf{W}}_{i,k} &\triangleq [W_{i,k}(1)\ W_{i,k}(2) \cdots W_{i,k}(n_y)], \\ \mathbf{exp}(\varpi_i) &\triangleq [\exp(\Gamma_i(1))\ \exp(\Gamma_i(2)) \cdots \exp(\Gamma_i(n_y))]^T \end{aligned}$$

where $W_{i,k}(t) > 0$ for $t = 1, 2, \ldots, n_y$.

Definition 6.1 Let the disturbance attenuation level $\gamma > 0$, the weighted matrices $U_{i1} > 0$, $U_{i2} > 0$, $R_{i,k} > 0$, $Q_{i,k} > 0$, $T_{i1,k} > 0$, $T_{i2,k} > 0$ and $\bar{\mathbf{W}}_{i,k}$ be given. The filtering error system (6.9) is said to satisfy the H_∞-consensus performance constraint over the finite horizon $\mathscr{H}$ if the following inequality

$$\mathbb{E}\left\{\sum_{k=0}^{n-1}\Phi_{\mathscr{G}}(z_k)\right\} \leq \gamma^2\sum_{i=1}^{l_0}\sum_{k=0}^{n-1}\bar{\mathbf{W}}_{i,k}\mathbf{exp}(\varpi_i) + \gamma^2\sum_{i=1}^{N}\left(\eta_{i,0}^T\mathscr{U}_i\eta_{i,0} + \sum_{k=0}^{n-1}\|v_{i,k}\|^2_{\mathscr{T}_{i,k}}\right) \tag{6.10}$$

holds, where

$$\Phi_{\mathscr{G}}(z_k) \triangleq \sum_{i=1}^{N}\left(\sum_{j\in\mathscr{N}_i}\|z_{j,k} - z_{i,k}\|^2_{R_{i,k}} + \|z_{i,k}\|^2_{Q_{i,k}}\right),$$

$$\mathscr{U}_i \triangleq \text{diag}\{U_{i1}, U_{i2}\}, \quad \mathscr{T}_{i,k} \triangleq \text{diag}\{T_{i1,k}, T_{i2,k}\}.$$

Remark 6.1 In comparison with the traditional H_∞-consensus performance index, the effect of the measurement censoring is reflected in inequality (6.10) by introducing the second term on its right-hand side. It is noted that such a term is characterized by a weighted sum with respect to an exponential function of the censoring threshold over the time horizon $\mathscr{H}$, which would increase as the censoring threshold grows. This implies that the H_∞-consensus performance would become worse as the censoring phenomena becomes more severe, which is in agreement with the engineering practice. As such, the proposed new performance index (6.10) accounts for the influence of the measurement censoring very well.

The purpose of this chapter is to determine the filter gains $L_{i,k}$ and $K_{ij,k}$ such that the dynamics of the system (6.9) satisfies the proposed H_∞-consensus performance constraint (6.10) over the finite horizon $\mathscr{H}$.

6.2 Main Results

In this section, by resorting to the vector dissipativity theory, the analysis and design problems of the distributed H_∞-consensus filter (6.7) will be discussed. To facilitate the subsequent development, we give the following notations

$$\begin{aligned}
\varpi &\triangleq [\varpi_1^T \ \varpi_2^T \ \cdots \ \varpi_{l_0}^T]^T, \\
S_i(z_k, v_{i,k}, \varpi_i) &\triangleq \gamma^2\left(\|v_{i,k}\|^2_{\mathscr{T}_{i,k}} + \bar{\mathbf{W}}_{i,k}\mathbf{exp}(\varpi_i)\right) - \sum_{j\in\mathscr{N}_i}\|z_{j,k} - z_{i,k}\|^2_{R_{i,k}} - \|z_{i,k}\|^2_{Q_{i,k}}, \\
& i = 1, 2, \ldots, l_0, \\
S_i(z_k, v_{i,k}) &\triangleq \gamma^2\|v_{i,k}\|^2_{\mathscr{T}_{i,k}} - \|z_{i,k}\|^2_{Q_{i,k}} - \sum_{j\in\mathscr{N}_i}\|z_{j,k} - z_{i,k}\|^2_{R_{i,k}}, \\
& i = l_0 + 1, l_0 + 2, \ldots, N.
\end{aligned} \tag{6.11}$$

Definition 6.2 The filtering error dynamics (6.9) is said to be strictly stochastic vector-dissipative over the finite horizon $\mathscr{H}$ with respect to the following vector-valued supply rate function $\mathbf{S}(z_k, v_k, \varpi) \triangleq [S_1(z_k, v_{1,k}, \varpi_1) \quad \cdots \quad S_{l_0}(z_k, v_{l_0,k}, \varpi_{l_0}) \quad S_{l_0+1}(z_k, v_{l_0+1,k}) \quad \cdots \quad S_N(z_k, v_{N,k})]^T$, if there exists a vector of nonnegative definite storage functions $\mathbf{V}(\eta_k) \triangleq [V_1(\eta_{1,k}) \quad \cdots \quad V_N(\eta_{N,k})]^T$ (with $\mathbf{V}(0) = 0$) and a nonsingular column substochastic dissipation matrix sequence $U_k \in \mathbb{R}^{N\times N}$ such that the following vector dissipation inequality is satisfied for any $k \in \mathscr{H}$:

$$\mathbb{E}\{\mathbf{V}(\eta_{k+1})\} \ll U_k\mathbb{E}\{\mathbf{V}(\eta_k)\} + \mathbb{E}\{\mathbf{S}(z_k, v_k, \varpi)\}. \tag{6.12}$$

In the following, a set of sufficient conditions is provided to guarantee that the filtering error dynamics (6.9) is stochastic vector-dissipative over the finite horizon $\mathscr{H}$. In order to construct a desirable dissipation matrix U_k, define an interval-valued function $\mathfrak{I}_{q_i}$ with respect to out-degree q_i as follows:

$$\mathfrak{I}_{q_i} = \begin{cases} (0, \frac{1+q_i}{2q_i}), & \text{if } q_i \neq 0; \\ (0, 1], & \text{if } q_i = 0. \end{cases} \tag{6.13}$$

Theorem 6.1 *Let the disturbance attenuation level $\gamma > 0$, the scalar sequence $\alpha_{i,k} \in \mathfrak{I}_{q_i}$, the weighted matrices $R_{i,k} > 0$, $Q_{i,k} > 0$, $\mathscr{T}_{i,k} > 0$, and $\bar{\mathbf{W}}_{i,k}$, and the filter parameters $L_{i,k}$ and $K_{ij,k}$ be given. The filtering error dynamics (6.9) is stochastic vector-dissipative over the finite horizon $\mathscr{H}$ with respect to the vector-valued supply rate $\mathbf{S}(z_k, v_k, \varpi)$ and also satisfies the H_∞-consensus performance constraint (6.10), if there exists a vector-valued storage function $\mathbf{V}(\eta_k)$ (whose element is $V_i(\eta_{i,k}) = \eta_{i,k}^T \mathscr{P}_{i,k}\eta_{i,k}$, where $\{\mathscr{P}_{i,k}\}_{k\in\mathscr{H}\cup\{n\}}$ is a sequence of positive definite matrices with the initial condition $\mathscr{P}_{i,0} \leq \gamma^2 U_i$), such that the following conditions hold for $k \in \mathscr{H}$, $\forall i = 1, 2, \ldots, N$:*

$$\begin{cases} \Xi_{i,k} \triangleq \begin{bmatrix} \Xi_{i,k}^{11} & \Xi_{i,k}^{12} & \Xi_{i,k}^{13} \\ * & \Xi_{i,k}^{22} & \Xi_{i,k}^{23} \\ * & * & \Xi_{i,k}^{33} \end{bmatrix} < 0, & (6.14a) \\ \Psi_{i,k} \triangleq -\frac{1}{\gamma^2}(\mathscr{B}_k + \mathscr{D}_{i,k})\mathscr{T}_{i,k}^{-1}(\mathscr{B}_k + \mathscr{D}_{i,k})^T + \mathscr{P}_{i,k+1}^{-1} > 0, & (6.14b) \\ \forall i = 1, 2, \ldots, l_0, \end{cases}$$

and

$$\begin{cases} \Upsilon_{i,k} \triangleq \begin{bmatrix} \Upsilon_{i,k}^{11} & \Upsilon_{i,k}^{12} \\ * & \Upsilon_{i,k}^{22} \end{bmatrix} < 0, & (6.15a) \\ \bar{\Psi}_{i,k} \triangleq \mathscr{P}_{i,k+1}^{-1} - \frac{1}{\gamma^2}\mathscr{B}_k\mathscr{T}_{i,k}^{-1}\mathscr{B}_k^T > 0, & (6.15b) \\ \forall i = l_0 + 1, l_0 + 2, \ldots, N \end{cases}$$

where

$$
\begin{aligned}
\Xi_{i,k}^{11} =& \sum_{t=1}^{r} \mathscr{A}_{t,k}^T \mathscr{P}_{i,k+1} \mathscr{A}_{t,k} + \left(\mathscr{A}_{0,k} + \mathscr{C}_{i,k} - \sum_{j \in \mathscr{N}_i} \mathscr{K}_{ij,k} \right)^T \Psi_{i,k}^{-1} \left(\mathscr{A}_{0,k} + \mathscr{C}_{i,k} - \sum_{j \in \mathscr{N}_i} \mathscr{K}_{ij,k} \right) \\
& + \mathscr{F}^T (Q_{i,k} + R_{i,k}) \mathscr{F} - \rho_{i,k} \mathscr{P}_{i,k}, \\
\Xi_{i,k}^{12} =& \left[\left(\mathscr{A}_{0,k} + \mathscr{C}_{i,k} - \sum_{j \in \mathscr{N}_i} \mathscr{K}_{ij,k} \right)^T \Psi_{i,k}^{-1} \mathscr{K}_{ij_1,k} - \mathscr{F}^T R_{i,k} \mathscr{F} \cdots \right. \\
& \left. \left(\mathscr{A}_{0,k} + \mathscr{C}_{i,k} - \sum_{j \in \mathscr{N}_i} \mathscr{K}_{ij,k} \right)^T \Psi_{i,k}^{-1} \mathscr{K}_{ij_{p_i},k} - \mathscr{F}^T R_{i,k} \mathscr{F} \right], \\
\Xi_{i,k}^{22} =& \left[\mathscr{K}_{ij_1,k} \cdots \mathscr{K}_{ij_{p_i},k} \right]^T \Psi_{i,k}^{-1} \left[\mathscr{K}_{ij_1,k} \cdots \mathscr{K}_{ij_{p_i},k} \right] \\
& - \mathrm{diag} \left\{ \frac{\alpha_{j_{i_1},k}}{1 + q_{j_{i_1}}} \mathscr{P}_{j_{i_1},k} - \mathscr{F}^T R_{i,k} \mathscr{F}, \ldots, \frac{\alpha_{j_{ip_i},k}}{1 + q_{j_{ip_i}}} \mathscr{P}_{j_{ip_i},k} - \mathscr{F}^T R_{i,k} \mathscr{F} \right\}, \\
\Xi_{i,k}^{13} =& - \left(\mathscr{A}_{0,k} + \mathscr{C}_{i,k} - \sum_{j \in \mathscr{N}_i} \mathscr{K}_{ij,k} \right)^T \Psi_{i,k}^{-1} \mathscr{E}_{i,k} \varpi_i, \\
\Xi_{i,k}^{23} =& - \left[(\mathscr{K}_{ij_1,k} \Psi_{i,k}^{-1} \mathscr{E}_{i,k} \varpi_i)^T \cdots (\mathscr{K}_{ij_{p_i},k} \Psi_{i,k}^{-1} \mathscr{E}_{i,k} \varpi_i)^T \right]^T, \\
\Xi_{i,k}^{33} =& \varpi_i^T \mathscr{E}_{i,k}^T \Psi_{i,k}^{-1} \mathscr{E}_{i,k} \varpi_i - \gamma^2 \bar{\mathbf{W}}_{i,k} \mathbf{exp}(\varpi_i), \\
\Upsilon_{i,k}^{11} =& \sum_{t=1}^{r} \mathscr{A}_{t,k}^T \mathscr{P}_{i,k+1} \mathscr{A}_{t,k} + \left(\mathscr{A}_{0,k} - \sum_{j \in \mathscr{N}_i} \mathscr{K}_{ij,k} \right)^T \bar{\Psi}_{i,k}^{-1} \left(\mathscr{A}_{0,k} - \sum_{j \in \mathscr{N}_i} \mathscr{K}_{ij,k} \right) \\
& + \mathscr{F}^T (Q_{i,k} + R_{i,k}) \mathscr{F} - \rho_{i,k} \mathscr{P}_{i,k}, \\
\Upsilon_{i,k}^{12} =& \left[\left(\mathscr{A}_{0,k} - \sum_{j \in \mathscr{N}_i} \mathscr{K}_{ij,k} \right)^T \bar{\Psi}_{i,k}^{-1} \mathscr{K}_{ij_1,k} - \mathscr{F}^T R_{i,k} \mathscr{F} \ \ldots \right. \\
& \left. \left(\mathscr{A}_{0,k} - \sum_{j \in \mathscr{N}_i} \mathscr{K}_{ij,k} \right)^T \bar{\Psi}_{i,k}^{-1} \mathscr{K}_{ij_{p_i},k} - \mathscr{F}^T R_{i,k} \mathscr{F} \right], \\
\Upsilon_{i,k}^{22} =& - \mathrm{diag} \left\{ \frac{\alpha_{j_{i_1},k}}{1 + q_{j_{i_1}}} \mathscr{P}_{j_{i_1},k} - \mathscr{F}^T R_{i,k} \mathscr{F}, \ldots, \frac{\alpha_{j_{ip_i},k}}{1 + q_{j_{ip_i}}} \mathscr{P}_{j_{ip_i},k} - \mathscr{F}^T R_{i,k} \mathscr{F} \right\} \\
& + \left[\mathscr{K}_{ij_1,k} \cdots \mathscr{K}_{ij_{p_i},k} \right]^T \bar{\Psi}_{i,k}^{-1} \left[\mathscr{K}_{ij_1,k} \cdots \mathscr{K}_{ij_{p_i},k} \right].
\end{aligned}
$$

Proof For presentation clarity, the proof is divided into the following four steps.

(1) *Proof of the stochastic vector-dissipativity for the nodes* i $(i = 1, 2, \ldots, l_0)$ *over the finite horizon* $\mathscr{H}$.

First, for node i $(i = 1, 2, \ldots, l_0)$, it follows that

$$
\begin{aligned}
&\mathbb{E}\{V_i(\eta_{i,k+1})|\eta_{i,k},\eta_{j,k}, j\in\mathcal{N}_i\}\\
=&\mathbb{E}\{\eta_{i,k+1}^T\mathcal{P}_{i,k+1}\eta_{i,k+1}|\eta_{i,k},\eta_{j,k}, j\in\mathcal{N}_i\}\\
=&\eta_{i,k}^T\left(\mathcal{A}_{0,k}+\mathcal{C}_{i,k}-\sum_{j\in\mathcal{N}_i}\mathcal{K}_{ij,k}\right)^T\mathcal{P}_{i,k+1}\left(\mathcal{A}_{0,k}+\mathcal{C}_{i,k}-\sum_{j\in\mathcal{N}_i}\mathcal{K}_{ij,k}\right)\eta_{i,k}\\
&+\eta_{i,k}^T\sum_{t=1}^{r}\mathcal{A}_{t,k}^T\mathcal{P}_{i,k+1}\mathcal{A}_{t,k}\eta_{i,k}+\nu_{i,k}^T(\mathcal{B}_k+\mathcal{D}_{i,k})^T\mathcal{P}_{i,k+1}(\mathcal{B}_k+\mathcal{D}_{i,k})\nu_{i,k}\\
&+\left(\sum_{j\in\mathcal{N}_i}\eta_{j,k}^T\mathcal{K}_{ij,k}^T\right)\mathcal{P}_{i,k+1}\left(\sum_{j\in\mathcal{N}_i}\mathcal{K}_{ij,k}\eta_{j,k}\right)+\varpi_i^T\mathcal{E}_{i,k}^T\mathcal{P}_{i,k+1}\mathcal{E}_{i,k}\varpi_i\\
&+2\eta_{i,k}^T\left(\mathcal{A}_{0,k}+\mathcal{C}_{i,k}-\sum_{j\in\mathcal{N}_i}\mathcal{K}_{ij,k}\right)^T\mathcal{P}_{i,k+1}(\mathcal{B}_k+\mathcal{D}_{i,k})\nu_{i,k}\\
&+2\eta_{i,k}^T\left(\mathcal{A}_{0,k}+\mathcal{C}_{i,k}-\sum_{j\in\mathcal{N}_i}\mathcal{K}_{ij,k}\right)^T\mathcal{P}_{i,k+1}\sum_{j\in\mathcal{N}_i}\mathcal{K}_{ij,k}\eta_{j,k}\\
&-2\eta_{i,k}^T\left(\mathcal{A}_{0,k}+\mathcal{C}_{i,k}-\sum_{j\in\mathcal{N}_i}\mathcal{K}_{ij,k}\right)^T\mathcal{P}_{i,k+1}\mathcal{E}_{i,k}\varpi_i\\
&+2\nu_{i,k}^T(\mathcal{B}_k+\mathcal{D}_{i,k})^T\mathcal{P}_{i,k+1}\sum_{j\in\mathcal{N}_i}\mathcal{K}_{ij,k}\eta_{j,k}\\
&-2\nu_{i,k}^T(\mathcal{B}_k+\mathcal{D}_{i,k})^T\mathcal{P}_{i,k+1}\mathcal{E}_{i,k}\varpi_i\\
&-2\varpi_i^T\mathcal{E}_{i,k}^T\mathcal{P}_{i,k+1}\sum_{j\in\mathcal{N}_i}\mathcal{K}_{ij,k}\eta_{j,k}\\
&+\gamma^2\nu_{i,k}^T\mathcal{T}_{i,k}\nu_{i,k}-\gamma^2\nu_{i,k}^T\mathcal{T}_{i,k}\nu_{i,k}.
\end{aligned}\tag{6.16}
$$

Next, it is readily obtained from (6.14b) that $\Psi_{i,k}>0$. Then, by applying the Schur Complement Lemma to (6.14b), one has

$$
\Phi_{i,k}^{-1}\triangleq\gamma^2\mathcal{T}_{i,k}-(\mathcal{B}_k+\mathcal{D}_{i,k})^T\mathcal{P}_{i,k+1}(\mathcal{B}_k+\mathcal{D}_{i,k})>0.
$$

On the other hand, it is easy to see that the following inequality is true:

$$\Bigg(\nu_{i,k} - \Phi_{i,k}(\mathscr{B}_k + \mathscr{D}_{i,k})^T \mathscr{P}_{i,k+1}(\mathscr{A}_{0,k} + \mathscr{C}_{i,k} - \sum_{j\in\mathscr{N}_i} \mathscr{K}_{ij,k})\eta_{i,k}$$

$$-\Phi_{i,k}(\mathscr{B}_k + \mathscr{D}_{i,k})^T \mathscr{P}_{i,k+1} \sum_{j\in\mathscr{N}_i} \mathscr{K}_{ij,k}\eta_{j,k} + \Phi_{i,k}(\mathscr{B}_k + \mathscr{D}_{i,k})^T \mathscr{P}_{i,k+1}\mathscr{E}_{i,k}\varpi_i\Bigg)^T$$

$$\times \Phi_{i,k}^{-1}\Bigg(\nu_{i,k} - \Phi_{i,k}(\mathscr{B}_k + \mathscr{D}_{i,k})^T \mathscr{P}_{i,k+1}(\mathscr{A}_{0,k} + \mathscr{C}_{i,k} - \sum_{j\in\mathscr{N}_i} \mathscr{K}_{ij,k})\eta_{i,k}$$

$$-\Phi_{i,k}(\mathscr{B}_k + \mathscr{D}_{i,k})^T \mathscr{P}_{i,k+1} \sum_{j\in\mathscr{N}_i} \mathscr{K}_{ij,k}\eta_{j,k} + \Phi_{i,k}(\mathscr{B}_k + \mathscr{D}_{i,k}) \mathscr{P}_{i,k+1}\mathscr{E}_{i,k}\varpi_i\Bigg) \geq 0,$$

and therefore

$$\begin{aligned}
&\nu_{i,k}^T((\mathscr{B}_k + \mathscr{D}_{i,k})^T \mathscr{P}_{i,k+1}(\mathscr{B}_k + \mathscr{D}_{i,k}) - \gamma^2 \mathscr{T}_{i,k})\nu_{i,k} \\
&+ 2\eta_{i,k}^T\left(\mathscr{A}_{0,k} + \mathscr{C}_{i,k} - \sum_{j\in\mathscr{N}_i} \mathscr{K}_{ij,k}\right)^T \mathscr{P}_{i,k+1}(\mathscr{B}_k + \mathscr{D}_{i,k})\nu_{i,k} \\
&+ 2\nu_{i,k}^T(\mathscr{B}_k + \mathscr{D}_{i,k})^T \mathscr{P}_{i,k+1} \sum_{j\in\mathscr{N}_i} \mathscr{K}_{ij,k}\eta_{j,k} \\
&- 2\nu_{i,k}^T(\mathscr{B}_k + \mathscr{D}_{i,k})^T \mathscr{P}_{i,k+1}\mathscr{E}_{i,k}\varpi_i \\
\leq&\eta_{i,k}^T\left(\mathscr{A}_{0,k} + \mathscr{C}_{i,k} - \sum_{j\in\mathscr{N}_i} \mathscr{K}_{ij,k}\right)^T \Lambda_{i,k}\left(\mathscr{A}_{0,k} + \mathscr{C}_{i,k} - \sum_{j\in\mathscr{N}_i} \mathscr{K}_{ij,k}\right)\eta_{i,k} \\
&+ 2\eta_{i,k}^T\left(\mathscr{A}_{0,k} + \mathscr{C}_{i,k} - \sum_{j\in\mathscr{N}_i} \mathscr{K}_{ij,k}\right)^T \Lambda_{i,k}\sum_{j\in\mathscr{N}_i} \mathscr{K}_{ij,k}\eta_{j,k} \\
&+ \varpi_i^T\mathscr{E}_{i,k}^T\Lambda_{i,k}\mathscr{E}_{i,k}\varpi_i - 2\sum_{j\in\mathscr{N}_i}\eta_{j,k}^T\mathscr{K}_{ij,k}^T\Lambda_{i,k}\mathscr{E}_{i,k}\varpi_i \\
&- 2\eta_{i,k}^T\left(\mathscr{A}_{0,k} + \mathscr{C}_{i,k} - \sum_{j\in\mathscr{N}_i} \mathscr{K}_{ij,k}\right)^T \Lambda_{i,k}\mathscr{E}_{i,k}\varpi_i \\
&+ \sum_{j\in\mathscr{N}_i}\eta_{j,k}^T\mathscr{K}_{ij,k}^T\Lambda_{i,k}\sum_{j\in\mathscr{N}_i}\mathscr{K}_{ij,k}\eta_{j,k}.
\end{aligned} \tag{6.17}$$

where

$$\Lambda_{i,k} \triangleq \mathscr{P}_{i,k+1}(\mathscr{B}_k + \mathscr{D}_{i,k})\Phi_{i,k}(\mathscr{B}_k + \mathscr{D}_{i,k})^T \mathscr{P}_{i,k+1}. \tag{6.18}$$

Moreover, with the help of the matrix inverse formula, we derive that

$$\begin{aligned}
&\Lambda_{i,k} + \mathscr{P}_{i,k+1} \\
&= \mathscr{P}_{i,k+1}(\mathscr{B}_k + \mathscr{D}_{i,k})\Phi_{i,k}(\mathscr{B}_k + \mathscr{D}_{i,k})^T \mathscr{P}_{i,k+1} + \mathscr{P}_{i,k+1} \\
&= \left(\mathscr{P}_{i,k+1}^{-1} - \frac{1}{\gamma^2}(\mathscr{B}_k + \mathscr{D}_{i,k})\mathscr{T}_{i,k}^{-1}(\mathscr{B}_k + \mathscr{D}_{i,k})^T \right)^{-1} \\
&\triangleq \Psi_{i,k}^{-1}.
\end{aligned} \tag{6.19}$$

In combination of (6.16)–(6.19), one has

$$\begin{aligned}
&\mathbb{E}\{V_i(\eta_{i,k+1})|\eta_{i,k}, \eta_{j,k}, j \in \mathscr{N}_i\} \\
\le&\eta_{i,k}^T \left(\mathscr{A}_{0,k} + \mathscr{C}_{i,k} - \sum_{j\in\mathscr{N}_i} \mathscr{K}_{ij,k} \right)^T \Psi_{i,k}^{-1} \left(\mathscr{A}_{0,k} + \mathscr{C}_{i,k} - \sum_{j\in\mathscr{N}_i} \mathscr{K}_{ij,k} \right) \eta_{i,k} \\
&- 2\eta_{i,k}^T \left(\mathscr{A}_{0,k} + \mathscr{C}_{i,k} - \sum_{j\in\mathscr{N}_i} \mathscr{K}_{ij,k} \right)^T \Psi_{i,k}^{-1} \mathscr{E}_{i,k}\varpi_i \\
&+ 2\eta_{i,k}^T \left(\mathscr{A}_{0,k} + \mathscr{C}_{i,k} - \sum_{j\in\mathscr{N}_i} \mathscr{K}_{ij,k} \right)^T \Psi_{i,k}^{-1} \sum_{j\in\mathscr{N}_i} \mathscr{K}_{ij,k}\eta_{j,k} \\
&- 2\sum_{j\in\mathscr{N}_i} \eta_{j,k}^T \mathscr{K}_{ij,k}^T \Psi_{i,k}^{-1} \mathscr{E}_{i,k}\varpi_i + \sum_{j\in\mathscr{N}_i} \eta_{j,k}^T \mathscr{K}_{ij,k}^T \Psi_{i,k}^{-1} \sum_{j\in\mathscr{N}_i} \mathscr{K}_{ij,k}\eta_{j,k} \\
&+ \eta_{i,k}^T \sum_{t=1}^{r} \mathscr{A}_{t,k}^T \mathscr{P}_{i,k+1}\mathscr{A}_{t,k}\eta_{i,k} + \varpi_i^T \mathscr{E}_{i,k}^T \Psi_{i,k}^{-1} \mathscr{E}_{i,k}\varpi_i + \gamma^2 v_{i,k}^T \mathscr{T}_{i,k} v_{i,k} \\
\triangleq& \chi_{i,k}^T \Pi_{i,k} \chi_{i,k} + \gamma^2 v_{i,k}^T \mathscr{T}_{i,k} v_{i,k}
\end{aligned} \tag{6.20}$$

where $\chi_{i,k} \triangleq [\eta_{i,k}^T \quad \eta_{\mathscr{N}_i,k}^T \quad 1]^T$, $\eta_{\mathscr{N}_i,k} \triangleq [\eta_{j_{i_1},k}^T \quad \cdots \quad \eta_{j_{i_{p_i}},k}^T]^T$ and

$$\Pi_{i,k} = \begin{bmatrix} \Pi_{i,k}^{11} & \Pi_{i,k}^{12} & \Pi_{i,k}^{13} \\ * & \Pi_{i,k}^{22} & \Pi_{i,k}^{23} \\ * & * & \Pi_{i,k}^{33} \end{bmatrix}$$

with

$$\Pi_{i,k}^{11} = \left(\mathscr{A}_{0,k} + \mathscr{C}_{i,k} - \sum_{j\in\mathscr{N}_i} \mathscr{K}_{ij,k} \right)^T \Psi_{i,k}^{-1} \left(\mathscr{A}_{0,k} + \mathscr{C}_{i,k} - \sum_{j\in\mathscr{N}_i} \mathscr{K}_{ij,k} \right) + \sum_{t=1}^{r} \mathscr{A}_{t,k}^T \mathscr{P}_{i,k+1}\mathscr{A}_{t,k},$$

$$\Pi_{i,k}^{12} = \left(\mathscr{A}_{0,k} + \mathscr{C}_{i,k} - \sum_{j\in\mathscr{N}_i} \mathscr{K}_{ij,k} \right)^T \Psi_{i,k}^{-1} \left[\mathscr{K}_{ij_1,k} \quad \cdots \quad \mathscr{K}_{ij_{p_i},k} \right],$$

$$
\begin{aligned}
\Pi_{i,k}^{13} &= -\left(\mathscr{A}_{0,k} + \mathscr{C}_{i,k} - \sum_{j\in\mathscr{N}_i} \mathscr{K}_{ij,k}\right)^T \Psi_{i,k}^{-1}\mathscr{E}_{i,k}\varpi_i,\\
\Pi_{i,k}^{22} &= \left[\mathscr{K}_{ij_1,k} \quad \cdots \quad \mathscr{K}_{ij_{p_i},k}\right]^T \Psi_{i,k}^{-1}\left[\mathscr{K}_{ij_1,k} \quad \cdots \quad \mathscr{K}_{ij_{p_i},k}\right],\\
\Pi_{i,k}^{23} &= -\left[(\mathscr{K}_{ij_1,k}\Psi_{i,k}^{-1}\mathscr{E}_{i,k}\varpi_i)^T \quad \cdots \quad (\mathscr{K}_{ij_{p_i},k}\Psi_{i,k}^{-1}\mathscr{E}_{i,k}\varpi_i)^T\right]^T,\\
\Pi_{i,k}^{33} &= \varpi_i^T\mathscr{E}_{i,k}^T\Psi_{i,k}^{-1}\mathscr{E}_{i,k}\varpi_i.
\end{aligned}
$$

By some straightforward algebraic manipulations, we can see that the following inequality is true:

$$
\begin{aligned}
&\chi_{i,k}^T\Pi_{i,k}\chi_{i,k} + \sum_{j\in\mathscr{N}_i}\|z_{j,k} - z_{i,k}\|_{R_{i,k}}^2 - \rho_{i,k}\eta_{i,k}^T\mathscr{P}_{i,k}\eta_{i,k} - \sum_{j\in\mathscr{N}_i}\frac{\alpha_{j,k}}{1+q_j}\eta_{j,k}^T\mathscr{P}_{j,k}\eta_{j,k}\\
&\quad + \|z_{i,k}\|_{Q_{i,k}}^2 - \gamma^2\bar{\mathbf{W}}_{i,k}\mathbf{exp}(\varpi_i)\\
&\triangleq \chi_{i,k}^T\Xi_{i,k}\chi_{i,k} < 0,
\end{aligned}
$$

which implies

$$
\begin{aligned}
\chi_{i,k}^T\Pi_{i,k}\chi_{i,k} < &-\sum_{j\in\mathscr{N}_i}\|z_{j,k} - z_{i,k}\|_{R_{i,k}}^2 - \|z_{i,k}\|_{Q_{i,k}}^2 + \gamma^2\bar{\mathbf{W}}_{i,k}\mathbf{exp}(\varpi_i)\\
&+\sum_{j\in\mathscr{N}_i}\frac{\alpha_{j,k}}{1+q_j}\eta_{j,k}^T\mathscr{P}_{j,k}\eta_{j,k} + \rho_{i,k}\eta_{i,k}^T\mathscr{P}_{i,k}\eta_{i,k}.
\end{aligned}
$$

Recalling the expression of $S_i(z_k, v_{i,k}, \varpi_i)$ and (6.14a), one has

$$
\begin{aligned}
&\mathbb{E}\{V_i(\eta_{i,k+1})|\eta_{i,k}, \eta_{j,k}, j\in\mathscr{N}_i\}\\
\leq&\chi_{i,k}^T\Pi_{i,k}\chi_{i,k} + \gamma^2 v_{i,k}^T\mathscr{T}_{i,k}v_{i,k}\\
<&\gamma^2 v_{i,k}^T\mathscr{T}_{i,k}v_{i,k} + \gamma^2\bar{\mathbf{W}}_{i,k}\mathbf{exp}(\varpi_i) - \sum_{j\in\mathscr{N}_i}\|z_{j,k} - z_{i,k}\|_{R_{i,k}}^2\\
&-\|z_{i,k}\|_{Q_{i,k}}^2 + \sum_{j\in\mathscr{N}_i}\frac{\alpha_{j,k}}{1+q_j}\eta_{j,k}^T\mathscr{P}_{j,k}\eta_{j,k} + \rho_{i,k}\eta_{i,k}^T\mathscr{P}_{i,k}\eta_{i,k}\\
=&\left[\frac{a_{i1}\alpha_{1,k}}{1+q_1} \quad \cdots \quad \rho_{i,k} \quad \cdots \quad \frac{a_{iN}\alpha_{N,k}}{1+q_N}\right]\mathbf{V}(\eta_k) + S_i(z_k, v_{i,k}, \varpi_i)\\
=&[U_k\mathbf{V}(\eta_k)]_i + S_i(z_k, v_{i,k}, \varpi_i)
\end{aligned}
$$

where $[U_k\mathbf{V}(\eta_k)]_i$ is the i-th element of vector $U_k\mathbf{V}(\eta_k)$ with

$$U_k \triangleq \begin{bmatrix} \rho_{1,k} & \cdots & \frac{a_{1i}\alpha_{i,k}}{1+q_i} & \cdots & \frac{a_{1N}\alpha_{N,k}}{1+q_N} \\ \vdots & \ddots & \vdots & \cdots & \vdots \\ \frac{a_{i1}\alpha_{1,k}}{1+q_1} & \cdots & \rho_{i,k} & \cdots & \frac{a_{iN}\alpha_{N,k}}{1+q_N} \\ \vdots & \cdots & \vdots & \ddots & \vdots \\ \frac{a_{N1}\alpha_{1,k}}{1+q_N} & \cdots & \frac{a_{Ni}\alpha_{i,k}}{1+q_i} & \cdots & \rho_{N,k} \end{bmatrix}. \tag{6.21}$$

Since $\alpha_{i,k} \in \mathfrak{I}_{q_i}$, it is inferred from [6] that U_k is strictly diagonally dominant and is thus nonsingular. Notice $\mathbf{1}^T U_k = \mathbf{1}^T$ and, therefore, U_k is the dissipation matrix. Moreover, by using the elementary properties of the conditional expectation, one can obtain the following inequality:

$$\mathbb{E}\{V_i(\eta_{i,k+1})\} < \mathbb{E}\{[U_k\mathbf{V}(\eta_k)]_i\} + \mathbb{E}\{S_i(z_k, v_{i,k}, \varpi_i)\}, i = 1, 2, \ldots, l_0. \tag{6.22}$$

(2) *Proof of the stochastic vector-dissipativity for the nodes* i $(i = l_0 + 1, l_0 + 2, \ldots, N)$ *over the finite horizon* $\mathscr{H}$.

For the node i $(i = l_0 + 1, \ldots, N)$, it follows immediately from the system dynamics (6.9) that

$$\begin{aligned} &\mathbb{E}\{V_i(\eta_{i,k+1})|\eta_{i,k}, \eta_{j,k}, j \in \mathcal{N}_i\} \\ =&\mathbb{E}\{\eta_{i,k+1}^T \mathscr{P}_{i,k+1}\eta_{i,k+1}|\eta_{i,k}, \eta_{j,k}, j \in \mathcal{N}_i\} \\ =&\eta_{i,k}^T\left(\mathscr{A}_{0,k} - \sum_{j\in\mathcal{N}_i}\mathscr{K}_{ij,k}\right)^T \mathscr{P}_{i,k+1}\left(\mathscr{A}_{0,k} - \sum_{j\in\mathcal{N}_i}\mathscr{K}_{ij,k}\right)\eta_{i,k} \\ &+ \eta_{i,k}^T\sum_{t=1}^{r}\mathscr{A}_{t,k}^T\mathscr{P}_{i,k+1}\mathscr{A}_{t,k}\eta_{i,k} + v_{i,k}^T\mathscr{B}_k^T\mathscr{P}_{i,k+1}\mathscr{B}_k v_{i,k} \\ &+ \left(\sum_{j\in\mathcal{N}_i}\eta_{j,k}^T\mathscr{K}_{ij,k}^T\right)\mathscr{P}_{i,k+1}\left(\sum_{j\in\mathcal{N}_i}\mathscr{K}_{ij,k}\eta_{j,k}\right) \\ &+ 2\eta_{i,k}^T\left(\mathscr{A}_{0,k} - \sum_{j\in\mathcal{N}_i}\mathscr{K}_{ij,k}\right)^T\mathscr{P}_{i,k+1}\mathscr{B}_k v_{i,k} \\ &+ 2\eta_{i,k}^T\left(\mathscr{A}_{0,k} - \sum_{j\in\mathcal{N}_i}\mathscr{K}_{ij,k}\right)^T\mathscr{P}_{i,k+1}\sum_{j\in\mathcal{N}_i}\mathscr{K}_{ij,k}\eta_{j,k} \\ &+ 2v_{i,k}^T\mathscr{B}_k^T\mathscr{P}_{i,k+1}\sum_{j\in\mathcal{N}_i}\mathscr{K}_{ij,k}\eta_{j,k} \\ &+ \gamma^2 v_{i,k}^T\mathscr{T}_{i,k}v_{i,k} - \gamma^2 v_{i,k}^T\mathscr{T}_{i,k}v_{i,k}. \end{aligned} \tag{6.23}$$

It is obvious that (6.15b) leads to $\bar{\Psi}_{i,k} > 0$. Then, by applying the Schur Complement Lemma again, one obtains $\bar{\Phi}_{i,k}^{-1} = \gamma^2 \mathscr{T}_{i,k} - \mathscr{B}_k^T \mathscr{P}_{i,k+1} \mathscr{B}_k > 0$. On the basis of the following inequality

$$\left(\nu_{i,k} - \bar{\Phi}_{i,k} \mathscr{B}_k^T \mathscr{P}_{i,k+1} \left(\mathscr{A}_{0,k} - \sum_{j \in \mathscr{N}_i} \mathscr{K}_{ij,k} \right) \eta_{i,k} - \bar{\Phi}_{i,k} \mathscr{B}_k^T \mathscr{P}_{i,k+1} \sum_{j \in \mathscr{N}_i} \mathscr{K}_{ij,k} \eta_{j,k} \right)^T \bar{\Phi}_{i,k}^{-1}$$
$$\times \left(\nu_{i,k} - \bar{\Phi}_{i,k} \mathscr{B}_k^T \mathscr{P}_{i,k+1} \left(\mathscr{A}_{0,k} - \sum_{j \in \mathscr{N}_i} \mathscr{K}_{ij,k} \right) \eta_{i,k} - \bar{\Phi}_{i,k} \mathscr{B}_k^T \mathscr{P}_{i,k+1} \sum_{j \in \mathscr{N}_i} \mathscr{K}_{ij,k} \eta_{j,k} \right)$$
$$\geq 0,$$

we have

$$\begin{aligned}
& \nu_{i,k}^T (\mathscr{B}_k^T \mathscr{P}_{i,k+1} \mathscr{B}_k - \gamma^2 \mathscr{T}_{i,k}) \nu_{i,k} \\
& + 2\eta_{i,k}^T \left(\mathscr{A}_{0,k} - \sum_{j \in \mathscr{N}_i} \mathscr{K}_{ij,k} \right)^T \mathscr{P}_{i,k+1} \mathscr{B}_k \nu_{i,k} \\
& + 2\nu_{i,k}^T \mathscr{B}_k^T \mathscr{P}_{i,k+1} \sum_{j \in \mathscr{N}_i} \mathscr{K}_{ij,k} \eta_{j,k} \\
< & \eta_{i,k}^T \left(\mathscr{A}_{0,k} - \sum_{j \in \mathscr{N}_i} \mathscr{K}_{ij,k} \right)^T \mathscr{P}_{i,k+1} \mathscr{B}_k \bar{\Phi}_{i,k} \mathscr{B}_k^T \mathscr{P}_{i,k+1} \left(\mathscr{A}_{0,k} - \sum_{j \in \mathscr{N}_i} \mathscr{K}_{ij,k} \right) \eta_{i,k} \\
& + 2\eta_{i,k}^T \left(\mathscr{A}_{0,k} - \sum_{j \in \mathscr{N}_i} \mathscr{K}_{ij,k} \right)^T \mathscr{P}_{i,k+1} \mathscr{B}_k \bar{\Phi}_{i,k} \mathscr{B}_k^T \mathscr{P}_{i,k+1} \sum_{j \in \mathscr{N}_i} \mathscr{K}_{ij,k} \eta_{j,k} \\
& + \sum_{j \in \mathscr{N}_i} \eta_{j,k}^T \mathscr{K}_{ij,k}^T \mathscr{P}_{i,k+1} \mathscr{B}_k \bar{\Phi}_{i,k} \mathscr{B}_k^T \mathscr{P}_{i,k+1} \sum_{j \in \mathscr{N}_i} \mathscr{K}_{ij,k} \eta_{j,k}.
\end{aligned} \tag{6.24}$$

According to the matrix inverse formula, one derives

$$\begin{aligned}
& \mathscr{P}_{i,k+1} \mathscr{B}_k \bar{\Phi}_{i,k} \mathscr{B}_k^T \mathscr{P}_{i,k+1} + \mathscr{P}_{i,k+1} \\
= & \left(\mathscr{P}_{i,k+1}^{-1} - \frac{1}{\gamma^2} \mathscr{B}_k \mathscr{T}_{i,k}^{-1} \mathscr{B}_k^T \right)^{-1} \\
\triangleq & \bar{\Psi}_{i,k}^{-1}.
\end{aligned} \tag{6.25}$$

Then, it is not difficult to verify from (6.23)–(6.25) that

$$
\begin{aligned}
&\mathbb{E}\{V_i(\eta_{i,k+1})|\eta_{i,k},\eta_{j,k},j\in\mathcal{N}_i\}\\
\leq&\eta_{i,k}^T\left(\mathscr{A}_{0,k}-\sum_{j\in\mathcal{N}_i}\mathscr{K}_{ij,k}\right)^T\bar{\Psi}_{i,k}^{-1}\left(\mathscr{A}_{0,k}-\sum_{j\in\mathcal{N}_i}\mathscr{K}_{ij,k}\right)\eta_{i,k}\\
&+2\eta_{i,k}^T\left(\mathscr{A}_{0,k}-\sum_{j\in\mathcal{N}_i}\mathscr{K}_{ij,k}\right)^T\bar{\Psi}_{i,k}^{-1}\sum_{j\in\mathcal{N}_i}\mathscr{K}_{ij,k}\eta_{j,k}\\
&+\sum_{j\in\mathcal{N}_i}\eta_{j,k}^T\mathscr{K}_{ij,k}^T\Psi_{i,k}^{-1}\sum_{j\in\mathcal{N}_i}\mathscr{K}_{ij,k}\eta_{j,k}\\
&+\eta_{i,k}^T\sum_{t=1}^{r}\mathscr{A}_{t,k}^T\mathscr{P}_{i,k+1}\mathscr{A}_{t,k}\eta_{i,k}+\gamma^2\nu_{i,k}^T\mathscr{T}_{i,k}\nu_{i,k}\\
\triangleq&\bar{\chi}_{i,k}^T\bar{\Pi}_{i,k}\bar{\chi}_{i,k}+\gamma^2\nu_{i,k}^T\mathscr{T}_{i,k}\nu_{i,k}
\end{aligned}
\tag{6.26}
$$

where

$$
\begin{aligned}
&\bar{\chi}_{i,k}\triangleq\begin{bmatrix}\eta_{i,k}^T & \eta_{\mathcal{N}_i,k}^T\end{bmatrix}^T,\quad \eta_{\mathcal{N}_i,k}\triangleq\begin{bmatrix}\eta_{j_{i_1},k}^T & \cdots & \eta_{j_{i_{p_i}},k}^T\end{bmatrix}^T,\\
&\bar{\Pi}_{i,k}=\begin{bmatrix}\bar{\Pi}_{i,k}^{11} & \bar{\Pi}_{i,k}^{12}\\ * & \bar{\Pi}_{i,k}^{22}\end{bmatrix},\\
&\bar{\Pi}_{i,k}^{11}=\left(\mathscr{A}_{0,k}-\sum_{j\in\mathcal{N}_i}\mathscr{K}_{ij,k}\right)^T\bar{\Psi}_{i,k}^{-1}\left(\mathscr{A}_{0,k}-\sum_{j\in\mathcal{N}_i}\mathscr{K}_{ij,k}\right)+\sum_{t=1}^{r}\mathscr{A}_{t,k}^T\mathscr{P}_{i,k+1}\mathscr{A}_{t,k},\\
&\bar{\Pi}_{i,k}^{12}=\begin{bmatrix}(\mathscr{A}_{0,k}-\sum_{j\in\mathcal{N}_i}\mathscr{K}_{ij,k})^T\bar{\Psi}_{i,k}^{-1}\mathscr{K}_{ij_1,k} & \ldots & (\mathscr{A}_{0,k}-\sum_{j\in\mathcal{N}_i}\mathscr{K}_{ij,k})^T\Psi_{i,k}^{-1}\mathscr{K}_{ij_{p_i},k}\end{bmatrix},\\
&\bar{\Pi}_{i,k}^{22}=\begin{bmatrix}\mathscr{K}_{ij_1,k} & \cdots & \mathscr{K}_{ij_{p_i},k}\end{bmatrix}^T\bar{\Psi}_{i,k}^{-1}\begin{bmatrix}\mathscr{K}_{ij_1,k} & \cdots & \mathscr{K}_{ij_{p_i},k}\end{bmatrix}.
\end{aligned}
$$

Moreover, (6.15a) implies

$$
\begin{aligned}
&\bar{\chi}_{i,k}^T\bar{\Pi}_{i,k}\bar{\chi}_{i,k}+\sum_{j\in\mathcal{N}_i}\|z_{j,k}-z_{i,k}\|_{R_{i,k}}^2+\|z_{i,k}\|_{Q_{i,k}}^2\\
&-\sum_{j\in\mathcal{N}_i}\frac{\alpha_{j,k}}{1+q_j}\eta_{j,k}^T\mathscr{P}_{j,k}\eta_{j,k}-\rho_{i,k}\eta_{i,k}^T\mathscr{P}_{i,k}\eta_{i,k}\\
=&\zeta_{i,k}^T\Upsilon_{i,k}\zeta_{i,k}<0,
\end{aligned}
\tag{6.27}
$$

namely,

$$\bar{\chi}_{i,k}^T \bar{\Pi}_{i,k} \bar{\chi}_{i,k}$$
$$< - \sum_{j \in \mathscr{N}_i} \|z_{j,k} - z_{i,k}\|^2_{R_{i,k}} - \|z_{i,k}\|^2_{Q_{i,k}} + \sum_{j \in \mathscr{N}_i} \frac{\alpha_{j,k}}{1+q_j} \eta_{j,k}^T \mathscr{P}_{j,k} \eta_{j,k} + \rho_{i,k} \eta_{i,k}^T \mathscr{P}_{i,k} \eta_{i,k}.$$

Noticing $S_i(z_k, \nu_{i,k})$ and (6.14a), one has the following conclusion:

$$\begin{aligned}
&\mathbb{E}\{V_i(\eta_{i,k+1})|\eta_{i,k}, \eta_{j,k}, j \in \mathscr{N}_i\} \\
\leq &\bar{\chi}_{i,k}^T \bar{\Pi}_{i,k} \bar{\chi}_{i,k} \\
< &\gamma^2 \nu_{i,k}^T \mathscr{T}_{i,k} \nu_{i,k} - \sum_{j \in \mathscr{N}_i} \|z_{j,k} - z_{i,k}\|^2_{R_{i,k}} - \|z_{i,k}\|^2_{Q_{i,k}} \\
&+ \sum_{j \in \mathscr{N}_i} \frac{\alpha_{j,k}}{1+q_j} \eta_{j,k}^T \mathscr{P}_{j,k} \eta_{j,k} + \rho_{i,k} \eta_{i,k}^T \mathscr{P}_{i,k} \eta_{i,k} \\
= &\begin{bmatrix} \frac{a_{i1}\alpha_{1,k}}{1+q_1} & \cdots & \rho_{i,k} & \cdots & \frac{a_{iN}\alpha_{N,k}}{1+q_N} \end{bmatrix} \mathbf{V}(\eta_k) + S_i(z_k, \nu_{i,k}) \\
= &[U_k \mathbf{V}(\eta_k)]_i + S_i(z_k, \nu_{i,k}).
\end{aligned} \tag{6.28}$$

By applying the property of the conditional expectation to (6.28), one further obtains the following result

$$\mathbb{E}\{V_i(\eta_{i,k+1})\} < \mathbb{E}\{[U_k \mathbf{V}(\eta_k)]_i\} + \mathbb{E}\{S_i(z_k, \nu_{i,k})\}, \quad i = l_0 + 1, \ldots, N. \tag{6.29}$$

(3) *Proof of the stochastic vector-dissipativity for the nodes* $i = 1, 2, \ldots, l_0, l_0 + 1, \ldots, N$.

Based on the above results, it is observed from (6.22) and (6.29) that the following vector inequality is satisfied:

$$\mathbb{E}\{\mathbf{V}(\eta_{k+1})\} \ll U_k \mathbb{E}\{\mathbf{V}(\eta_k)\} + \mathbb{E}\{\mathbf{S}(z_k, \nu_k, \varpi)\}. \tag{6.30}$$

Consequently, it is readily seen from Definition 6.2 that the dynamics (6.9) is stochastic vector-dissipative over the finite horizon $\mathscr{H}$.

(4) *Proof of the* H_∞*-consensus performance index.*

In this step, we aim to prove the guaranteed H_∞-consensus performance of the distributed filtering scheme. First, left-multiplying $\mathbf{1}^T$ on both sides of (6.30) yields

$$\mathbb{E}\{\mathbf{1}^T \mathbf{V}(\eta_{k+1})\} < \mathbb{E}\{\mathbf{1}^T \mathbf{S}(z_k, \nu_k, \varpi)\} + \mathbb{E}\{\mathbf{1}^T U_k \mathbf{V}(\eta_k)\}. \tag{6.31}$$

Denoting $v(\eta_k) \triangleq \mathbf{1}^T \mathbf{V}(\eta_k)$, (6.31) is further reformulated as follows:

$$\mathbb{E}\{v(\eta_{k+1})\} < \sum_{i=1}^{l_0} \mathbb{E}\{S_i(z_k, \nu_{i,k}, \varpi_i)\} + \sum_{i=l_0+1}^{N} \mathbb{E}\{S_i(z_k, \nu_{i,k})\} + \mathbb{E}\{v(\eta_k)\}. \tag{6.32}$$

Keeping the notations of $S_i(z_k, \nu_{i,k}, \varpi_i)(i = 1, 2, \ldots, l_0)$ and $S_i(z_k, \nu_{i,k})(i = l_0 + 1, \ldots, N)$ in mind, we arrive at

$$\mathbb{E}\{v(\eta_{k+1})\} - \mathbb{E}\{v(\eta_k)\} + \Phi_{\mathscr{G}}(z_k) < \sum_{i=1}^{N} \gamma^2 \nu_{i,k}^T \mathscr{T}_{i,k} \nu_{i,k} + \sum_{i=1}^{l_0} \gamma^2 \bar{\mathbf{W}}_{i,k} \mathbf{exp}(\varpi_i). \tag{6.33}$$

Summing up both sides of (6.33) from 0 to $n-1$ with respect to k yields

$$\mathbb{E}\{v(\eta_n)\} - \mathbb{E}\{v(\eta_0)\} + \sum_{k=0}^{n-1} \Phi_{\mathscr{G}}(z_k) < \gamma^2 \sum_{k=0}^{n-1}\sum_{i=1}^{N} \nu_{i,k}^T \mathscr{T}_{i,k} \nu_{i,k} + \sum_{k=0}^{n-1}\sum_{i=1}^{l_0} \gamma^2 \bar{\mathbf{W}}_{i,k} \mathbf{exp}(\varpi_i). \tag{6.34}$$

Taking the fact $\mathbb{E}\{v(\eta_n)\} > 0$ and the initial condition $\mathscr{P}_{i,0} \le \gamma^2 U_i$ into account, the H_∞-consensus performance index (6.10) is satisfied. Thus, the proof of this theorem is now complete.

Theorem 6.2 *Let the disturbance attenuation level $\gamma > 0$, the scalar sequence $\alpha_{i,k} \in \mathbb{J}_{q_i}$, and the matrices $R_{i,k}$, $Q_{i,k}$, $\mathscr{T}_{i,k}$ and $\bar{\mathbf{W}}_{i,k}$ be given. The filtering error system (6.9) achieves the H_∞-consensus performance index over the finite horizon $\mathscr{H}$, if there exists a sequence of positive definite matrices $\{\mathscr{P}_{i,k}\}_{k\in\mathscr{H}\cup\{n\}} \triangleq \mathrm{diag}\{P_{i,k}^1, P_{i,k}^2\}$, and matrices $E_{i,k}$ and $F_{ij,k}$ satisfying the initial condition $\mathscr{P}_{i,0} \le \gamma^2 U_i$ ($\forall i = 1, 2, \ldots, N$), such that the following recursive linear matrix inequalities are satisfied for all $k \in \mathscr{H}$:*

$$\begin{cases} \Omega_{i,k} \triangleq \begin{bmatrix} \Omega_{i,k}^{11} & \Omega_{i,k}^{12} & 0 & \Omega_{i,k}^{14} & 0 & \Omega_{i,k}^{16} \\ * & \Omega_{i,k}^{22} & 0 & \Omega_{i,k}^{24} & 0 & 0 \\ * & * & \Omega_{i,k}^{33} & \Omega_{i,k}^{34} & 0 & 0 \\ * & * & * & \Omega_{i,k}^{44} & \Omega_{i,k}^{45} & 0 \\ * & * & * & * & \Omega_{i,k}^{55} & 0 \\ * & * & * & * & * & \Omega_{i,k}^{66} \end{bmatrix} < 0, & \text{(6.35a)} \\ \Phi_{i,k}^{-1} \triangleq -(\mathscr{B}_k + \mathscr{D}_{i,k})^T \mathscr{P}_{i,k+1}(\mathscr{B}_k + \mathscr{D}_{i,k}) + \gamma^2 \mathscr{T}_{i,k} > 0, & \text{(6.35b)} \\ \forall i = 1, 2, \ldots, l_0, \end{cases}$$

and

$$\begin{cases} \Delta_{i,k} \triangleq \begin{bmatrix} \Delta_{i,k}^{11} & \Delta_{i,k}^{12} & \Delta_{i,k}^{13} & 0 & \Delta_{i,k}^{15} \\ * & \Delta_{i,k}^{22} & \Delta_{i,k}^{23} & 0 & 0 \\ * & * & \Delta_{i,k}^{33} & \Delta_{i,k}^{34} & 0 \\ * & * & * & \Delta_{i,k}^{44} & 0 \\ * & * & * & * & \Delta_{i,k}^{55} \end{bmatrix} < 0, & \text{(6.36a)} \\ \bar{\Phi}_{i,k}^{-1} \triangleq \gamma^2 \mathscr{T}_{i,k} - \mathscr{B}_k^T \mathscr{P}_{i,k+1} \mathscr{B}_k > 0, & \text{(6.36b)} \\ \forall i = l_0 + 1, l_0 + 2, \ldots, N \end{cases}$$

where

$$\Omega_{i,k}^{11} = \mathscr{F}^T Q_{i,k}\mathscr{F} + \mathscr{F}^T R_{i,k}\mathscr{F} - \rho_{i,k}\mathscr{P}_{i,k}, \quad \Omega_{i,k}^{12} = -[\underbrace{\mathscr{F}^T R_{i,k}\mathscr{F} \ \ldots \ \mathscr{F}^T R_{i,k}\mathscr{F}}_{p_i}],$$

$$\Omega_{i,k}^{22} = -\operatorname{diag}\left\{\frac{\alpha_{j_{i_1},k}}{1+q_{j_{i_1}}}\mathscr{P}_{j_{i_1},k} - \mathscr{F}^T R_{i,k}\mathscr{F}, \ldots, \frac{\alpha_{j_{ip_i},k}}{1+q_{j_{ip_i}}}\mathscr{P}_{j_{ip_i},k} - \mathscr{F}^T R_{i,k}\mathscr{F}\right\},$$

$$\Omega_{i,k}^{33} = -\gamma^2 \bar{\mathbf{W}}_{i,k}\mathbf{exp}(\varpi_i), \quad \Omega_{i,k}^{14} = \left(\bar{\mathscr{A}}_{0,k} + \bar{\mathscr{C}}_{i,k} - \sum_{j\in\mathscr{N}_i}\bar{\mathscr{K}}_{ij,k}\right)^T, \quad \Omega_{i,k}^{34} = \bar{\mathscr{E}}_{i,k}^T,$$

$$\Omega_{i,k}^{24} = \left[\bar{\mathscr{K}}_{ij_1,k} \cdots \bar{\mathscr{K}}_{ij_{p_i},k}\right]^T, \quad \bar{\mathscr{K}}_{ij,k} = \operatorname{diag}\{0, F_{ij,k}\},$$

$$\Omega_{i,k}^{44} = -\mathscr{P}_{i,k+1}, \quad \Omega_{i,k}^{55} = -\gamma^2\mathscr{T}_{i,k}, \quad \Omega_{i,k}^{45} = \bar{\mathscr{B}}_k + \bar{\mathscr{D}}_{i,k},$$

$$\Omega_{i,k}^{16} = \left[\bar{\mathscr{A}}_{1,k}^T \cdots \bar{\mathscr{A}}_{r,k}^T\right], \quad \Omega_{i,k}^{66} = -\operatorname{diag}_r\{\mathscr{P}_{i,k+1}\},$$

$$\bar{\mathscr{A}}_{0,k} = \operatorname{diag}\left\{P_{i,k+1}^1 A_{0,k}, P_{i,k+1}^2 A_{0,k}\right\}, \quad \bar{\mathscr{D}}_{i,k} = \operatorname{diag}\left\{0, -E_{i,k}p_{i,k}D_{i,k}\right\},$$

$$\bar{\mathscr{A}}_{t,k} = \begin{bmatrix} P_{i,k+1}^1 A_{t,k} & 0 \\ P_{i,k+1}^2 A_{t,k} & 0 \end{bmatrix}, \quad \bar{\mathscr{B}}_k = \begin{bmatrix} P_{i,k+1}^1 B_k & 0 \\ P_{i,k+1}^2 B_k & 0 \end{bmatrix},$$

$$\bar{\mathscr{C}}_{i,k} = \begin{bmatrix} 0 & 0 \\ E_{i,k}(1-p_{i,k})C_{i,k} & -E_{i,k}C_{i,k} \end{bmatrix}, \quad \bar{\mathscr{E}}_{i,k} = \begin{bmatrix} 0 & 0 \\ 0 & E_{i,k}(1-p_{i,k})\varpi_i \end{bmatrix},$$

$$\Delta_{i,k}^{11} = \mathscr{F}^T Q_{i,k}\mathscr{F} + \mathscr{F}^T R_{i,k}\mathscr{F} - \rho_{i,k}\mathscr{P}_{i,k}, \quad \Delta_{i,k}^{34} = \bar{\mathscr{B}}_k,$$

$$\Delta_{i,k}^{12} = -[\underbrace{\mathscr{F}^T R_{i,k}\mathscr{F} \ \cdots \ \mathscr{F}^T R_{i,k}\mathscr{F}}_{p_i}], \quad \Delta_{i,k}^{13} = \left(\bar{\mathscr{A}}_{0,k} - \sum_{j\in\mathscr{N}_i}\bar{\mathscr{K}}_{ij,k}\right)^T,$$

$$\Delta_{i,k}^{22} = -\operatorname{diag}\left\{\frac{\alpha_{j_{i_1},k}}{1+q_{j_{i_1}}}\mathscr{P}_{j_{i_1},k} - \mathscr{F}^T R_{i,k}\mathscr{F}, \ldots, \frac{\alpha_{j_{ip_i},k}}{1+q_{j_{ip_i}}}\mathscr{P}_{j_{ip_i},k} - \mathscr{F}^T R_{i,k}\mathscr{F}\right\},$$

$$\Delta_{i,k}^{23} = \left[\bar{\mathscr{K}}_{ij_1,k} \ \cdots \ \bar{\mathscr{K}}_{ij_{p_i},k}\right]^T, \quad \Delta_{i,k}^{33} = -\mathscr{P}_{i,k+1}, \quad \Delta_{i,k}^{44} = -\gamma^2\mathscr{T}_{i,k},$$

$$\Delta_{i,k}^{55} = -\operatorname{diag}_r\{\mathscr{P}_{i,k+1}\}, \quad \Delta_{i,k}^{15} = \left[\bar{\mathscr{A}}_{1,k}^T \ \cdots \ \bar{\mathscr{A}}_{r,k}^T\right].$$

Moreover, the desired filter gains are given by

$$L_{i,k} = \left(P_{i,k+1}^2\right)^{-1} E_{i,k} \quad \text{and} \quad K_{ij,k} = \left(P_{i,k+1}^2\right)^{-1} F_{ij,k}. \tag{6.37}$$

Proof Performing the congruence transformation

$$\operatorname{diag}\left\{I, I, I, \mathscr{P}_{i,k+1}^{-1}, I, \underbrace{\mathscr{P}_{i,k+1}^{-1}, \ldots, \mathscr{P}_{i,k+1}^{-1}}_{r}\right\}$$

to (6.35a) leads to

$$\begin{bmatrix} \Omega_{i,k}^{11} & \Omega_{i,k}^{12} & 0 & \bar{\Omega}_{i,k}^{14} & 0 & \bar{\Omega}_{i,k}^{16} \\ * & \Omega_{i,k}^{22} & 0 & \bar{\Omega}_{i,k}^{24} & 0 & 0 \\ * & * & \Omega_{i,k}^{33} & \bar{\Omega}_{i,k}^{34} & 0 & 0 \\ * & * & * & \bar{\Omega}_{i,k}^{44} & \bar{\Omega}_{i,k}^{45} & 0 \\ * & * & * & * & \bar{\Omega}_{i,k}^{55} & 0 \\ * & * & * & * & * & \bar{\Omega}_{i,k}^{66} \end{bmatrix} < 0, \tag{6.38}$$

and, similarly, using the congruence transformation

$$\operatorname{diag}\left\{ I,\ I,\ \mathscr{P}_{i,k+1}^{-1},\ I,\ \underbrace{\mathscr{P}_{i,k+1}^{-1}, \ldots, \mathscr{P}_{i,k+1}^{-1}}_{r} \right\}$$

to (6.36a), one has

$$\begin{bmatrix} \Delta_{i,k}^{11} & \Delta_{i,k}^{12} & \bar{\Delta}_{i,k}^{13} & 0 & \bar{\Delta}_{i,k}^{15} \\ * & \Delta_{i,k}^{22} & \bar{\Delta}_{i,k}^{23} & 0 & 0 \\ * & * & \bar{\Delta}_{i,k}^{33} & \bar{\Delta}_{i,k}^{34} & 0 \\ * & * & * & \Delta_{i,k}^{44} & 0 \\ * & * & * & * & \bar{\Delta}_{i,k}^{55} \end{bmatrix} < 0 \tag{6.39}$$

where

$$\begin{aligned}
\bar{\Omega}_{i,k}^{14} =& \left(\mathscr{A}_{0,k} + \mathscr{C}_{i,k} - \sum_{j \in \mathscr{N}_i} \mathscr{K}_{ij,k} \right)^T, \quad \bar{\Omega}_{i,k}^{34} = \mathscr{E}_{i,k}^T, \\
\bar{\Omega}_{i,k}^{24} =& \begin{bmatrix} \mathscr{K}_{ij_1,k} & \cdots & \mathscr{K}_{ij_{p_i},k} \end{bmatrix}^T, \quad \bar{\Omega}_{i,k}^{44} = -\mathscr{P}_{i,k+1}^{-1}, \\
\bar{\Omega}_{i,k}^{45} =& \mathscr{B}_k + \mathscr{D}_{i,k}, \quad \bar{\Omega}_{i,k}^{16} = \begin{bmatrix} \mathscr{A}_{1,k} & \mathscr{A}_{2,k} & \cdots & \mathscr{A}_{r,k} \end{bmatrix}, \\
\bar{\Omega}_{i,k}^{55} =& -\gamma^2 \mathscr{T}_{i,k}, \quad \bar{\Omega}_{i,k}^{66} = -\operatorname{diag}_r\{\mathscr{P}_{i,k+1}^{-1}\}, \\
\bar{\Delta}_{i,k}^{13} =& \left(\mathscr{A}_{0,k} - \sum_{j \in \mathscr{N}_i} \mathscr{K}_{ij,k} \right)^T, \quad \bar{\Delta}_{i,k}^{33} = -\mathscr{P}_{i,k+1}^{-1}, \\
\bar{\Delta}_{i,k}^{23} =& \begin{bmatrix} \mathscr{K}_{ij_1,k} & \cdots & \mathscr{K}_{ij_{p_i},k} \end{bmatrix}^T, \quad \bar{\Delta}_{i,k}^{34} = \mathscr{B}_k, \\
\bar{\Delta}_{i,k}^{15} =& \begin{bmatrix} \mathscr{A}_{1,k} \ \mathscr{A}_{2,k} & \cdots & \mathscr{A}_{r,k} \end{bmatrix}, \quad \bar{\Delta}_{i,k}^{55} = -\operatorname{diag}_r\{\mathscr{P}_{i,k+1}^{-1}\}.
\end{aligned}$$

Then, it is easy to verify the following relationships

$$
\begin{aligned}
&\Omega_{i,k}^{14} = \bar{\Omega}_{i,k}^{14}\mathscr{P}_{i,k+1}, \quad \Omega_{i,k}^{24} = \bar{\Omega}_{i,k}^{24}\mathscr{P}_{i,k+1},\\
&\Omega_{i,k}^{34} = \bar{\Omega}_{i,k}^{34}\mathscr{P}_{i,k+1}, \quad \Omega_{i,k}^{44} = \mathscr{P}_{i,k+1}\bar{\Omega}_{i,k}^{44}\mathscr{P}_{i,k+1},\\
&\Omega_{i,k}^{45} = \mathscr{P}_{i,k+1}\bar{\Omega}_{i,k}^{45}, \quad \Omega_{i,k}^{16} = \mathscr{P}_{i,k+1}\bar{\Omega}_{i,k}^{16},\\
&\Omega_{i,k}^{66} = \mathrm{diag}_r\{\mathscr{P}_{i,k+1}\}\bar{\Omega}_{i,k}^{66}\mathrm{diag}_r\{\mathscr{P}_{i,k+1}\},\\
&\Delta_{i,k}^{13} = \bar{\Delta}_{i,k}^{13}\mathscr{P}_{i,k+1}, \quad \Delta_{i,k}^{23} = \bar{\Delta}_{i,k}^{23}\mathscr{P}_{i,k+1},\\
&\Delta_{i,k}^{33} = \mathscr{P}_{i,k+1}\bar{\Delta}_{i,k}^{33}\mathscr{P}_{i,k+1}, \quad \Delta_{i,k}^{34} = \mathscr{P}_{i,k+1}\bar{\Delta}_{i,k}^{34},\\
&\Delta_{i,k}^{55} = \mathrm{diag}_r\{\mathscr{P}_{i,k+1}\}\bar{\Delta}_{i,k}^{55}\mathrm{diag}_r\{\mathscr{P}_{i,k+1}\},\\
&\Delta_{i,k}^{15} = \bar{\Delta}_{i,k}^{15}\mathscr{P}_{i,k+1}, \quad E_{i,k} = P_{i,k+1}^2 L_{i,k},\\
&F_{ij,k} = P_{i,k+1}^2 K_{ij,k}.
\end{aligned}
$$

In terms of Schur Complement Lemma, one immediately draws the conclusion from (6.38) and (6.39) that the conditions (6.14a) and (6.15a) are satisfied. In addition, (6.35b) and (6.36b) hold if and only if (6.14b) and (6.15b) are true, and the rest of the proof follows directly from Theorem 6.1. The proof is thus complete.

Remark 6.2 This chapter investigates the partial-nodes-based distributed filtering problem for a class of discrete-time systems with censored measurements. A set of deterministic indicative variables is introduced to describe the censored measurements and a new H_∞ performance index is proposed to better reflect the impact of censored measurements on the filtering performance. With the aid of the local performance analysis method, sufficient conditions are established for two kinds of nodes (i.e., having or not having available measurement outputs) such that the prescribed H_∞ performance index is achieved. Compared with the existing results, the developed filtering scheme has the following distinguished features: (1) a censored-related term is involved in the H_∞ performance index in an exponential function form; (2) the designed algorithm is valid for the nodes with or without the sensing capabilities, respectively; and (3) a local design method is applied to deal with the filtering issue in the distributed sense.

6.3 An Illustrative Example

In this section, a simulation example is carried out to illustrate the effectiveness of the proposed distributed filter design scheme.

Consider an SN with 5 nodes, whose topology is represented by a digraph $\mathscr{G} = (\mathscr{V}, \mathscr{E}, \mathcal{A})$ with the set of nodes $\mathscr{V} = \{1, 2, 3, 4, 5\}$ and the set of edges $\mathscr{E} = \{(1,4), (2,1), (3,2), (4,3), (5,4), (5,1), (2,4), (5,3)\}$, as shown in Fig.6.1. Obviously, the in-degree and out-degree of every node can be given as $p_1 = 1$, $p_2 = 2$, $p_3 = p_4 = 1$, $p_5 = 3$, and $q_1 = q_3 = q_4 = 2$, $q_2 = 1$, $q_5 = 0$. In this example, the measurement output of node 5 is not available.

Fig. 6.1 Topology of a sensor network

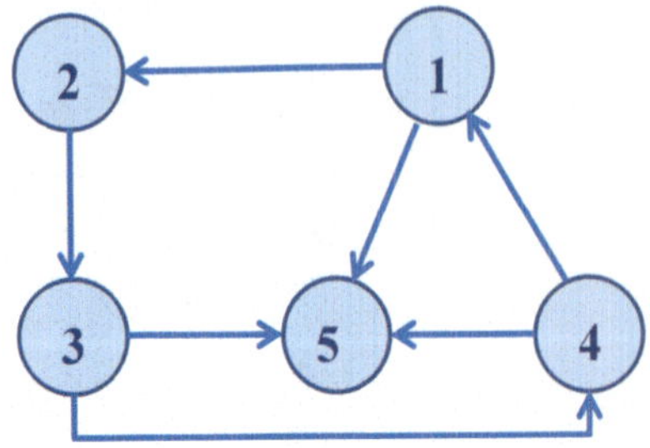

The parameters of the target plant are set as:

$$A_k = \begin{bmatrix} -0.80 & 0.15 \\ 0.15 & 0.21\sin(6k) \end{bmatrix}, \quad B_k = \begin{bmatrix} 0.1 \\ 0.1 \end{bmatrix},$$
$$A_{1,k} = \begin{bmatrix} -0.80 & 0.15 \\ 0.15 & 0.21\sin(6k) \end{bmatrix},$$
$$C_{i,k} = \begin{bmatrix} 0.8 \\ 1.2\sin((i+1)k) \end{bmatrix}^T, \quad D_{i,k} = 1$$

where $\phi_{1,k}$ is a Gaussian white noise sequence with zero mean and unity variance. The time horizon $n-1$, H_∞ attenuation level γ, scalar sequence $\alpha_{i,k}$, initial condition $\mathscr{P}_{i,0}$, initial weight matrix U_i, disturbance weight matrix $\mathscr{T}_{i,k}$, censored threshold Γ_i and censored weighted matrix $\bar{\mathbf{W}}_{i,k}$ are, respectively, given by $n = 21$, $\gamma = 0.16$, $\alpha_{i,k} \equiv 0.5$, $\mathscr{P}_{i,0} = \mathrm{diag}\{2I, 6I\}$, $U_i = 100\mathscr{P}_{i,0}$, $\mathscr{T}_{i,k} \equiv \mathrm{diag}\{1, 1\}$, $\Gamma_i = -0.1$ and $\bar{\mathbf{W}}_{i,k} = 1$.

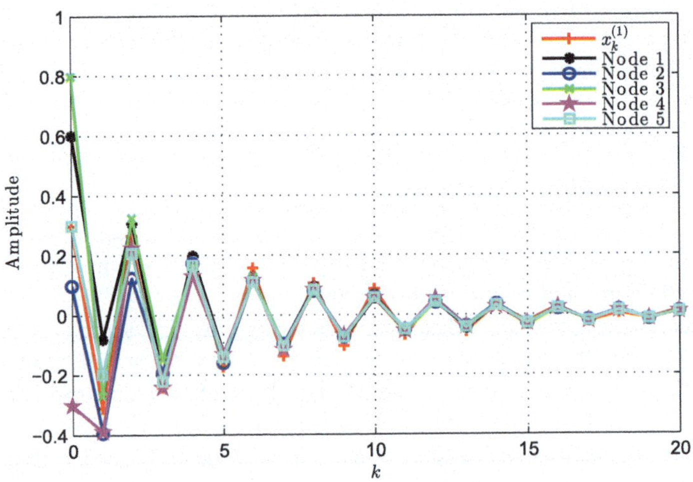

Fig. 6.2 The trajectories of $x_k^{(1)}$ and its estimates $\hat{x}_{i,k}^{(1)}$

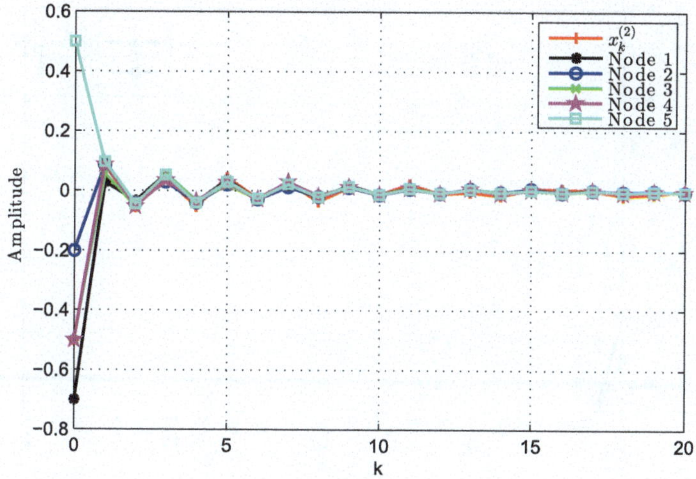

Fig. 6.3 The trajectories of $x_k^{(2)}$ and its estimates $\hat{x}_{i,k}^{(2)}$

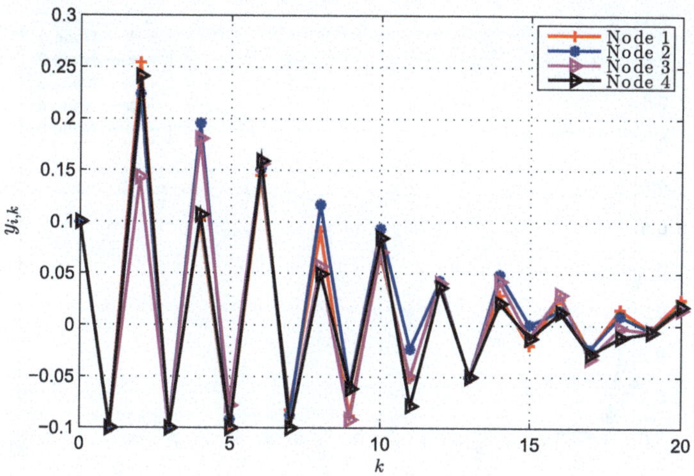

Fig. 6.4 The censored measurements $y_{i,k}$

By using the YALMIP toolbox in MATLAB software [7], all the desired filter gains are recursively calculated according to Theorem 6.2. The initial values are set as follows: $x_0 = [0.3\ \ -0.5]^T$, $\hat{x}_{1,0} = [0.6\ \ -0.7]^T$, $\hat{x}_{2,0} = [0.1\ \ -0.2]^T$, $\hat{x}_{3,0} = [0.8\ \ -0.5]^T$, $\hat{x}_{4,0} = [-0.3\ \ -0.5]^T$ and $\hat{x}_{5,0} = [0.3\ 0.5]^T$. The disturbances are chosen as $w_k = 0.1\cos(5k)$ and $\xi_{i,k} = 0.1\cos(5k)$, respectively.

The simulation results are shown in Figs. 6.2, 6.3, 6.4, 6.5 and 6.6, where the state trajectories of the plant and their estimates are displayed in Figs. 6.2 and 6.3, the censored measurements $y_{i,k}$ $(i = 1, \ldots, 4)$ are depicted in Fig. 6.4, the norms of the consensus errors $\sum_{j\in\mathcal{N}_i} \|e_{j,k} - e_{i,k}\|^2$ are given in Fig. 6.5 and the norms of the

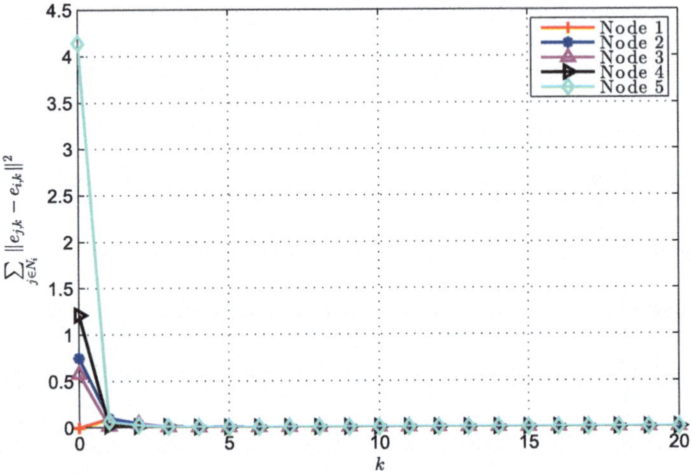

Fig. 6.5 The sum of the norms of consensus errors $\sum_{j\in\mathcal{N}_i}\|e_{j,k}-e_{i,k}\|^2$

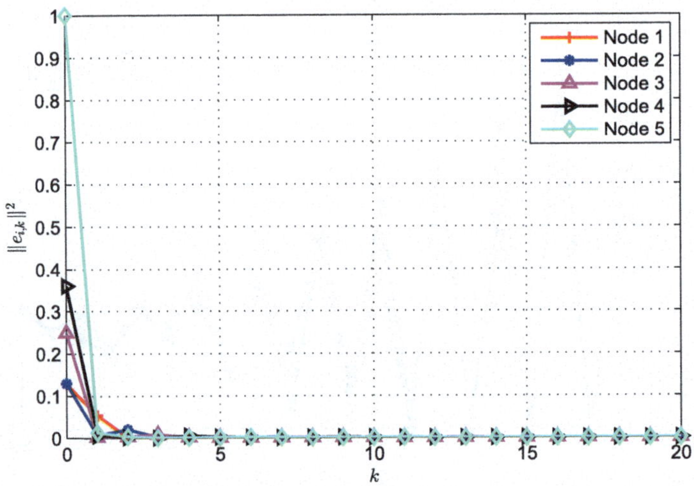

Fig. 6.6 The norms of the filtering errors $\|e_{i,k}\|^2$

filtering errors $\|e_{i,k}\|^2$ are plotted in Fig. 6.6. It is observed from all the simulation results that the proposed distributed filtering scheme performs very well.

6.4 Summary

In this chapter, we have investigated the scalable distributed H_∞-consensus filtering problem for a class of discrete time-varying systems with multiplicative noises and censored measurements over sensor networks where only partial nodes have the ability to conduct the measurement task. A set of deterministic indicative variables has been employed to characterize the censored measurements. A novel H_∞-consensus performance index has been established to reflect the impacts of the censored measurements on the filtering error dynamics. By using the local performance analysis method, a set of sufficient conditions has been established so as to meet the prescribed H_∞-consensus performance requirement. Furthermore, the desired filter gains have been recursively calculated in a distributed manner. Finally, the effectiveness of the main results has been illustrated by a simulation example.

References

1. Allik, B., Miller, C., Piovoso, M., Zurakowski, R.: Estimation of saturated data using the Tobit Kalman filter. In: Proceedings of 2014 American Control Conference, Portland, Oregon, USA, pp. 4151–4156 (2014)
2. Allik, B., Miller, C., Piovoso, M., Zurakowski, R.: Nonlinear estimators for censored data: a comparison of the EKF, the UKF and the Tobit Kalman filter. In: Procedings of the American Control Conference, Chicago, IL, USA, pp. 5146–5151 (2015)
3. Allik, B., Miller, C., Piovoso, M., Zurakowski, R.: The Tobit Kalman filter: an estimator for censored measurements. IEEE Trans. Control Syst. Technol. **24**(1), 365–371 (2016)
4. Han, F., Dong, H., Wang, Z., Li, G., Alsaadi, F.: Improved Tobit Kalman filtering for systems with random parameters via conditional expectation. Signal Process. **147**, 35–45 (2018)
5. Li, J., Wei, G., Ding, D., Li, Y.: Set-membership filtering for discrete time-varying nonlinear systems with censored measurements under Round-Robin protocol. Neurocomputing **281**, 20–26 (2018)
6. Horn, R., Johnson, C.: Matrix Analysis, 2nd edn. Cambridge University Press (2013)
7. Löfberg, J.: YALMIP: A toolbox for modeling and optimization in MATLAB. In: Proceedings of the CACSD Conference, Taipei, Taiwan (2004)

Chapter 7
Distributed H_∞-Consensus Filtering over Sensor Networks Under Deception Attacks

Deception attacks are a common kind of cyber-attacks. How to characterize such this attack in a reasonable manner? How to measure its impact on the desired performance? How to design the distributed filtering under this attack? We will seek the solutions to these problems in this chapter.

This chapter is concerned with the local design of the distributed H_∞-consensus filtering problem for a class of discrete time-varying systems subject to both multiplicative noises and deception attacks over sensor networks. The target plant and the measuring sensors are disturbed by multiplicative noises with known statistics. The malicious signal involved in deception attacks is constrained by a specific sector-like bounded condition, which reflects a certain tolerable bound on the difference between the attack signal and the true signal. Attention is paid to the design of filter gains for guaranteeing a desirable filtering performance that simultaneously characterizes the filtering accuracy and the consensus requirement. To handle the proposed filtering problem, the supply rate function is firstly chosen for each node and then the dissipation matrix is constructed as a column sub-stochastic matrix based on the stochastic vector dissipativity theory. Subsequently, sufficient conditions by means of recursive linear matrix inequalities are presented for each node such that the filtering error and the consensus error satisfy the desirable H_∞-consensus performance index over a finite horizon. Finally, an illustrative simulation is presented to demonstrate the effectiveness of the proposed filter strategy.

F. Han et al., *Distributed Filtering, Control and Synchronization*,
Studies in Systems, Decision and Control 428,
https://doi.org/10.1007/978-3-030-97075-8_7

7.1 Problem Formulation

Consider the following discrete-time stochastic system defined on the finite horizon $\mathscr{H} \triangleq \{0, 1, \ldots, n-1\}$:

$$x_{k+1} = \left(A_{0,k} + \sum_{t=1}^{r_1} \phi_{t,k} A_{t,k}\right) x_k + B_k w_k \tag{7.1}$$

and the measurement is modeled by

$$y_{i,k} = \left(C_{i0,k} + \sum_{s=1}^{r_2} \theta_{is,k} C_{is,k}\right) x_k + D_{i,k} v_{i,k} \tag{7.2}$$

where $x_k \in \mathbb{R}^{n_x}$ is the system state, $y_{i,k} \in \mathbb{R}^{n_y}$ is the measurement of the sensor node i, $w_k \in \mathbb{R}^{n_w}$ and $v_{i,k} \in \mathbb{R}^{n_v}$ are the external disturbances belonging to $l_2[0, n-1]$. For $t = 1, 2, \ldots, r_1$ and $s = 1, 2, \ldots, r_2$, $\phi_{t,k} \in \mathbb{R}$ and $\theta_{is,k} \in \mathbb{R}$ are mutually uncorrelated Gaussian white noises with zero mean and unity variance. $A_{t,k}$, B_k, $C_{is,k}$, and $D_{i,k}$ are known and time-varying matrices with compatible dimensions.

The traditional distributed filter is constructed as follows:

$$\hat{x}_{i,k+1} = A_{0,k}\hat{x}_{i,k} + L_{i,k}(y_{i,k} - C_{i0,k}\hat{x}_{i,k}) + \sum_{j \in \mathscr{N}_i} K_{ij,k}(\hat{x}_{j,k} - \hat{x}_{i,k}) \tag{7.3}$$

where, for the ith sensor node, $\hat{x}_{i,k} \in \mathbb{R}^{n_x}$ is the estimate of the state x_k, matrices $L_{i,k}$ and $K_{ij,k}$ are the gain parameters.

In this chapter, some potential attackers would launch attacks to the signals transmitted to the filters through the shared network. For the ith filter, the ideal estimate $\hat{x}_{j,k}$ $(j \in \mathscr{N}_i)$ subject to the deception attacks are modeled by:

$$\bar{x}_{j,k} = \hat{x}_{j,k} + \alpha_{j,k}\varphi_{j,k} \tag{7.4}$$

where $\bar{x}_{j,k} \in \mathbb{R}^{n_x}$ is the actual estimate (the updated signal) received from the neighboring node j, $\varphi_{j,k} \in \mathbb{R}^{n_x}$ is the deception signal, and $\alpha_{j,k} \in \mathbb{R}$ is a random variable modeled by a Bernoulli distributed white sequence to verify whether the deception attacks happen or not. The occurrence probability of the attack can be obtained via numerous statistical tests. Let's set $\text{Prob}\{\alpha_{j,k} = 1\} = \bar{\alpha}_j$ with $\bar{\alpha}_j$ being a known constant. Moreover, the deception signal $\varphi_{j,k}$ is formulated as:

$$\varphi_{j,k} = \xi_{j,k} - \hat{x}_{j,k}$$

where $\xi_{j,k}$ is a limited magnitude signal satisfying the following sector-like bounded condition:

$$(\xi_{j,k} - \Upsilon_{j,k}\hat{x}_{j,k})^T(\xi_{j,k} - \Upsilon_{j,k}\hat{x}_{j,k}) \leq \delta_j^2 \hat{x}_{j,k}^T \hat{x}_{j,k}, \tag{7.5}$$

in which $\Upsilon_{j,k}$ is a known real matrix and $\delta_j \geq 0$ is a known real constant.

Remark 7.1 According to our literature review, the signal $\xi_{j,k}$ is usually assumed to satisfy the constraint $\xi_{j,k}^T \xi_{j,k} \leq \delta_j^2$, which indicates the energy constraint on the deception signal via the parameter δ_j. Unfortunately, this constraint could result in certain conservatism because the information of $\hat{x}_{j,k}$ is not considered. An improved constraint is $\xi_{j,k}^T \xi_{j,k} \leq \delta_j^2 \hat{x}_{j,k}^T \hat{x}_{j,k}$, which is equivalent to the sector-bounded condition $(\xi_{j,k} - \delta_j \hat{x}_{j,k})^T(\xi_{j,k} + \delta_j \hat{x}_{j,k}) \leq 0$. As is well known, a more general sector-bounded condition is $(\xi_{j,k} - \Lambda_{1j,k}\hat{x}_{j,k})^T(\xi_{j,k} - \Lambda_{2j,k}\hat{x}_{j,k}) \leq 0$ with known matrices $\Lambda_{1j,k}$ and $\Lambda_{2j,k}$. It can be seen that $(\xi_{j,k} - \Lambda_{1j,k}\hat{x}_{j,k})^T(\xi_{j,k} - \Lambda_{2j,k}\hat{x}_{j,k}) \leq 0$ is equivalent to $(\xi_{j,k} - \bar{\Lambda}_{j,k}\hat{x}_{j,k})^T(\xi_{j,k} - \bar{\Lambda}_{j,k}\hat{x}_{j,k}) \leq \hat{x}_{j,k}^T \tilde{\Lambda}_{j,k}^T \tilde{\Lambda}_{j,k}\hat{x}_{j,k}$, where $\bar{\Lambda}_{j,k} = \frac{\Lambda_{1j,k}+\Lambda_{2j,k}}{2}$ and $\tilde{\Lambda}_{j,k} = \frac{\Lambda_{1j,k}-\Lambda_{2j,k}}{2}$. It is noticeable that, if $\Lambda_{1j,k} = \Lambda_{2j,k} = 0$, then the constraint is reduced to $\xi_{j,k}^T \xi_{j,k} \leq 0$, which implies that the standard sector-bounded condition cannot include $\xi_{j,k}^T \xi_{j,k} \leq \delta_j^2 \hat{x}_{j,k}^T \hat{x}_{j,k}$. Therefore, a new constraint (7.5) is proposed to unify all the discussed constraints.

Remark 7.2 In practical engineering, an attacker usually likes to send the attack signal without being detected or monitored. In this regard, the novel sector-like bounded condition (7.5) is introduced to characterize the malicious signal. Actually, $\Upsilon_{j,k}\hat{x}_{j,k}$ in (7.5) can be viewed as the predicted attack signal and $\delta_j^2 \hat{x}_{j,k}^T \hat{x}_{j,k}$ characterizes a tolerable bound on the difference between the attack signal and the predicted signal, where the parameters $\Upsilon_{j,k}$ and δ_j involved in (7.5) are estimated through statistical tests or specified by certain security requirements. It is easy to see that, if $\Upsilon_{j,k}^T = \Upsilon_{j,k}$, then (7.5) is equivalent to the conventional sector-bounded condition $(\xi_{j,k} - (\Upsilon_{j,k} - \delta_j I)\hat{x}_{j,k})^T(\xi_{j,k} - (\Upsilon_{j,k} + \delta_j I)\hat{x}_{j,k}) \leq 0$. Moreover, if $\Upsilon_{j,k} = 0$, then (7.5) is reduced to $\xi_{j,k}^T \xi_{j,k} \leq \delta_j^2 \hat{x}_{j,k}^T \hat{x}_{j,k}$.

According to (7.3)–(7.4), for the ith sensor node, the distributed filter subject to deception attacks is reconstructed as:

$$\hat{x}_{i,k+1} = A_{0,k}\hat{x}_{i,k} + L_{i,k}(y_{i,k} - C_{i0,k}\hat{x}_{i,k}) + \sum_{j \in \mathcal{N}_i} K_{ij,k}(\bar{x}_{j,k} - \hat{x}_{i,k}) \tag{7.6}$$

where $\bar{x}_{j,k} \in \mathbb{R}^{n_x}$ is the estimate of state x_k, matrices $L_{i,k}$ and $K_{ij,k}$ are the filter gains to be determined later.

Set $\tilde{\alpha}_{j,k} \triangleq \alpha_{j,k} - \bar{\alpha}_j$ and $e_{i,k} \triangleq x_k - \hat{x}_{i,k}$. From (7.1)–(7.4) and (7.6), the consensus error and the filtering error are respectively described as:

$$\begin{cases} \bar{x}_{j,k} - \hat{x}_{i,k} = e_{i,k} - e_{j,k} + \bar{\alpha}_j(e_{j,k} + \xi_{j,k} - x_k) + \tilde{\alpha}_{j,k}(e_{j,k} + \xi_{j,k} - x_k), \\ e_{i,k+1} = (A_{0,k} - L_{i,k}C_{i0,k})e_{i,k} + \sum_{t=1}^{r_1} \phi_{t,k}A_{t,k}x_k - \sum_{s=1}^{r_2} \theta_{is,k}L_{i,k}C_{is,k}x_k \\ \qquad - L_{i,k}D_{i,k}v_{i,k} + B_k w_k - \sum_{j\in\mathcal{N}_i} K_{ij,k}(\bar{x}_{j,k} - \hat{x}_{i,k}), \\ e_{i,0} = x_0 - \hat{x}_{i,0}. \end{cases} \tag{7.7}$$

Remarkably, the filtering error $e_{i,k}$ measures the difference between the estimate of node i and the real state, which depicts the filtering accuracy. The consensus error $\bar{x}_{j,k} - \hat{x}_{i,k}$ measures the difference between the estimate of node i and the estimate of its neighbor node j possibly suffering from deception attacks, which characterizes the consensus requirement. Accordingly, the following definition is given to evaluate the distributed consensus filtering performance.

Definition 7.1 For a given disturbance attenuation level $\gamma > 0$, positive definite matrices U_{1i}, U_{2i}, $R_{i,k}$, $Q_{i,k}$, $T_{1i,k}$, and $T_{2i,k}$, the error dynamics (7.7) is said to satisfy the H_∞-consensus performance constraint over the finite horizon $\mathscr{H}$ if the following inequality holds:

$$\begin{aligned} &\sum_{k=0}^{n-1}\sum_{i=1}^{N} \mathbb{E}\left\{\sum_{j\in\mathcal{N}_i} \|\bar{x}_{j,k} - \hat{x}_{i,k}\|^2_{R_{i,k}} + \|e_{i,k}\|^2_{Q_{i,k}}\right\} \\ &\leq \gamma^2 \sum_{i=1}^{N}\left[\|x_0\|^2_{U_{1i}} + \|e_{i,0}\|^2_{U_{2i}} + \sum_{k=0}^{n-1}\left(\|w_k\|^2_{T_{i1,k}} + \|v_{i,k}\|^2_{T_{i2,k}}\right)\right]. \end{aligned} \tag{7.8}$$

Remark 7.3 Actually, (7.8) can be rewritten as

$$\begin{aligned} &\sum_{k=0}^{n-1}\sum_{i=1}^{N} \mathbb{E}\Bigg\{\sum_{j\in\mathcal{N}_i} \|e_{i,k} - e_{j,k}\|^2_{R_{i,k}} + \|e_{i,k}\|^2_{Q_{i,k}} + \sum_{j\in\mathcal{N}_i} \bar{\alpha}_j \|\xi_{j,k} - \hat{x}_{j,k}\|^2_{R_{i,k}} \\ &\qquad + 2\sum_{j\in\mathcal{N}_i} \bar{\alpha}_j (\hat{x}_{j,k} - \hat{x}_{i,k})^T R_{i,k}(\xi_{j,k} - \hat{x}_{j,k})\Bigg\} \\ &\leq \gamma^2 \sum_{i=1}^{N}\left[\|x_0\|^2_{U_{1i}} + \|e_{i,0}\|^2_{U_{2i}} + \sum_{k=0}^{n-1}\left(\|w_k\|^2_{T_{i1,k}} + \|v_{i,k}\|^2_{T_{i2,k}}\right)\right] \end{aligned} \tag{7.9}$$

where the traditional consensus error $e_{i,k} - e_{j,k}$ and the filtering error $e_{i,k}$ are both constrained. Moreover, due to the introduction of the deception attacks , the signal changes before and after attacks $\xi_{j,k} - \hat{x}_{j,k}$ are also constrained, and this is different from existing references [1, 2]. Clearly, if there is no attack, i.e., $\bar{\alpha}_j = 0$, then (7.9) can be reduced to the traditional distributed consensus filtering performance index [1, 2].

Remark 7.4 As is well known, the performance index is critical in investigating the filtering problem. Different performance indices would lead to different filtering problems. The finite-time boundedness and input-output finite-time stability in [3] emphasize the transient behaviors of the filtering error dynamics at each time instant over a finite time. The exponential stability and the H_∞ performance in [4] reflect, respectively, the convergence speed and the noise attenuation level over an infinite time horizon. Different from these performance indices, in this chapter, we consider the boundedness of the norm sums of the filtering error and the consensus error for a period of time.

For analysis convenience, denote $\eta_{i,k} \triangleq [x_k^T \quad e_{i,k}^T]^T$ and $\nu_{i,k} \triangleq [w_k^T \quad v_{i,k}^T]^T$. Then, the following concise form is obtained

$$\begin{aligned}\eta_{i,k+1} = &\left(\mathscr{A}_{i0,k} - \sum_{j\in\mathscr{N}_i} \mathscr{K}_{ij,k}(I - \bar{\alpha}_j \mathscr{I}_1)\right)\eta_{i,k} + \sum_{t=1}^{r_1} \phi_{t,k}\mathscr{A}_{t,k}\eta_{i,k} - \sum_{s=1}^{r_2} \theta_{is,k}\mathscr{C}_{is,k}\eta_{i,k} \\ &+ \sum_{j\in\mathscr{N}_i} (1-\bar{\alpha}_j)\mathscr{K}_{ij,k}\eta_{j,k} - \sum_{j\in\mathscr{N}_i} \tilde{\alpha}_{j,k}\mathscr{K}_{ij,k}\eta_{j,k} + \sum_{j\in\mathscr{N}_i} \tilde{\alpha}_{j,k}\mathscr{K}_{ij,k}\mathscr{I}_1\eta_{j,k} \\ &- \sum_{j\in\mathscr{N}_i} \tilde{\alpha}_{j,k}\mathscr{K}_{ij,k}\mathscr{I}_2\xi_{j,k} - \sum_{j\in\mathscr{N}_i} \bar{\alpha}_j\mathscr{K}_{ij,k}\mathscr{I}_2\xi_{j,k} + \mathscr{B}_{i,k}\nu_{i,k},\end{aligned} \tag{7.10}$$

where

$$\begin{aligned}&\mathscr{A}_{i0,k} = \mathrm{diag}\{A_{0,k},\, A_{0,k} - L_{i,k}C_{i0,k}\}, \quad \mathscr{K}_{ij,k} = \mathrm{diag}\{0,\, K_{ij,k}\}, \\ &\mathscr{A}_{t,k} = \begin{bmatrix} A_{t,k} & 0 \\ A_{t,k} & 0 \end{bmatrix}, \quad \mathscr{C}_{is,k} = \begin{bmatrix} 0 & 0 \\ L_{i,k}C_{is,k} & 0 \end{bmatrix}, \quad \mathscr{I}_2 = \begin{bmatrix} 0 \\ I \end{bmatrix}, \\ &\mathscr{B}_{i,k} = \begin{bmatrix} B_k & 0 \\ B_k & -L_{i,k}D_{i,k} \end{bmatrix}, \quad \mathscr{I}_1 = \begin{bmatrix} 0 & 0 \\ I & 0 \end{bmatrix}.\end{aligned}$$

For the locally augmented dynamics (7.10), the following auxiliary index is introduced:

$$\begin{aligned}\sum_{k=0}^{n-1}\sum_{i=1}^{N} \mathbb{E}\Bigg\{ &\sum_{j\in\mathscr{N}_i} \|\mathscr{I}_3\eta_{i,k} + \mathscr{I}_{5j}\eta_{j,k} + \bar{\alpha}_j\xi_{j,k} + \tilde{\alpha}_{j,k}(\xi_{j,k} + (\mathscr{I}_3 - \mathscr{I}_4)\eta_{j,k})\|_{R_{i,k}}^2 \\ &+ \|\mathscr{I}_3\eta_{i,k}\|_{Q_{i,k}}^2 \Bigg\} \le \gamma^2 \sum_{i=1}^{N}\left(\|\eta_{i,0}\|_{\mathscr{U}_i}^2 + \sum_{k=0}^{n-1}\|\nu_k\|_{\mathscr{T}_{i,k}}^2\right)\end{aligned} \tag{7.11}$$

where

$$\begin{aligned}&\mathscr{U}_i = \mathrm{diag}\{U_{1i},\, U_{2i}\}, \quad \mathscr{T}_{i,k} = \mathrm{diag}\{T_{i1,k},\, T_{i2,k}\}, \\ &\mathscr{I}_3 = [0 \quad I], \quad \mathscr{I}_4 = [I \quad 0], \quad \mathscr{I}_{5j} = (\bar{\alpha}_j - 1)\mathscr{I}_3 - \bar{\alpha}_j\mathscr{I}_4.\end{aligned}$$

It is worth emphasizing that (7.8) is satisfied if (7.11) holds.

The purpose of the problem addressed in this chapter is to design the distributed filter (7.6) such that the consensus error and the filtering error of system (7.7) satisfy the performance index (7.8). In other words, the desired object is to determine the filter gains $L_{i,k}$ and $K_{ij,k}$ ($i \in \mathscr{V}$, $j \in \mathscr{N}_i$) such that the augmented dynamics (7.10) satisfies the auxiliary index (7.11).

7.2 Main Results

In this section, by means of the vector dissipativity theory [1, 2, 5], sufficient conditions will be established for guaranteeing that the error dynamics (7.7) satisfies the H_∞-consensus performance criterion (7.8).

Denote $\eta_k \triangleq [\eta_{1,k}^T \ \cdots \ \eta_{N,k}^T]^T$, $\nu_k \triangleq [\nu_{1,k}^T \ \cdots \ \nu_{N,k}^T]^T$, $\xi_k \triangleq [\xi_{1,k}^T \ \cdots \ \xi_{N,k}^T]^T$, and

$$
\begin{aligned}
& S_i\left(\eta_k, \xi_k, \nu_{i,k}\right) \\
\triangleq & -\sum_{j \in \mathscr{N}_i} \|\mathscr{I}_3\eta_{i,k} + \mathscr{I}_{5j}\eta_{j,k} + \bar{\alpha}_j\xi_{j,k} + \tilde{\alpha}_{j,k}\xi_{j,k} - \tilde{\alpha}_{j,k}\mathscr{I}_4\eta_{j,k} + \tilde{\alpha}_{j,k}\mathscr{I}_3\eta_{j,k}\|^2_{R_{i,k}} \\
& + \gamma^2\|\nu_{i,k}\|^2_{\mathscr{T}_{i,k}} - \|\mathscr{I}_3\eta_{i,k}\|^2_{Q_{i,k}}.
\end{aligned}
$$

The definition of the stochastic vector dissipativity is given in what follows.

Definition 7.2 ([1, 2]) The discrete time-varying locally augmented dynamics (7.10) is stochastic vector dissipative-like over the finite horizon $\mathscr{H}$ with respect to a vector of supply rate functions

$$
\mathbf{S}(\eta_k, \xi_k, \nu_k) \triangleq [S_1(\eta_k, \xi_k, \nu_{1,k}) \quad \cdots \quad S_N(\eta_k, \xi_k, \nu_{N,k})]^T,
$$

if there exists a vector of nonnegative definite storage functions $\mathbf{V}(\eta_k) \triangleq [V_1(\eta_{1,k}) \quad \cdots \quad V_N(\eta_{N,k})]^T$ with $\mathbf{V}(0) = 0$, and a sequence of nonsingular column sub-stochastic dissipation matrices $W_k \in \mathbb{R}^{N\times N}$ such that the following vector dissipation inequality is satisfied for all $k \in \mathscr{H}$:

$$
\mathbb{E}\{\mathbf{V}(\eta_{k+1})\} \leq\leq W_k\mathbb{E}\{\mathbf{V}(\eta_k)\} + \mathbb{E}\{\mathbf{S}(\eta_k, \xi_k, \nu_k)\}. \tag{7.12}
$$

Noting that (7.12) is of a vector form, one rewrites (7.12) in a scalar form related to node i ($\forall i \in \mathscr{V}$) as:

$$
\mathbb{E}\{\mathbf{V}_i(\eta_{i,k+1})\} \leq [W_k\mathbb{E}\{\mathbf{V}(\eta_k)\}]_i + \mathbb{E}\{\mathbf{S}_i(\eta_k\xi_k, \nu_{i,k})\} \tag{7.13}
$$

where $[W_k\mathbb{E}\{\mathbf{V}(\eta_k)\}]_i$ is the ith element of the vector $W_k\mathbb{E}\{\mathbf{V}(\eta_k)\}$. Additionally, for constructing a desirable dissipation matrix W_k, define an interval-valued function with respect to the out-degree q_i as follows:

$$\mathfrak{I}_{q_i} = \begin{cases} (0, \frac{1+q_i}{2q_i}), & \text{if } q_i \neq 0; \\ (0, 1], & \text{if } q_i = 0. \end{cases} \tag{7.14}$$

The following theorem provides a sufficient criterion to guarantee the stochastic vector dissipative-like property of system (7.10), which further infers the desired performance (7.8) for the filtering error dynamics (7.7).

Theorem 7.1 *For $i \in \mathscr{V}$, $j \in \mathscr{N}_i$ and $k \in \mathscr{H}$, let the disturbance attenuation level $\gamma > 0$, the scalar sequence $\rho_{i,k} \in \mathfrak{I}_{q_i}$, the scalars δ_j and $\bar{\alpha}_j$, the matrices $R_{i,k}$, $Q_{i,k}$, $\mathscr{T}_{i,k}$, $\mathscr{U}_i$, $L_{i,k}$ and $K_{ij,k}$ be given. Assume that there is scalar sequence $\lambda_{ij,k}$ ($j \in \mathscr{N}_i$) and a sequence of positive definite matrices $\{\mathscr{P}_{i,k}\}_{k \in \mathscr{H} \cup \{n\}}$ with $\mathscr{P}_{i,0} \leq \gamma^2 U_i$ satisfying the following inequality:*

$$\Xi_{i,k} \triangleq \begin{bmatrix} \Xi_{i,k}^{11} & \Xi_{i,k}^{12} & \Xi_{i,k}^{13} & \Xi_{i,k}^{14} \\ * & \Xi_{i,k}^{22} & \Xi_{i,k}^{23} & \Xi_{i,k}^{24} \\ * & * & \Xi_{i,k}^{33} & \Xi_{i,k}^{34} \\ * & * & * & -\Psi_{i,k} \end{bmatrix} < 0 \tag{7.15}$$

where

$$\Xi_{i,k}^{11} = \sum_{t=1}^{r_1} \mathscr{A}_{t,k}^T \mathscr{P}_{i,k+1} \mathscr{A}_{t,k} + \sum_{s=1}^{r_2} \mathscr{C}_{is,k}^T \mathscr{P}_{i,k+1} \mathscr{C}_{is,k} + \mathscr{I}_3^T Q_{i,k} \mathscr{I}_3 + p_i \mathscr{I}_3^T R_{i,k} \mathscr{I}_3,$$

$$- \rho_{i,k} \mathscr{P}_{i,k}, \quad \rho_{i,k} = \frac{1 + q_i(1 - \rho_{i,k})}{1 + q_i}, \quad \Xi_{i,k}^{12} = \mathrm{vec}_{p_i}^j \left[\mathscr{I}_3^T R_{i,k} \mathscr{I}_{5j} \right],$$

$$\begin{aligned} \Theta_{ij,k} = {} & \beta_j (I - \mathscr{I}_1)^T \mathscr{K}_{ij,k}^T \mathscr{P}_{i,k+1} \mathscr{K}_{ij,k} (I - \mathscr{I}_1) + \lambda_{ij,k} \delta_j^2 (\mathscr{I}_4 - \mathscr{I}_3)^T (\mathscr{I}_4 - \mathscr{I}_3) \\ & - \frac{\rho_{j,k}}{1 + q_j} \mathscr{P}_{j,k} - \lambda_{ij,k} \bar{\Upsilon}_{j,k}^T \bar{\Upsilon}_{j,k} + \mathscr{I}_{5j}^T R_{i,k} \mathscr{I}_{5j} + \beta_j (\mathscr{I}_4 - \mathscr{I}_3)^T R_{i,k} (\mathscr{I}_4 - \mathscr{I}_3), \end{aligned}$$

$$\begin{aligned} \Xi_{i,k}^{23} = {} & \mathrm{diag}_{p_i}^j \Big\{ \lambda_{ij,k} \bar{\Upsilon}_{j,k}^T + \beta_j (I - \mathscr{I}_1)^T \mathscr{K}_{ij,k}^T \mathscr{P}_{i,k+1} \mathscr{K}_{ij,k} + \bar{\alpha}_j \mathscr{I}_{5j}^T R_{i,k} - \beta_j \mathscr{I}_4^T R_{i,k} \\ & + \beta_j \mathscr{I}_3^T R_{i,k} \Big\}, \quad \Xi_{i,k}^{13} = \mathrm{vec}_{p_i}^j \left[\bar{\alpha}_j \mathscr{I}_3^T R_{i,k} \right], \quad \Xi_{i,k}^{22} = \mathrm{diag}_{p_i}^j \left\{ \Theta_{ij,k} \right\}, \end{aligned}$$

$$\Xi_{i,k}^{33} = \mathrm{diag}_{p_i}^j \left\{ \beta_j \mathscr{I}_2^T \mathscr{K}_{ij,k}^T \mathscr{P}_{i,k+1} \mathscr{K}_{ij,k} \mathscr{I}_2 - \lambda_{ij,k} I + \bar{\alpha}_j R_{i,k} \right\},$$

$$\Xi_{i,k}^{14} = \Big(\mathscr{A}_{i0,k} - \sum_{j \in \mathscr{N}_i} \mathscr{K}_{ij,k} (I - \bar{\alpha}_j \mathscr{I}_1) \Big)^T, \quad \Xi_{i,k}^{24} = \left(\mathrm{vec}_{p_i}^j \left[(1 - \bar{\alpha}_j) \mathscr{K}_{ij,k} \right] \right)^T,$$

$$\Xi_{i,k}^{34} = - \left(\mathrm{vec}_{p_i}^j [\bar{\alpha}_j \mathscr{K}_{ij,k} \mathscr{I}_2] \right)^T, \quad \Psi_{i,k} = \mathscr{P}_{i,k+1}^{-1} - \frac{1}{\gamma^2} \mathscr{B}_{i,k} \mathscr{T}_{i,k}^{-1} \mathscr{B}_{i,k}^T,$$

$$\bar{\Upsilon}_{j,k} = \left[\Upsilon_{j,k} \; -\Upsilon_{j,k} \right], \quad \beta_j = \bar{\alpha}_j - \bar{\alpha}_j^2.$$

Then, the locally augmented dynamics (7.10) is stochastic vector dissipative-like over the finite horizon $\mathscr{H}$ with respect to the vector supply rate $\mathbf{S}(\eta_k, \xi_k, \nu_k)$. Moreover, the filtering error dynamics (7.7) satisfies the H_∞-consensus performance criterion (7.8).

Proof Via the expectation formula, one obtains

$$\mathbb{E}\{\tilde{\alpha}_{j,k}\} = \mathbb{E}\{\alpha_{j,k} - \bar{\alpha}_j\} = 0, \quad \mathbb{E}\{\tilde{\alpha}_{j,k}^2\} = \bar{\alpha}_j(1 - \bar{\alpha}_j). \tag{7.16}$$

Set the storage function as $V_i(\eta_{i,k}) = \eta_{i,k}^T \mathscr{P}_{i,k}\eta_{i,k}$. In order to verify inequality (7.13), the storage function concerning the ith node is calculated along the trajectory of system (7.10) with the aid of (7.16) as:

$$\begin{aligned}
&\mathbb{E}\{V_i(\eta_{i,k+1})|\eta_{i,k}, \eta_{j,k}, j \in \mathscr{N}_i\} \\
&= \mathbb{E}\{\eta_{i,k+1}^T \mathscr{P}_{i,k+1}\eta_{i,k+1}|\eta_{i,k}, \eta_{j,k}, j \in \mathscr{N}_i\} \\
&= \eta_{i,k}^T \left(\mathscr{A}_{i0,k} - \sum_{j\in\mathscr{N}_i} \mathscr{K}_{ij,k}(I - \bar{\alpha}_j\mathscr{I}_1)\right)^T \mathscr{P}_{i,k+1} \left(\mathscr{A}_{i0,k} - \sum_{j\in\mathscr{N}_i} \mathscr{K}_{ij,k}(I - \bar{\alpha}_j\mathscr{I}_1)\right) \eta_{i,k} \\
&\quad + \sum_{s=1}^{r_2} \eta_{i,k}^T \mathscr{C}_{is,k}^T \mathscr{P}_{i,k+1}\mathscr{C}_{is,k}\eta_{i,k} + \sum_{j\in\mathscr{N}_i} \eta_{j,k}^T(1 - \bar{\alpha}_j)\mathscr{K}_{ij,k}^T \mathscr{P}_{i,k+1} \sum_{j\in\mathscr{N}_i} (1 - \bar{\alpha}_j)\mathscr{K}_{ij,k}\eta_{j,k} \\
&\quad + \sum_{j\in\mathscr{N}_i} \beta_j\eta_{j,k}^T \mathscr{K}_{ij,k}^T \mathscr{P}_{i,k+1}\mathscr{K}_{ij,k}\eta_{j,k} + \sum_{j\in\mathscr{N}_i} \beta_j\eta_{j,k}^T \mathscr{I}_1^T \mathscr{K}_{ij,k}^T \mathscr{P}_{i,k+1}\mathscr{K}_{ij,k}\mathscr{I}_1\eta_{j,k} \\
&\quad + \sum_{j\in\mathscr{N}_i} \beta_j\xi_{j,k}^T \mathscr{I}_2^T \mathscr{K}_{ij,k}^T \mathscr{P}_{i,k+1}\mathscr{K}_{ij,k}\mathscr{I}_2\xi_{j,k} - 2\sum_{j\in\mathscr{N}_i} \beta_j\eta_{j,k}^T \mathscr{K}_{ij,k}^T \mathscr{P}_{i,k+1}\mathscr{K}_{ij,k}\mathscr{I}_1\eta_{j,k} \\
&\quad + 2\sum_{j\in\mathscr{N}_i} \beta_j\eta_{j,k}^T \mathscr{K}_{ij,k}^T \mathscr{P}_{i,k+1}\mathscr{K}_{ij,k}\mathscr{I}_2\xi_{j,k} - 2\sum_{j\in\mathscr{N}_i} \beta_j\eta_{j,k}^T \mathscr{I}_1^T \mathscr{K}_{ij,k}^T \mathscr{P}_{i,k+1}\mathscr{K}_{ij,k}\mathscr{I}_2\xi_{j,k} \\
&\quad - 2v_{i,k}^T \mathscr{B}_{i,k}^T \mathscr{P}_{i,k+1} \sum_{j\in\mathscr{N}_i} \bar{\alpha}_j\mathscr{K}_{ij,k}\mathscr{I}_2\xi_{j,k} + 2v_{i,k}^T \mathscr{B}_{i,k}^T \mathscr{P}_{i,k+1} \sum_{j\in\mathscr{N}_i} (1 - \bar{\alpha}_j)\mathscr{K}_{ij,k}\eta_{j,k} \\
&\quad - 2\sum_{j\in\mathscr{N}_i} (1 - \bar{\alpha}_j)\eta_{j,k}^T \mathscr{K}_{ij,k}^T \mathscr{P}_{i,k+1} \sum_{j\in\mathscr{N}_i} \bar{\alpha}_j\mathscr{K}_{ij,k}\mathscr{I}_2\xi_{j,k} + \sum_{t=1}^{r_1} \eta_{i,k}^T \mathscr{A}_{t,k}^T \mathscr{P}_{i,k+1}\mathscr{A}_{t,k}\eta_{i,k} \\
&\quad + 2\eta_{i,k}^T \left(\mathscr{A}_{i0,k} - \sum_{j\in\mathscr{N}_i} \mathscr{K}_{ij,k}(I - \bar{\alpha}_j\mathscr{I}_1)\right)^T \mathscr{P}_{i,k+1} \sum_{j\in\mathscr{N}_i} (1 - \bar{\alpha}_j)\mathscr{K}_{ij,k}\eta_{j,k} \\
&\quad - 2\eta_{i,k}^T \left(\mathscr{A}_{i0,k} - \sum_{j\in\mathscr{N}_i} \mathscr{K}_{ij,k}(I - \bar{\alpha}_j\mathscr{I}_1)\right)^T \mathscr{P}_{i,k+1} \sum_{j\in\mathscr{N}_i} \bar{\alpha}_j\mathscr{K}_{ij,k}\mathscr{I}_2\xi_{j,k} \\
&\quad + 2\eta_{i,k}^T \left(\mathscr{A}_{i0,k} - \sum_{j\in\mathscr{N}_i} \mathscr{K}_{ij,k}(I - \bar{\alpha}_j\mathscr{I}_1)\right)^T \mathscr{P}_{i,k+1}\mathscr{B}_{i,k}v_{i,k} \\
&\quad + \sum_{j\in\mathscr{N}_i} \bar{\alpha}_j\xi_{j,k}^T \mathscr{I}_2^T \mathscr{K}_{ij,k}^T \mathscr{P}_{i,k+1} \sum_{j\in\mathscr{N}_i} \bar{\alpha}_j\mathscr{K}_{ij,k}\mathscr{I}_2\xi_{j,k} \\
&\quad + \gamma^2 v_{i,k}^T \mathscr{T}_{i,k}v_{i,k} + v_{i,k}^T(\mathscr{B}_{i,k}^T \mathscr{P}_{i,k+1}\mathscr{B}_{i,k} - \gamma^2\mathscr{T}_{i,k})v_{i,k}.
\end{aligned} \tag{7.17}$$

Denote $\Phi_{i,k} = \gamma^2 \mathscr{T}_{i,k} - \mathscr{B}_{i,k}^T \mathscr{P}_{i,k+1} \mathscr{B}_{i,k} > 0$. By using the completing-the-square technique, we have

$$\begin{aligned}
&2\nu_{i,k}^T \mathscr{B}_{i,k}^T \mathscr{P}_{i,k+1} \sum_{j \in \mathscr{N}_i} (1-\bar{\alpha}_j) \mathscr{K}_{ij,k} \eta_{j,k} \\
&+ 2\eta_{i,k}^T \left(\mathscr{A}_{i0,k} - \sum_{j \in \mathscr{N}_i} \mathscr{K}_{ij,k} (I - \bar{\alpha}_j \mathscr{I}_1) \right)^T \mathscr{P}_{i,k+1} \mathscr{B}_{i,k} \nu_{i,k} \\
&- 2\nu_{i,k}^T \mathscr{B}_{i,k}^T \mathscr{P}_{i,k+1} \sum_{j \in \mathscr{N}_i} \bar{\alpha}_j \mathscr{K}_{ij,k} \mathscr{I}_2 \xi_{j,k} + \nu_{i,k}^T (\mathscr{B}_{i,k}^T \mathscr{P}_{i,k+1} \mathscr{B}_{i,k} - \gamma^2 \mathscr{T}_{i,k}) \nu_{i,k} \\
\leq & \sum_{j \in \mathscr{N}_i} \bar{\alpha}_j \xi_{j,k}^T \mathscr{I}_2^T \mathscr{K}_{ij,k}^T \Gamma_{i,k} \sum_{j \in \mathscr{N}_i} \bar{\alpha}_j \mathscr{K}_{ij,k} \mathscr{I}_2 \xi_{j,k} \\
&- 2 \sum_{j \in \mathscr{N}_i} \bar{\alpha}_j \xi_{j,k}^T \mathscr{I}_2^T \mathscr{K}_{ij,k}^T \Gamma_{i,k} \sum_{j \in \mathscr{N}_i} (1-\bar{\alpha}_j) \mathscr{K}_{ij,k} \eta_{j,k} \\
&+ \sum_{j \in \mathscr{N}_i} (1-\bar{\alpha}_j) \eta_{j,k}^T \mathscr{K}_{ij,k}^T \Gamma_{i,k} \sum_{j \in \mathscr{N}_i} (1-\bar{\alpha}_j) \mathscr{K}_{ij,k} \eta_{j,k} \\
&- 2\eta_{i,k}^T \left(\mathscr{A}_{i0,k} - \sum_{j \in \mathscr{N}_i} \mathscr{K}_{ij,k} (I - \bar{\alpha}_j \mathscr{I}_1) \right)^T \Gamma_{i,k} \sum_{j \in \mathscr{N}_i} \bar{\alpha}_j \mathscr{K}_{ij,k} \mathscr{I}_2 \xi_{j,k} \\
&+ 2\eta_{i,k}^T \left(\mathscr{A}_{i0,k} - \sum_{j \in \mathscr{N}_i} \mathscr{K}_{ij,k} (I - \bar{\alpha}_j \mathscr{I}_1) \right)^T \Gamma_{i,k} \sum_{j \in \mathscr{N}_i} (1-\bar{\alpha}_j) \mathscr{K}_{ij,k} \eta_{j,k} \\
&+ \eta_{i,k}^T \left(\mathscr{A}_{i0,k} - \sum_{j \in \mathscr{N}_i} \mathscr{K}_{ij,k} (I - \bar{\alpha}_j \mathscr{I}_1) \right)^T \Gamma_{i,k} \left(\mathscr{A}_{i0,k} - \sum_{j \in \mathscr{N}_i} \mathscr{K}_{ij,k} (I - \bar{\alpha}_j \mathscr{I}_1) \right) \eta_{i,k}
\end{aligned} \tag{7.18}$$

where

$$\Gamma_{i,k} \triangleq \mathscr{P}_{i,k+1} \mathscr{B}_{i,k} \Phi_{i,k}^{-1} \mathscr{P}_{i,k+1} \mathscr{B}_{i,k}^T \mathscr{P}_{i,k+1}.$$

It follows from the matrix inverse formula that

$$\Psi_{i,k}^{-1} \triangleq \Gamma_{i,k} + \mathscr{P}_{i,k+1} = \left(\mathscr{P}_{i,k+1}^{-1} - \frac{1}{\gamma^2} \mathscr{B}_{i,k} \mathscr{T}_{i,k}^{-1} \mathscr{B}_{i,k}^T \right)^{-1}. \tag{7.19}$$

In addition, (7.5) is transformed into an augmented form:

$$\left(\xi_{j,k} - \bar{\Upsilon}_{j,k} \eta_{j,k} \right)^T \left(\xi_{j,k} - \bar{\Upsilon}_{j,k} \eta_{j,k} \right) \leq \delta_j^2 \eta_{j,k}^T (\mathscr{I}_4 - \mathscr{I}_3)^T (\mathscr{I}_4 - \mathscr{I}_3) \eta_{j,k}. \tag{7.20}$$

Based on (7.18)–(7.20), (7.17) is further derived as:

$$
\begin{aligned}
&\mathbb{E}\{V_i(\eta_{i,k+1})|\eta_{i,k},\eta_{j,k},j\in\mathscr{N}_i\}\\
\leq&\sum_{t=1}^{r_1}\eta_{i,k}^T\mathscr{A}_{t,k}^T\mathscr{P}_{i,k+1}\mathscr{A}_{t,k}\eta_{i,k}+\sum_{s=1}^{r_2}\eta_{i,k}^T\mathscr{C}_{is,k}^T\mathscr{P}_{i,k+1}\mathscr{C}_{is,k}\eta_{i,k}\\
&+\sum_{j\in\mathscr{N}_i}\beta_j\eta_{j,k}^T\mathscr{I}_1^T\mathscr{K}_{ij,k}^T\mathscr{P}_{i,k+1}\mathscr{K}_{ij,k}\mathscr{I}_1\eta_{j,k}\\
&+\sum_{j\in\mathscr{N}_i}\beta_j\xi_{j,k}^T\mathscr{I}_2^T\mathscr{K}_{ij,k}^T\mathscr{P}_{i,k+1}\mathscr{K}_{ij,k}\mathscr{I}_2\xi_{j,k}\\
&+\sum_{j\in\mathscr{N}_i}\eta_{j,k}^T(1-\bar{\alpha}_j)\mathscr{K}_{ij,k}^T\Psi_{i,k}^{-1}\sum_{j\in\mathscr{N}_i}(1-\bar{\alpha}_j)\mathscr{K}_{ij,k}\eta_{j,k}\\
&+\sum_{j\in\mathscr{N}_i}\beta_j\eta_{j,k}^T\mathscr{K}_{ij,k}^T\mathscr{P}_{i,k+1}\mathscr{K}_{ij,k}\eta_{j,k}\\
&-2\sum_{j\in\mathscr{N}_i}\beta_j\eta_{j,k}^T\mathscr{I}_1^T\mathscr{K}_{ij,k}^T\mathscr{P}_{i,k+1}\mathscr{K}_{ij,k}\mathscr{I}_2\xi_{j,k}\\
&+\sum_{j\in\mathscr{N}_i}\bar{\alpha}_j\xi_{j,k}^T\mathscr{I}_2^T\mathscr{K}_{ij,k}^T\Psi_{i,k}^{-1}\sum_{j\in\mathscr{N}_i}\bar{\alpha}_j\mathscr{K}_{ij,k}\mathscr{I}_2\xi_{j,k}\\
&+\gamma^2v_{i,k}^T\mathscr{T}_{i,k}v_{i,k}-2\sum_{j\in\mathscr{N}_i}(1-\bar{\alpha}_j)\eta_{j,k}^T\mathscr{K}_{ij,k}^T\Psi_{i,k}^{-1}\sum_{j\in\mathscr{N}_i}\bar{\alpha}_j\mathscr{K}_{ij,k}\mathscr{I}_2\xi_{j,k}\\
&-2\eta_{i,k}^T\left(\mathscr{A}_{i0,k}-\sum_{j\in\mathscr{N}_i}\mathscr{K}_{ij,k}(I-\bar{\alpha}_j\mathscr{I}_1)\right)^T\Psi_{i,k}^{-1}\sum_{j\in\mathscr{N}_i}\bar{\alpha}_j\mathscr{K}_{ij,k}\mathscr{I}_2\xi_{j,k}\\
&+2\eta_{i,k}^T\left(\mathscr{A}_{i0,k}-\sum_{j\in\mathscr{N}_i}\mathscr{K}_{ij,k}(I-\bar{\alpha}_j\mathscr{I}_1)\right)^T\Psi_{i,k}^{-1}\sum_{j\in\mathscr{N}_i}(1-\bar{\alpha}_j)\mathscr{K}_{ij,k}\eta_{j,k}\\
&-\sum_{j\in\mathscr{N}_i}\lambda_{ij,k}\Big[(\xi_{j,k}-\bar{\Upsilon}_{j,k}\eta_{j,k})^T(\xi_{j,k}-\bar{\Upsilon}_{j,k}\eta_{j,k})-\delta_j^2\eta_{j,k}^T(\mathscr{I}_4-\mathscr{I}_3)^T(\mathscr{I}_4-\mathscr{I}_3)\eta_{j,k}\Big]\\
&-2\sum_{j\in\mathscr{N}_i}\beta_j\eta_{j,k}^T\mathscr{K}_{ij,k}^T\mathscr{P}_{i,k+1}\mathscr{K}_{ij,k}\mathscr{I}_1\eta_{j,k}+2\sum_{j\in\mathscr{N}_i}\beta_j\eta_{j,k}^T\mathscr{K}_{ij,k}^T\mathscr{P}_{i,k+1}\mathscr{K}_{ij,k}\mathscr{I}_2\xi_{j,k}\\
&+\eta_{i,k}^T\left(\mathscr{A}_{i0,k}-\sum_{j\in\mathscr{N}_i}\mathscr{K}_{ij,k}(I-\bar{\alpha}_j\mathscr{I}_1)\right)^T\Psi_{i,k}^{-1}\left(\mathscr{A}_{i0,k}-\sum_{j\in\mathscr{N}_i}\mathscr{K}_{ij,k}(I-\bar{\alpha}_j\mathscr{I}_1)\right)\eta_{i,k}\\
\triangleq&\varpi_{i,k}^T\Pi_{i,k}\varpi_{i,k}+\gamma^2v_{i,k}^T\mathscr{T}_{i,k}v_{i,k}
\end{aligned}
\tag{7.21}
$$

where

$$\varpi_{i,k} = \begin{bmatrix} \eta_{i,k}^T & \eta_{\mathcal{N}_i,k}^T & \xi_{\mathcal{N}_i,k}^T \end{bmatrix}^T, \quad \eta_{\mathcal{N}_i,k} = \begin{bmatrix} \eta_{j_{i_1},k}^T & \cdots & \eta_{j_{ip_i},k}^T \end{bmatrix}^T,$$

$$\Pi_{i,k} = \begin{bmatrix} \Pi_{i,k}^{11} & \Pi_{i,k}^{12} & \Pi_{i,k}^{13} \\ * & \Pi_{i,k}^{22} & \Pi_{i,k}^{23} \\ * & * & \Pi_{i,k}^{33} \end{bmatrix}, \quad \xi_{\mathcal{N}_i,k} = \begin{bmatrix} \xi_{j_{i_1},k}^T & \cdots & \xi_{j_{ip_i},k}^T \end{bmatrix}^T,$$

$$\begin{aligned}
\Pi_{i,k}^{11} = {} & \left(\mathscr{A}_{i0,k} - \sum_{j\in\mathcal{N}_i} \mathscr{K}_{ij,k}(I - \bar{\alpha}_j \mathscr{I}_1) \right)^T \Psi_{i,k}^{-1} \left(\mathscr{A}_{i0,k} - \sum_{j\in\mathcal{N}_i} \mathscr{K}_{ij,k}(I - \bar{\alpha}_j \mathscr{I}_1) \right) \\
& + \sum_{t=1}^{r_1} \mathscr{A}_{t,k}^T \mathscr{P}_{i,k+1} \mathscr{A}_{t,k} + \sum_{s=1}^{r_2} \mathscr{C}_{is,k}^T \mathscr{P}_{i,k+1} \mathscr{C}_{is,k}, \\
\Pi_{i,k}^{22} = {} & \mathrm{vec}_{p_i}^j \left[(1 - \bar{\alpha}_j)\mathscr{K}_{ij,k} \right]^T \Psi_{i,k}^{-1} \mathrm{vec}_{p_i}^j \left[(1 - \bar{\alpha}_j)\mathscr{K}_{ij,k} \right] \\
& + \mathrm{diag}_{p_i}^j \left\{ \beta_j (\mathscr{K}_{ij,k} - \mathscr{K}_{ij,k}\mathscr{I}_1)^T \mathscr{P}_{i,k+1} (\mathscr{K}_{ij,k} - \mathscr{K}_{ij,k}\mathscr{I}_1) \right. \\
& \left. + \lambda_{ij,k}\delta_j^2 (\mathscr{I}_4 - \mathscr{I}_3)^T (\mathscr{I}_4 - \mathscr{I}_3) - \lambda_{ij,k} \bar{\Upsilon}_{j,k}^T \bar{\Upsilon}_{j,k} \right\}, \\
\Pi_{i,k}^{33} = {} & \mathrm{diag}_{p_i}^j \left\{ \beta_j \mathscr{I}_2^T \mathscr{K}_{ij,k}^T \mathscr{P}_{i,k+1} \mathscr{K}_{ij,k} \mathscr{I}_2 - \lambda_{ij,k} I \right\} \\
& + \mathrm{vec}_{p_i}^j \left[\bar{\alpha}_j \mathscr{K}_{ij,k} \mathscr{I}_2 \right]^T \Psi_{i,k}^{-1} \mathrm{vec}_{p_i}^j \left[\bar{\alpha}_j \mathscr{K}_{ij,k} \mathscr{I}_2 \right], \\
\Pi_{i,k}^{12} = {} & \mathrm{vec}_{p_i}^j \left[\left(\mathscr{A}_{i0,k} - \sum_{j\in\mathcal{N}_i} \mathscr{K}_{ij,k}(I - \bar{\alpha}_j \mathscr{I}_1) \right)^T \Psi_{i,k}^{-1} (1 - \bar{\alpha}_j)\mathscr{K}_{ij,k} \right], \\
\Pi_{i,k}^{13} = {} & \mathrm{vec}_{p_i}^j \left[- \left(\mathscr{A}_{i0,k} - \sum_{j\in\mathcal{N}_i} \mathscr{K}_{ij,k}(I - \bar{\alpha}_j \mathscr{I}_1) \right)^T \Psi_{i,k}^{-1} \bar{\alpha}_j \mathscr{K}_{ij,k} \mathscr{I}_2 \right], \\
\Pi_{i,k}^{23} = {} & - \left(\mathrm{vec}_{p_i}^j \left[(1 - \bar{\alpha}_j)\mathscr{K}_{ij,k} \right] \right)^T \Psi_{i,k}^{-1} \mathrm{vec}_{p_i}^j \left[\bar{\alpha}_j \mathscr{K}_{ij,k} \mathscr{I}_2 \right] + \mathrm{diag}_{p_i}^j \left[\lambda_{ij,k} \bar{\Upsilon}_{j,k}^T \right] \\
& + \mathrm{diag}_{p_i}^j \left[\beta_j (\mathscr{K}_{ij,k} - \mathscr{K}_{ij,k}\mathscr{I}_1)^T \mathscr{P}_{i,k+1} \mathscr{K}_{ij,k} \mathscr{I}_2 \right].
\end{aligned}$$

From the denotations of (7.15), one further has

$$\begin{aligned}
& \varpi_{i,k}^T \Pi_{i,k} \varpi_{i,k} - \sum_{j\in\mathcal{N}_i} \frac{\rho_{j,k}}{1+q_j} \eta_{j,k}^T \mathscr{P}_{j,k} \eta_{j,k} - \rho_{i,k} \eta_{i,k}^T \mathscr{P}_{i,k} \eta_{i,k} + \mathbb{E}\| \mathscr{I}_3 \eta_{i,k} \|_{Q_{i,k}}^2 \\
& + \mathbb{E} \left\{ \sum_{j\in\mathcal{N}_i} \| \mathscr{I}_3 \eta_{i,k} + \mathscr{I}_{5j} \eta_{j,k} + \bar{\alpha}_j \xi_{j,k} + \tilde{\alpha}_{j,k} \xi_{j,k} - \tilde{\alpha}_{j,k} \mathscr{I}_4 \eta_{j,k} + \tilde{\alpha}_{j,k} \mathscr{I}_3 \eta_{j,k} \|_{R_{i,k}}^2 \right\} \\
= {} & \varpi_{i,k}^T \left(\$_{i,k} + \begin{bmatrix} (\Xi_{i,k}^{14})^T & (\Xi_{i,k}^{24})^T & (\Xi_{i,k}^{34})^T \end{bmatrix}^T \Psi_{i,k}^{-1} \begin{bmatrix} \Xi_{i,k}^{41} & \Xi_{i,k}^{42} & \Xi_{i,k}^{43} \end{bmatrix} \right) \varpi_{i,k}
\end{aligned}$$

where

$$
\$_{i,k} = \begin{bmatrix} \Xi_{i,k}^{11} & \Xi_{i,k}^{12} & \Xi_{i,k}^{13} \\ * & \Xi_{i,k}^{22} & \Xi_{i,k}^{23} \\ * & * & \Xi_{i,k}^{33} \end{bmatrix}.
$$

According to the Schur Complement Lemma, it follows from $\Xi_{i,k} \leq 0$ that

$$
\$_{i,k} + \left[(\Xi_{i,k}^{14})^T \quad (\Xi_{i,k}^{24})^T \quad (\Xi_{i,k}^{34})^T\right]^T \Psi_{i,k}^{-1} \left[\Xi_{i,k}^{41} \quad \Xi_{i,k}^{42} \quad \Xi_{i,k}^{43}\right] \leq 0,
$$

which further results in

$$
\begin{aligned}
&\varpi_{i,k}^T \Pi_{i,k} \varpi_{i,k} \\
&\leq -\mathbb{E}\left\{ \sum_{j\in\mathcal{N}_i} \|\mathscr{I}_3\eta_{i,k} + \mathscr{I}_{5j}\eta_{j,k} + \bar{\alpha}_j\xi_{j,k} + \tilde{\alpha}_{j,k}\xi_{j,k} - \tilde{\alpha}_{j,k}\mathscr{I}_4\eta_{j,k} + \tilde{\alpha}_{j,k}\mathscr{I}_3\eta_{j,k}\|_{R_{i,k}}^2 \right\} \\
&\quad - \mathbb{E}\|\mathscr{I}_3\eta_{i,k}\|_{Q_{i,k}}^2 + \sum_{j\in\mathcal{N}_i} \frac{\rho_{j,k}}{1+q_j}\eta_{j,k}^T \mathscr{P}_{j,k}\eta_{j,k} + \rho_{i,k}\eta_{i,k}^T\mathscr{P}_{i,k}\eta_{i,k}.
\end{aligned} \tag{7.22}
$$

According to (7.21) and (7.22), one has

$$
\begin{aligned}
&\mathbb{E}\{V_i(\eta_{i,k+1})|\eta_{i,k}, \eta_{j,k}, j \in \mathcal{N}_i\} \\
&\leq \varpi_{i,k}^T \Pi_{i,k} \varpi_{i,k} + \gamma^2 v_{i,k}^T \mathscr{T}_{i,k} v_{i,k} \\
&\leq -\mathbb{E}\left\{ \sum_{j\in\mathcal{N}_i} \|\mathscr{I}_3\eta_{i,k} + \mathscr{I}_{5j}\eta_{j,k} + \bar{\alpha}_j\xi_{j,k} + \tilde{\alpha}_{j,k}\xi_{j,k} - \tilde{\alpha}_{j,k}\mathscr{I}_4\eta_{j,k} + \tilde{\alpha}_{j,k}\mathscr{I}_3\eta_{j,k}\|_{R_{i,k}}^2 \right\} \\
&\quad - \mathbb{E}\|\mathscr{I}_3\eta_{i,k}\|_{Q_{i,k}}^2 + \sum_{j\in\mathcal{N}_i} \frac{\rho_{j,k}}{1+q_j}\eta_{j,k}^T \mathscr{P}_{j,k}\eta_{j,k} + \rho_{i,k}\eta_{i,k}^T\mathscr{P}_{i,k}\eta_{i,k} + \gamma^2 v_{i,k}^T \mathscr{T}_{i,k} v_{i,k}.
\end{aligned} \tag{7.23}
$$

From the expression of $S_i(\eta_k, \xi_k, v_{i,k})$ and $\mathbf{V}(\eta_k)$, it follows from (7.23) that

$$
\begin{aligned}
&\mathbb{E}\{V_i(\eta_{i,k+1})|\eta_{i,k}, \eta_{j,k}, j \in \mathcal{N}_i\} \\
&\leq \sum_{j\in\mathcal{N}_i} \frac{\rho_{j,k}}{1+q_j}\eta_{j,k}^T \mathscr{P}_{j,k}\eta_{j,k} + \rho_{i,k}\eta_{i,k}^T\mathscr{P}_{i,k}\eta_{i,k} + S_i(\eta_k, \xi_k, v_{i,k}) \\
&= \left[\frac{a_{i1}\rho_{1,k}}{1+q_1} \quad \cdots \quad \rho_{i,k} \quad \cdots \quad \frac{a_{iN}\rho_{N,k}}{1+q_N}\right] \mathbf{V}(\eta_k) + S_i(\eta_k, \xi_k, v_{i,k}).
\end{aligned} \tag{7.24}
$$

By defining a new matrix W_k, (7.24) can be further rewritten as:

$$
\mathbb{E}\{V_i(\eta_{i,k+1})|\eta_{i,k}, \eta_{j,k}, j \in \mathcal{N}_i\} \leq [W_k\mathbf{V}(\eta_k)]_i + S_i(\eta_k, \xi_k, v_{i,k}) \tag{7.25}
$$

where

$$W_k = \begin{bmatrix} \rho_{1,k} & \cdots & \frac{a_{1i}\rho_{i,k}}{1+q_i} & \cdots & \frac{a_{1N}\rho_{N,k}}{1+q_N} \\ \vdots & \ddots & \vdots & \vdots & \vdots \\ \frac{a_{i1}\rho_{1,k}}{1+q_1} & \cdots & \rho_{i,k} & \cdots & \frac{a_{iN}\rho_{N,k}}{1+q_N} \\ \vdots & \vdots & \vdots & \ddots & \vdots \\ \frac{a_{N1}\rho_{1,k}}{1+q_1} & \cdots & \frac{a_{Ni}\rho_{i,k}}{1+q_i} & \cdots & \rho_{N,k} \end{bmatrix}.$$

In view of $\rho_{i,k} \in \mathbb{J}_{q_i}, a_{ij} \geq 0$ and $q_i \geq 0$, one has $\mathbf{1}^T W_k \leq\leq \mathbf{1}^T$ and $W_k^{(ij)} \geq 0$, where $W_k^{(ij)}$ is the (i, j)th element of W_k. As a result, W_k is column sub-stochastic (p. 550, [6]) and thus is a dissipation matrix based on Definition 7.2.

By using the elementary properties of the conditional expectation, one obtains the following inequality:

$$\mathbb{E}\{V_i(\eta_{i,k+1})\} \leq \mathbb{E}\{[W_k\mathbf{V}(\eta_k)]_i\} + \mathbb{E}\{S_i(\eta_k, \xi_k, \nu_{i,k})\}.$$

Since the above inequality holds for all $i \in \mathscr{V}$, one has the following vector form:

$$\mathbb{E}\{\mathbf{V}(\eta_{k+1})\} \leq\leq W_k\mathbb{E}\{\mathbf{V}(\eta_k)\} + \mathbb{E}\{\mathbf{S}(\eta_k, \xi_k, \nu_k)\}. \tag{7.26}$$

Therefore, it follows from Definition 7.2 that (7.10) is stochastic vector dissipative-like over the finite horizon $\mathscr{H}$ with respect to the vector supply rate function $\mathbf{S}(\eta_k, \xi_k, \nu_k)$ and the dissipation matrix W_k.

Next, denote $v(\eta_k) \triangleq \mathbf{1}^T\mathbf{V}(\eta_k)$. By using the expression of $S_i(\eta_k, \xi_k, \nu_{i,k})$ and left-multiplying by $\mathbf{1}^T$ on both sides of (7.26), one arrives at

$$\begin{aligned} &\sum_{i=1}^{N}\sum_{j\in\mathscr{N}_i}\mathbb{E}\{\|\mathscr{I}_3\eta_{i,k} + \mathscr{I}_{5j}\eta_{j,k} + \bar{\alpha}_j\mathscr{I}_3\mathscr{I}_2\xi_{j,k} + \tilde{\alpha}_{j,k}\mathscr{I}_3\mathscr{I}_2\xi_{j,k} - \tilde{\alpha}_{j,k}\mathscr{I}_4\eta_{j,k} \\ &\qquad + \tilde{\alpha}_{j,k}\mathscr{I}_3\eta_{j,k}\|^2_{R_{i,k}}\} + \sum_{i=1}^{N}\mathbb{E}\|\mathscr{I}_3\eta_{i,k}\|^2_{Q_{i,k}} + \mathbb{E}\{v(\eta_{k+1})\} - \mathbb{E}\{v(\eta_k)\} \\ &\leq \gamma^2\sum_{i=1}^{N}\nu_{i,k}^T\mathscr{T}_{i,k}\nu_{i,k}. \end{aligned} \tag{7.27}$$

Summing up both sides of (7.27) from 0 to $n-1$ with respect to k, one has

$$\begin{aligned} &\sum_{k=0}^{n-1}\sum_{i=1}^{N}\sum_{j\in\mathscr{N}_i}\mathbb{E}\left\{\|\mathscr{I}_3\eta_{i,k} + \mathscr{I}_{5j}\eta_{j,k} + \bar{\alpha}_j\xi_{j,k} + \tilde{\alpha}_{j,k}\xi_{j,k} - \tilde{\alpha}_{j,k}\mathscr{I}_4\eta_{j,k} + \tilde{\alpha}_{j,k}\mathscr{I}_3\eta_{j,k}\|^2_{R_{i,k}}\right\} \\ &+ \sum_{k=0}^{n-1}\sum_{i=1}^{N}\mathbb{E}\|\mathscr{I}_3\eta_{i,k}\|^2_{Q_{i,k}} + \mathbb{E}\{v(\eta_n)\} - \mathbb{E}\{v(\eta_0)\} \leq \gamma^2\sum_{k=0}^{n-1}\sum_{i=1}^{N}\nu_{i,k}^T\mathscr{T}_{i,k}\nu_{i,k}. \end{aligned}$$

It follows from $v(\eta_n) \geq 0$ and $v(\eta_0) = \sum_{i=1}^{N} \eta_{i,0}^T \mathscr{P}_{i,0}\eta_{i,0}$ that

$$\sum_{k=0}^{n-1}\sum_{i=1}^{N}\mathbb{E}\left\{\sum_{j\in\mathscr{N}_i}\|\mathscr{I}_3\eta_{i,k}+\mathscr{I}_{5j}\eta_{j,k}+\bar{\alpha}_j\xi_{j,k}+\tilde{\alpha}_{j,k}\xi_{j,k}-\tilde{\alpha}_{j,k}\mathscr{I}_4\eta_{j,k}+\tilde{\alpha}_{j,k}\mathscr{I}_3\eta_{j,k}\|^2_{R_{i,k}}\right\}$$
$$+\sum_{k=0}^{n-1}\sum_{i=1}^{N}\mathbb{E}\|\mathscr{I}_3\eta_{i,k}\|^2_{Q_{i,k}} \leq \sum_{i=1}^{N}\left(\eta_{i,0}^T\mathscr{P}_{i,0}\eta_{i,0}+\gamma^2\sum_{k=0}^{n-1}\|v_{i,k}\|^2_{\mathscr{T}_{i,k}}\right).$$

Combining with the initial condition $\mathscr{P}_{i,0} \leq \gamma^2 U_i$, one obtains that (7.11) is true and thus confirms the validity of (7.8). Therefore, the proof of this theorem is completed.

Up to now, we have developed the locally sufficient condition (7.15) for ensuring the H_∞-consensus performance (7.8). In the following, the distributed filter gains $L_{i,k}$ and $K_{ij,k}$ will be designed for each node based on Theorem 7.1.

Theorem 7.2 *For $i \in \mathscr{V}$, $j \in \mathscr{N}_i$ and $k \in \mathscr{H}$, let the disturbance attenuation level $\gamma > 0$, the scalar sequence $\rho_{i,k} \in \mathbb{J}_{q_i}$, the scalars δ_j and $\bar{\alpha}_j$, the matrices U_{i1}, U_{i2}, $R_{i,k}$, $Q_{i,k}$, $T_{i1,k}$ and $T_{i2,k}$ be given. Assume that there are matrix sequences $E_{ij,k}$, $F_{i,k}$, scalar sequence $\lambda_{ij,k}$, and positive definite matrix sequences $\{P_{i1,k}\}_{k\in\mathscr{H}\cup\{n\}}$ and $\{P_{i2,k}\}_{k\in\mathscr{H}\cup\{n\}}$ with $P_{i1,0} \leq \gamma^2 U_{i1}$ and $P_{i2,0} \leq \gamma^2 U_{i2}$ satisfying the following inequality:*

$$\begin{bmatrix} \Omega_{i,k}^a & \Omega_{i,k}^b \\ * & \Omega_{i,k}^c \end{bmatrix} < 0 \tag{7.28}$$

where

$$\Omega_{i,k}^a = \begin{bmatrix} \Omega_{i,k}^{11} & \Omega_{i,k}^{12} & \Omega_{i,k}^{13} & \Omega_{i,k}^{14} \\ * & \Omega_{i,k}^{22} & \Omega_{i,k}^{23} & \Omega_{i,k}^{24} \\ * & * & \Omega_{i,k}^{33} & \Omega_{i,k}^{34} \\ * & * & * & -\mathscr{P}_{i,k+1} \end{bmatrix}, \quad \Omega_{i,k}^b = \begin{bmatrix} \Omega_{i,k}^{15} & 0 & 0 \\ 0 & \Omega_{i,k}^{26} & 0 \\ 0 & \Omega_{i,k}^{36} & 0 \\ 0 & 0 & \bar{\mathscr{B}}_{i,k} \end{bmatrix},$$

$$\Omega_{i,k}^c = \mathrm{diag}\left\{\Omega_{i,k}^{55},\ \Omega_{i,k}^{66},\ -\gamma^2\mathscr{T}_{i,k}\right\}, \quad \Omega_{i,k}^{11} = -\rho_{i,k}\mathscr{P}_{i,k} + \mathscr{I}_3^T Q_{i,k}\mathscr{I}_3 + p_i\mathscr{I}_3^T R_{i,k}\mathscr{I}_3,$$
$$\Omega_{i,k}^{12} = \mathrm{vec}_{p_i}^{j}\left[\mathscr{I}_3^T R_{i,k}\mathscr{I}_{5j}\right], \quad \Omega_{i,k}^{13} = \mathrm{vec}_{p_i}^{j}\left[\bar{\alpha}_j\mathscr{I}_3^T R_{i,k}\right], \quad \Omega_{i,k}^{22} = \mathrm{diag}_{p_i}^{j}\left\{\bar{\Theta}_{ij,k}\right\},$$
$$\Omega_{i,k}^{23} = \mathrm{diag}_{p_i}^{j}\left\{\lambda_{ij,k}\bar{\Upsilon}_{j,k}^T + \bar{\alpha}_j\mathscr{I}_{5j}^T R_{i,k} - \beta_j\mathscr{I}_4^T R_{i,k} + \beta_j\mathscr{I}_3^T R_{i,k}\right\}, \quad \bar{\beta}_j = \sqrt{\beta_j},$$
$$\begin{aligned}\bar{\Theta}_{ij,k} = &-\frac{\rho_{j,k}}{1+q_j}\mathscr{P}_{j,k} + \lambda_{ij,k}\left(\delta_j^2(\mathscr{I}_4-\mathscr{I}_3)^T(\mathscr{I}_4-\mathscr{I}_3) - \bar{\Upsilon}_{j,k}^T\bar{\Upsilon}_{j,k}\right)\\ &+ \mathscr{I}_{5j}^T R_{i,k}\mathscr{I}_{5j} + \beta_j(\mathscr{I}_3-\mathscr{I}_4)^T R_{i,k}(\mathscr{I}_3-\mathscr{I}_4),\end{aligned}$$
$$\Omega_{i,k}^{33} = \mathrm{diag}_{p_i}^{j}\left\{-\lambda_{ij,k}I + \bar{\alpha}_j R_{i,k}\right\}, \quad \Omega_{i,k}^{14} = \left(\bar{\mathscr{A}}_{i0,k} - \sum_{j\in\mathscr{N}_i}\bar{\mathscr{K}}_{ij,k}(I-\bar{\alpha}_j\mathscr{I}_1)\right)^T,$$
$$\Omega_{i,k}^{24} = \left(\mathrm{vec}_{p_i}^{j}\left[(1-\bar{\alpha}_j)\bar{\mathscr{K}}_{ij,k}\right]\right)^T, \quad \Omega_{i,k}^{15} = \left[\bar{\mathscr{C}}_{i1,k}^T\ \cdots\ \bar{\mathscr{C}}_{ir_2,k}^T\ \bar{\mathscr{A}}_{i1,k}^T\ \cdots\ \bar{\mathscr{A}}_{ir_1,k}^T\right],$$
$$\Omega_{i,k}^{55} = \mathrm{diag}_{r_1+r_2}^{i}\left\{-\mathscr{P}_{i,k+1}\right\}, \quad \bar{\mathscr{A}}_{i0,k} = \mathrm{diag}\left\{P_{i1,k+1}A_{0,k},\ P_{i2,k+1}A_{0,k} - F_{i,k}C_{i0,k}\right\},$$
$$\Omega_{i,k}^{36} = \mathrm{diag}_{p_i}^{j}\left\{\bar{\beta}_j\mathscr{I}_2^T\bar{\mathscr{K}}_{ij,k}^T\right\}, \quad \Omega_{i,k}^{34} = -\left(\mathrm{vec}_{p_i}^{j}\left[\bar{\alpha}_j\bar{\mathscr{K}}_{ij,k}\mathscr{I}_2\right]\right)^T,$$

$$\Omega_{i,k}^{66} = \mathrm{diag}_{p_i}^{i}\left\{-\mathscr{P}_{i,k+1}\right\}, \quad \Omega_{i,k}^{26} = \mathrm{diag}_{p_i}^{j}\left\{\bar{\beta}_j(I-\mathscr{I}_1)^T \bar{\mathscr{K}}_{ij,k}^T\right\},$$
$$\mathscr{P}_{i,k} = \mathrm{diag}\{P_{i1,k}, P_{i2,k}\}, \quad \bar{\mathscr{K}}_{ij,k} = \mathrm{diag}\{0, E_{ij,k}\},$$
$$\bar{\mathscr{C}}_{is,k} = \begin{bmatrix} 0 & 0 \\ F_{i,k}C_{is,k} & 0 \end{bmatrix}, \quad \bar{\mathscr{A}}_{it,k} = \begin{bmatrix} P_{i1,k+1}A_{t,k} & 0 \\ P_{i2,k+1}A_{t,k} & 0 \end{bmatrix}, \quad \bar{\mathscr{B}}_{i,k} = \begin{bmatrix} P_{i1,k+1}B_k & 0 \\ P_{i2,k+1}B_k & -F_{i,k}D_{i,k} \end{bmatrix}.$$

Then, the error dynamics (7.7) satisfies the H_∞-consensus performance constraint (7.8). Moreover, the filter gains $L_{i,k}$ and $K_{ij,k}$ are given as follows:

$$L_{i,k} = P_{i2,k+1}^{-1}F_{i,k}, \quad K_{ij,k} = P_{i2,k+1}^{-1}E_{ij,k}. \tag{7.29}$$

Proof From Theorem 7.1 and the Schur complement Lemma, the inequality (7.15) holds if and only if

$$\begin{bmatrix} \Omega_{i,k}^{11} & \Omega_{i,k}^{12} & \Omega_{i,k}^{13} & \Xi_{i,k}^{14} & \Xi_{i,k}^{15} & 0 & 0 \\ * & \Omega_{i,k}^{22} & \Omega_{i,k}^{23} & \Xi_{i,k}^{24} & 0 & \Xi_{i,k}^{26} & 0 \\ * & * & \Omega_{i,k}^{33} & \Xi_{i,k}^{34} & 0 & \Xi_{i,k}^{36} & 0 \\ * & * & * & -\mathscr{P}_{i,k+1}^{-1} & 0 & 0 & \mathscr{B}_{i,k} \\ * & * & * & * & \Xi_{i,k}^{55} & 0 & 0 \\ * & * & * & * & * & \Xi_{i,k}^{66} & 0 \\ * & * & * & * & * & * & -\gamma^2\mathscr{T}_{i,k} \end{bmatrix} < 0 \tag{7.30}$$

where

$$\Xi_{i,k}^{15} = \left[\mathscr{C}_{i1,k}^T \quad \cdots \quad \mathscr{C}_{ir_2,k}^T \quad \mathscr{A}_{1,k}^T \quad \cdots \quad \mathscr{A}_{r_1,k}^T\right],$$
$$\Xi_{i,k}^{36} = \mathrm{diag}_{p_i}^{j}\left\{\sqrt{\beta_j}\mathscr{I}_2^T\mathscr{K}_{ij,k}^T\right\}, \quad \Xi_{i,k}^{66} = \mathrm{diag}_{p_i}^{i}\left\{-\mathscr{P}_{i,k+1}^{-1}\right\},$$
$$\Xi_{i,k}^{55} = \mathrm{diag}_{r_1+r_2}^{i}\left\{-\mathscr{P}_{i,k+1}^{-1}\right\}, \quad \Xi_{i,k}^{26} = \mathrm{diag}_{p_i}^{j}\left\{\sqrt{\beta_j}(I-\mathscr{I}_1)^T\mathscr{K}_{ij,k}^T\right\}.$$

Let's partition the matrix $\mathscr{P}_{i,k}$ as $\mathscr{P}_{i,k} = \mathrm{diag}\{P_{i1,k}, P_{i2,k}\}$, perform the congruence transformation to (7.30) by $\mathrm{diag}\left\{I, I, I, \mathscr{P}_{i,k+1}, \mathrm{diag}_{r_1+r_2}\{\mathscr{P}_{i,k+1}\},\right.$ $\left.\mathrm{diag}_{p_i}\{\mathscr{P}_{i,k+1}\}, I\right\}$ and denote

$$\bar{\mathscr{A}}_{it,k} = \mathscr{P}_{i,k+1}\mathscr{A}_{t,k}\ (t = 1, 2, \ldots, r_1), \quad \bar{\mathscr{C}}_{is,k} = \mathscr{P}_{i,k+1}\mathscr{C}_{is,k}\ (s = 1, 2, \ldots, r_2),$$
$$\Omega_{i,k}^{14} = \Xi_{i,k}^{14}\mathscr{P}_{i,k+1}, \quad \Omega_{i,k}^{24} = \Xi_{i,k}^{24}\mathscr{P}_{i,k+1}, \quad \Omega_{i,k}^{34} = \Xi_{i,k}^{34}\mathscr{P}_{i,k+1}, \quad \Omega_{i,k}^{15} = \Xi_{i,k}^{15}\mathscr{P}_{i,k+1},$$
$$\Omega_{i,k}^{26} = \Xi_{i,k}^{26}\mathscr{P}_{i,k+1}, \quad \bar{\mathscr{B}}_{i,k} = \mathscr{P}_{i,k+1}\mathscr{B}_{i,k}, \quad \Omega_{i,k}^{36} = \Xi_{i,k}^{36}\mathscr{P}_{i,k+1}, \quad \bar{\mathscr{A}}_{i0,k} = \mathscr{P}_{i,k+1}\mathscr{A}_{i0,k},$$
$$E_{ij,k} = P_{i2,k+1}K_{ij,k}, \quad F_{i,k} = P_{i2,k+1}L_{i,k}, \quad \bar{\mathscr{K}}_{ij,k} = \mathscr{P}_{i,k+1}\mathscr{K}_{ij,k},$$
$$\Omega_{i,k}^{55} = \mathscr{P}_{i,k+1}\Xi_{i,k}^{55}\mathscr{P}_{i,k+1}, \quad \Omega_{i,k}^{66} = \mathscr{P}_{i,k+1}\Xi_{i,k}^{66}\mathscr{P}_{i,k+1}.$$

It is not difficult to check that the condition (7.28) ensures the validity of (7.30). The proof is therefore complete from Theorem 7.1 and Definition 7.1.

Remark 7.5 Actually, the deception attack can be viewed as a kind of sensor failures and some interesting results concerning sensor failures can be used to address the deception attack problem [7–9]. A fault-detection method based on a higher-order sliding mode observer has been proposed in [7] to judge whether a failure occurs or not, that is, if a failure occurs, then the estimate from the observer can replace the original measurement, otherwise the measurement is still used. In [8], an observer-based fault detection filter has been constructed to make the residual error robust to unknown inputs while sensitive to faults. These observer-based fault detection methods are still efficient when confronting the possible deception attacks . Therefore, in the future, we would investigate the distributed robust filtering subject to deception attacks by borrowing the fault detection techniques.

Remark 7.6 In this chapter, the local design of the distributed H_∞-consensus filter with deception attacks and multiplicative noises has been studied via the stochastic vector dissipativity theory. Theorem 7.2 presents the distributed filtering algorithm regarding node i ($\forall i \in \mathscr{V}$) that includes the information of node i and its neighbors j. As all the information from neighbors is known, the algorithm can be implemented for each node i. Comparing with the existing literature, our main results have the following features: (1) the distributed H_∞-consensus filtering problem subject to deception attacks and multiplicative noises has been addressed within the framework of the local design; (2) the sector-like bounded condition under consideration is employed to describe the malicious signal for better formulating the practical engineering; and (3) the designed algorithm is executed for every node with the updated information received from the neighbors.

According to Theorem 7.2, the algorithm executed for node i is summarized in Table 7.1.

Table 7.1 Distributed H_∞-consensus filtering with deception attacks

Filtering Algorithm Executed on Node i	
Step 1.	Set the average H_∞ performance index γ, the matrices U_{1i}, U_{2i}, $R_{i,k}$, $Q_{i,k}$, $T_{1i,k}$, and $T_{2i,k}$, the initial states x_0 and $\hat{x}_{i,0}$, in-degree p_i and out-degree q_i, and scalar sequence $\alpha_{i,k}$; determine matrices $P^1_{i,0}$ and $P^2_{i,0}$ according to the initial constraints $P^1_{i,0} \leq \gamma^2 U_{1i}$ and $P^2_{i,0} \leq \gamma^2 U_{2i}$; set $k = 0$
Step 2.	Compute the consensus error $\bar{x}_{j,k} - \hat{x}_{i,k}$
Step 3.	Compute the positive definite matrices $P^1_{i,k+1}$ and $P^2_{i,k+1}$, and the filter matrices $F_{i,k}$ and $E_{ij,k}$ at time instant k by solving the RLMIs (7.28)
Step 4.	Calculate the parameter matrices $L_{i,k}$ and $K_{ij,k}$ of the distributed filter of node i from (7.30), and compute $\hat{x}_{i,k+1}$ according to (7.6)
Step 5.	If $k < n$, then return to Step 2, else proceed to Step 6
Step 6.	Stop

7.3 An Illustrative Example

In this section, a simulation example is presented to illustrate the effectiveness of the proposed design scheme.

Consider an SN with 10 nodes, whose topology is represented by a digraph $\mathscr{G} = (\mathscr{V}, \mathscr{E}, \mathcal{A})$ with the set of nodes $\mathscr{V} = \{1, 2, 3, 4, 5, 6, 7, 8, 9, 10\}$ and the set of edges $\mathscr{E} = \{(2, 1), (3, 2), (4, 3), (5, 4), (6, 5), (7, 6), (8, 7), (9, 8), (10, 9), (1, 10)\}$. Obviously, $p_i = q_i = 1$ for $i = 1, 2, \ldots, 10$.

The system parameters are given as follows:

$$A_{0,k} = \begin{bmatrix} -0.87 & 0.4 \\ 0.15 & 0.87\sin(6k) \end{bmatrix}, \quad A_{1,k} = \begin{bmatrix} 0.1 & 0 \\ 0 & 0 \end{bmatrix},$$
$$B_k = \begin{bmatrix} 0.1 & 0.1 \end{bmatrix}^T, \quad C_{i0,k} = \begin{bmatrix} 0.3 & 0.2\sin(ik) \end{bmatrix}, \quad C_{i1,k} = \begin{bmatrix} 0.03 & 0 \end{bmatrix}.$$

The noise signals are chosen as $w_k = 0.1\cos(5k)$ and $v_{i,k} = 0.1\cos(5k)$. For other parameters, the time horizon, the H_∞ attenuation level γ, the scalar sequence $\rho_{i,k}$, and the weighted matrices U_i, $\mathscr{T}_{i,k}$, $Q_{i,k}$, and $R_{i,k}$ are given as $n = 30$, $\gamma = 0.3$, $\rho_{i,k} \equiv 0.9$, $U_i = \mathscr{P}_{i,0}$ and $\mathscr{T}_{i,k} = Q_{i,k} = R_{i,k} = \mathrm{diag}\{1, 1\}$.

The malicious signal $\xi_{j,k}$ is bounded by the following constraint:

$$(\xi_{j,k} - 3\hat{x}_{j,k})^T(\xi_{j,k} - 3\hat{x}_{j,k}) \le \hat{x}_{j,k}^T\hat{x}_{j,k}.$$

In this simulation, choose $\xi_{j,k} = 3\hat{x}_{j,k}$, which obviously satisfies the constraint (4.27). In order to clearly reflect the impact of deception attack and to verify the effectiveness of the proposed distributed filtering scheme, we consider the following two cases.

Case 1: no attack occurs (i.e., $\bar{\alpha}_i \equiv 0$) with $\mathscr{P}_{i,0} \equiv \mathrm{diag}\{8I, 8I\}$; **Case 2:** attacks may occur on Node 6 (i.e., $\bar{\alpha}_6 = 0.2, \bar{\alpha}_i \equiv 0\,(i \neq 6)$) with $\mathscr{P}_{i,0} \equiv \mathrm{diag}\{25I, 25I\}$. By means of the YALMIP toolbox in MATLAB software, the desired results are recursively calculated from Theorem 7.2. The initial values are chosen as $x_0 = [0.3 \;\; -0.5]^T$, $\hat{x}_{1,0} = [0.1 \;\; -0.2]^T$, $\hat{x}_{2,0} = [0.5 \;\; -0.7]^T$, $\hat{x}_{3,0} = [0.2 \;\; -0.4]^T$, $\hat{x}_{4,0} = [0 \;\; -0.6]^T$, $\hat{x}_{5,0} = [0.6 \;\; -0.3]^T$, $\hat{x}_{6,0} = [0.7 \;\; -0.1]^T$, $\hat{x}_{7,0} = [-0.1 \;\; -0.2]^T$, $\hat{x}_{8,0} = [-0.5 \;\; 0.7]^T$, $\hat{x}_{9,0} = [-0.2 \;\; 0.4]^T$ and $\hat{x}_{10,0} = [-0.7 \;\; 0.1]^T$. For these two cases, the corresponding simulation results are presented in Figs. 7.1a–e and 7.2a–d.

Figure 7.1a and b depict the norm sum of the consensus errors (NCEs) of each node (i.e., $\sum_{j\in\mathscr{N}_i}\|\bar{x}_{j,k} - \hat{x}_{i,k}\|^2$); Fig. 7.1c and d plot the norm of the filtering errors (NFEs) of each node (i.e., $\|e_{i,k}\|^2$); Fig. 7.1e plots the attack instants when $\bar{\alpha}_6 = 0.2$. In comparison with Fig. 7.1a and b, multiple sudden peaks can be observed from the trajectory of $\sum_{j\in\mathscr{N}_7}\|\bar{x}_{6,k} - \hat{x}_{7,k}\|^2$ at instants $k = 1, 5, 20, 26, 30$, which are induced by the possible deception attacks during the signal transmission from Nodes 6 to 7 as shown in Fig. 7.1e. By comparing with Fig. 7.1c and d, it can be observed that NFEs have substantial changes when taking attacks into account. On the other hand, the attack impact would be gradually reduced through the information exchange over the network topology.

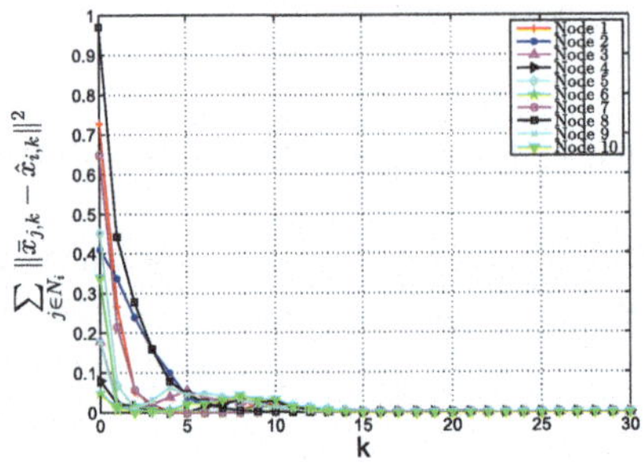

(a) NCE with $\bar{\alpha}_6 = 0$.

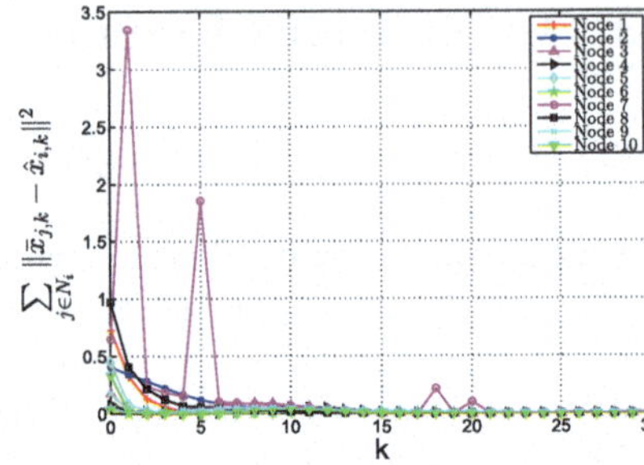

(b) NCE with $\bar{\alpha}_6 = 0.2$.

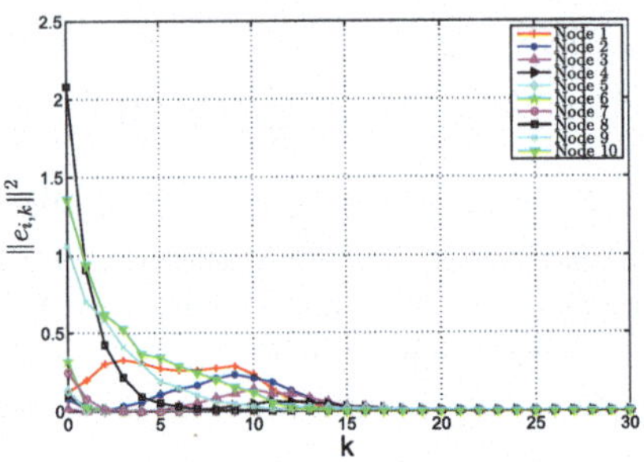

(c) NFE with $\bar{\alpha}_6 = 0$.

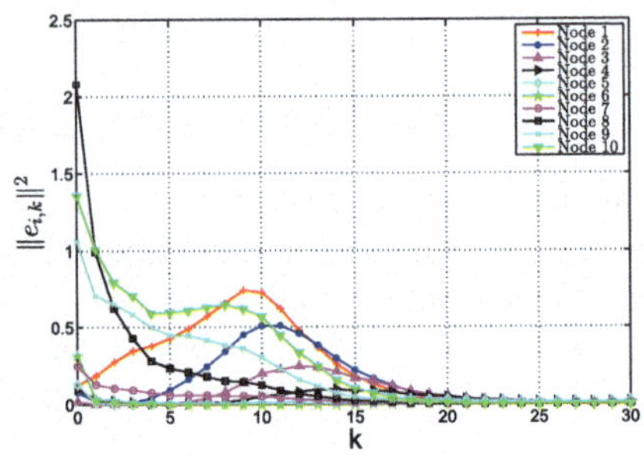

(d) NFE with $\bar{\alpha}_6 = 0.2$.

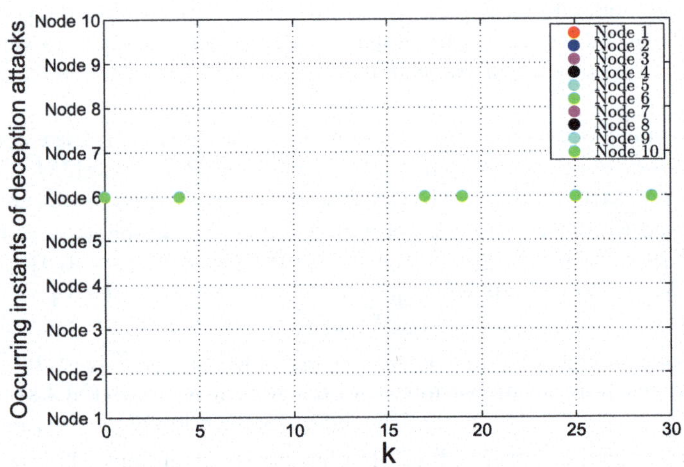

(e) Attack instant with $\bar{\alpha}_6 = 0.2$.

Fig. 7.1 NCEs and NFEs with $\bar{\alpha}_6 = 0$ and 0.2

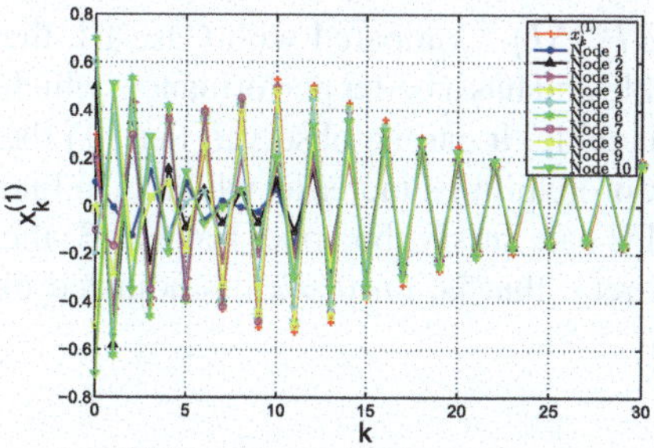

(a) $x_k^{(1)}$ and its estimates with $\bar{\alpha}_6 = 0$.

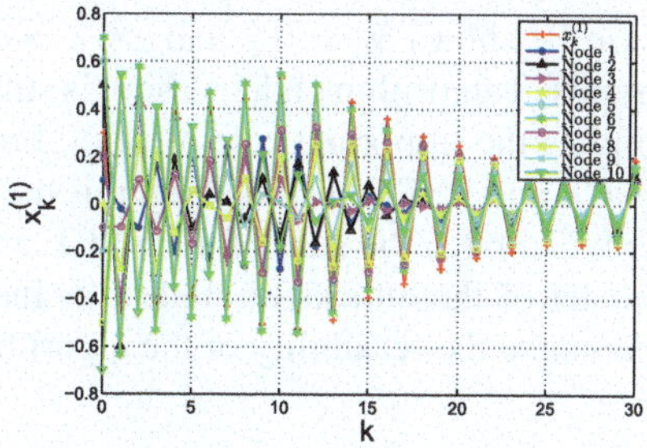

(b) $x_k^{(1)}$ and its estimates with $\bar{\alpha}_6 = 0.2$.

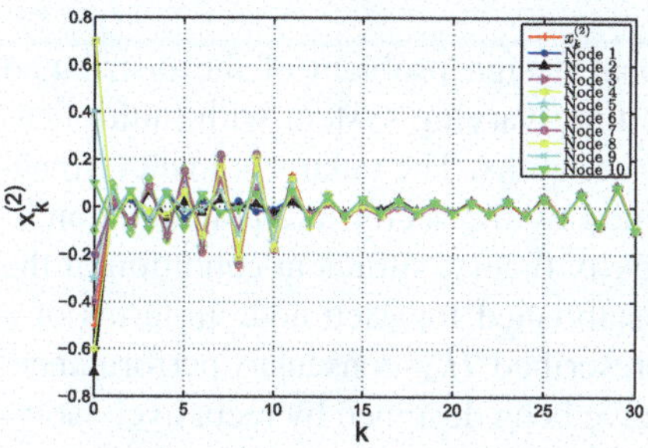

(c) $x_k^{(2)}$ and its estimates with $\bar{\alpha}_6 = 0$.

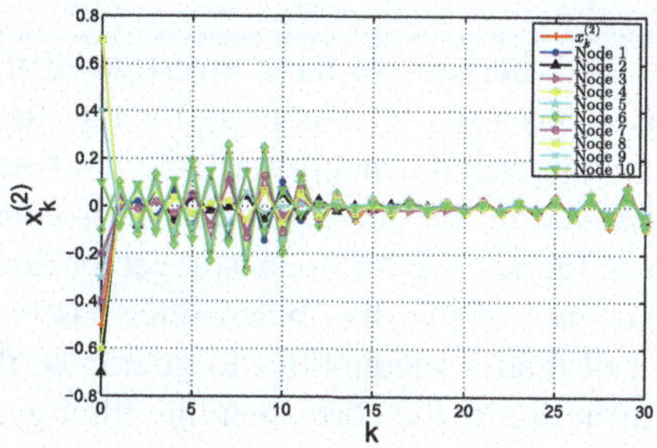

(d) $x_k^{(2)}$ and its estimates with $\bar{\alpha}_6 = 0.2$.

Fig. 7.2 x_k and its estimates under different attack probabilities

Table 7.2 Performance behaviors under different cases ($\bar{\alpha}_6$, $\mathscr{P}_{i,0}$)

($\bar{\alpha}_6$, $\mathscr{P}_{i,0}$)	Case I	Case II	Case III
SNCE	7.7215	13.5601	15.6150
SNFE	20.3683	29.9582	39.5037
Upper bound	71.6791	82.8991	223.4891

Figure 7.2a–d depict the plant state and its estimates, respectively. In comparison with Fig. 7.2a, b, and c, d, the possible attacks deteriorate the performance of the distributed filtering scheme: the time reaching consensus and the estimate approaching the real state are significantly extended under the attacks. From Figs. 7.1a–d and 7.2a–d, it is concluded that the designed filtering scheme is effective in presence of deception attacks and multiplicative noises.

Finally, in order to quantitatively compare these results, Table 7.2 summarizes the sum of norm of consensus error (SNCE), the sum of norm of filtering error (SNFE) and the upper bound (i.e., $\sum_{i=1}^{N}[\|x_0\|^2_{P_{i1,0}} + \|e_{i,0}\|^2_{P_{i2,0}} + \sum_{k=0}^{n-1}\gamma^2(\|w_k\|^2 + \|v_{i,k}\|^2)]$) in the performance index (7.8) for $\bar{\alpha}_6 = 0$ and 0.2. More specifically, three cases are considered, where Case I refers to $\bar{\alpha}_6 = 0$ and $\mathscr{P}_{i,0} = \text{diag}\{8I, 8I\}$, Case II indicates $\bar{\alpha}_6 = 0.2$, $\mathscr{P}_{6,0} = \text{diag}\{25I, 25I\}$, and $\mathscr{P}_{i,0} = \text{diag}\{8I, 8I\}(i \neq 6)$, and

Case III shows $\bar{\alpha}_6 = 0.2$ and $\mathscr{P}_{i,0} = \text{diag}\{25I, 25I\}$. Compared with Case III, the proposed algorithm under Case II is still feasible and has a better performance, which implies the important role of $\mathscr{P}_{i,0}$. From this table, it can be observed that (1) the SNCE and SNFE become larger as the attack probabilities increase, which is in line with (7.9); (2) the sum of the SNCE and SNFE can satisfy the upper bound; (3) the impact of the attacks on node 6 is indeed direct, that is, only $\mathscr{P}_{6,0}$ is adjusted to guarantee the feasibility of the algorithm.

7.4 Summary

In this chapter, we have investigated the local design problem of the distributed H_∞-consensus filteringissue for the discrete time-varying system with multiplicative noises and deception attacks over sensor networks. The unknown malice signal involved in the deception attacks is constrained by the sector-like bounded condition. In view of the stochastic vector dissipativity theory, sufficient condition of the existence of the distributed filter has been established for each node in terms of a set of matrix inequalities to guarantee the prescribed H_∞-consensus performance. Furthermore, the corresponding filter gains have been designed by recursively solving certain matrix inequalities. Finally, a simulation example has also been given to verify the effectiveness of the proposed filter algorithm. Some future research directions would be focused on the extension of the established approach to switched systems [10], nonlinear systems [11] and the application of the obtained results on fault detection [12] and image processing [13, 14].

References

1. Han, F., Wei, G., Ding, D., Song, Y.: Local condition-based consensus filtering with stochastic nonlinearities and multiple missing measurements. IEEE Trans. Autom. Control **62**(9), 4784–4790 (2017)
2. Han, F., Song, Y., Zhang, S., Li, W.: Local condition-based finite-horizon distributed H_∞-consensus filtering for random parameter system with event-triggering protocols. Neurocomputing **219**, 221–231 (2017)
3. Ren, H., Zong, G., Karimi, R.: Asynchronous finite-time filtering of networked switched systems and its application: an event-driven method. IEEE Trans. Circuits Syst. I Regul. Pap. **66**(1), 391–402 (2019). Jan.
4. Zhang, D., Xu, Z., Karimi, H., Wang, Q.: Distributed filtering for switched linear systems with sensor networks in presence of packet dropouts and quantization. IEEE Trans. Circ. Syst.-Part I **64**(10), 2783–2796 (2017)
5. Haddad, W., Chellabolna, V., Nersesov, S.: Vector dissipativity theory and stability of feedback interconnections for large-scale non-linear dynamical systems. Int. J. Control **77**(10), 907–919 (2004)
6. Horn, R., Johnson, C.: Matrix Analysis, 2nd edn. Cambridge University Press, Cambridge (2013)

7. Kommuri, S.K., Defoort, M., Karimi, H.R., Veluvolu, K.C.: A robust observer-based sensor fault-tolerant control for PMSM in electric electric vehicles. IEEE Trans. Ind. Electron. **63**(12), 7671–7681 (2016)
8. Li, Y., Karimi, H.R., Zhao, D., Xu, Y., Zhao, P.: H_∞ fault detection filter design for discrete-time nonlinear Markovian jump systems with missing measurements. Eur. J. Control. **44**, 27–39 (2018)
9. Dong, H., Wang, Z., Shen, B., Ding, D.: Variance-constrianed H_∞ control for a class of nonlinear stochastic discrete time-varying systems: the event-triggered design. Automatica **72**, 28–36 (2014)
10. Wu, L., Zheng, W., Gao, H.: Dissipativity-based sliding mode control of switched stochastic systems. IEEE Trans. Autom. Control **58**(3), 785–791 (2013)
11. Lv̈, H., Wei, G., Ding, Z., Ding, X.: Sensorless control for the brushless DC motor: an unscented Kalman filter algorithm. Syst. Sci. Control Eng.: An Open Access J. **3**(1), 8–13 (2015)
12. Li, J., Dong, H., Han, F., Hou, N., Li, X.: Filter design, fault estimation and reliable control for networked time-varying systems: a survey. Syst. Sci. Control Eng.: An Open Access J. **5**(1), 331–341 (2017)
13. Chen, D., Li, G., Sun, Y., Kong, J., Jiang, G., Tang, H., Ju, Z., Yu, H., Liu, H.: An interactive image segmentation method in hand gesture recognition. Sensors **17**(2), 253 (2017)
14. Liao, Y., Sun, Y., Li, G., Kong, J., Jiang, G., Jiang, D., Cai, H., Ju, Z., Yu, H., Liu, H.: Simultaneous calibration: a joint optimization approach for multiple kinect and external cameras. Sensors **17**(7), 1491 (2017)

Chapter 8
Distributed Resilient Filtering for Time-Delayed Systems with Stochastic Perturbations

In essence, compared to the traditional filtering, the significant difference of distributed filtering is the utilization of the space information collected by an SN. According to the previous chapters, such space information of a node from its neighboring nodes can be formulated by the dissipation matrix in the vector dissipativity theory. It is worth noting that, the state of the time-delayed system is influenced by the historical information from the past time. Moreover, there exist essential differences between the space information and the historical information. As a result, it is critical problem to construct a new dissipation matrix to reflect the historical information.

This chapter handles the distributed resilient filtering problem for a class of nonlinear time-delayed systems subject to stochastic perturbations. The plant and the measurements are disturbed by two Gaussian white stochastic processes with known statistical information, respectively. Also, a resilient filter is designed for each node by means of the parameter uncertainties and Bernoulli-distributed random variables. Then, a novel exponential-bounded performance index is put forward to measure the disturbance rejection level of the distributed filters against the external disturbances and the impact of the initial values. A new vector dissipation definition including multiple vectors of energy storage functions is established to deal with the time-delay filtering error dynamics. Within the framework of local performance analysis inspired by this new definition of vector dissipation, sufficient conditions in terms of recursive linear matrix inequalities are constructed for each node to guarantee the desirable performance index. Next, a local optimization problem subject to a set of recursive linear matrix inequalities is presented for each node to minimize the upper bound in the performance index, where the calculations can be conducted on every node in a distributed manner and the filter gains are also calculated. Finally, an illustrative simulation example is provided to verify the applicability of the proposed filters.

F. Han et al., *Distributed Filtering, Control and Synchronization*,
Studies in Systems, Decision and Control 428,
https://doi.org/10.1007/978-3-030-97075-8_8

8.1 Problem Formulation

Consider the following discrete-time nonlinear time-delayed plant with stochastic perturbations:

$$\begin{cases} x_{k+1} = A_k x_k + f(x_k) + g(x_{k-\tau_k}) + B_k w_k, \\ y_{i,k} = C_{i,k} x_k + D_{i,k} v_{i,k}, \\ x_s = \phi_s, s \in [-\tau_M, -\tau_M + 1, \ldots, 0] \end{cases} \tag{8.1}$$

where $x_k \in \mathbb{R}^{n_x}$ and $y_{i,k} \in \mathbb{R}^{n_y}$ are the plant state and the measurement output of node i, respectively; $f(\cdot)$ and $g(\cdot)$ are nonlinear vector-valued functions; ϕ_s denote the real-valued initial states; A_k, B_k, $C_{i,k}$ and $D_{i,k}$ are known time-varying matrices with appropriate dimensions; $w_k \in \mathbb{R}^{n_w}$ and $v_{i,k} \in \mathbb{R}^{n_v}$ are the Gaussian random vectors with zero means and covariance matrices $Q = \text{diag}\{\sigma_1^2, \sigma_2^2, \ldots, \sigma_{n_w}^2\}$ and $R_i = \text{diag}\{\sigma_{i1}^2, \sigma_{i2}^2, \ldots, \sigma_{in_v}^2\}$ ($\sigma_s > 0, s = 1, \ldots, n_w$; $\sigma_{it} > 0, i = 1, \ldots, N, t = 1, \ldots, n_v$), where the elements of w_k and $v_{i,k}$ are all mutually independent of each other.

The positive integer τ_k denotes the time-varying delay satisfying

$$0 \le \tau_m \le \tau_k \le \tau_M \tag{8.2}$$

where τ_m and τ_M are known positive integers.

The nonlinear vector-valued functions $f(\cdot)$ and $g(\cdot)$: $\mathbb{R}^{n_x} \mapsto \mathbb{R}^{n_x}$ with $f(0) = 0$ and $g(0) = 0$ are assumed to be continuous and satisfy the sector-bounded conditions:

$$\begin{aligned} &[f(z_1) - f(z_2) - \Phi_{1i}(z_1 - z_2)]^T [f(z_1) - f(z_2) - \Phi_{2i}(z_1 - z_2)] \le 0, \\ &[g(z_1) - g(z_2) - \Psi_{1i}(z_1 - z_2)]^T [g(z_1) - g(z_2) - \Psi_{2i}(z_1 - z_2)] \le 0 \end{aligned} \tag{8.3}$$

for all $z_1, z_2 \in \mathbb{R}^{n_x}$, where Φ_{1i}, Φ_{2i}, Ψ_{1i} and Ψ_{2i} are all known real matrices with appropriate dimensions.

In this chapter, inspired by [1], the following distributed resilient filter is constructed by considering both the randomly occurring gain fluctuations and the consensus requirement:

$$\begin{aligned} \hat{x}_{i,k+1} &= A_k \hat{x}_{i,k} + (L_{i,k} + \beta_{1i,k} \Delta L_{i,k})(y_{i,k} - C_{i,k} \hat{x}_{i,k}) + f(\hat{x}_{i,k}) \\ &\quad + g(\hat{x}_{i,k-\tau_k}) + \sum_{j \in \mathcal{N}_i} (K_{ij,k} + \beta_{2i,k} \Delta K_{ij,k})(\hat{x}_{j,k} - \hat{x}_{i,k}), \\ \hat{x}_{i,s} &= \hat{\phi}_{i,s}, s \le 0 \end{aligned} \tag{8.4}$$

where $\hat{x}_{i,k} \in \mathbb{R}^{n_x}$ is the estimate of x_k on node i, $L_{i,k}$ and $K_{ij,k}$ are the filter gains to be determined later. $\hat{\phi}_{i,s}$ is the initial condition of filter (8.4). The real-valued matrices $\Delta L_{i,k}$ and $\Delta K_{ij,k}$ stand for the norm-bounded parameter uncertainties satisfying

$$\Delta L_{i,k} = M_{1i,k} F_{1i,k} N_{1i,k}, \quad \Delta K_{ij,k} = M_{2i,k} F_{2ij,k} N_{2i,k} \tag{8.5}$$

where $M_{1i,k}$, $N_{1i,k}$, $M_{2i,k}$, and $N_{2i,k}$ are known real matrices, and $F_{1i,k}$ and $F_{2ij,k}$ are unknown matrices satisfying

$$F_{1i,k}F_{1i,k}^T \leq I, \quad F_{2ij,k}F_{2ij,k}^T \leq I. \tag{8.6}$$

The stochastic variables $\beta_{1i,k}$ and $\beta_{2i,k}$ taking values in a finite set $L = \{0, 1\}$ are mutually independent obeying the Bernoulli distribution with the following probabilities:

$$\begin{aligned} &\text{Prob}\{\beta_{1i,k} = 1\} = \bar{\beta}_{1i}, \quad \text{Prob}\{\beta_{1i,k} = 0\} = 1 - \bar{\beta}_{1i}, \\ &\text{Prob}\{\beta_{2i,k} = 1\} = \bar{\beta}_{2i}, \quad \text{Prob}\{\beta_{2i,k} = 0\} = 1 - \bar{\beta}_{2i} \end{aligned} \tag{8.7}$$

where $\bar{\beta}_{1i}$ and $\bar{\beta}_{2i}$ are known real constants satisfying $0 \leq \bar{\beta}_{1i} \leq 1$ and $0 \leq \bar{\beta}_{2i} \leq 1$.

Denote $\tilde{\beta}_{1i,k} \triangleq \beta_{1i,k} - \bar{\beta}_{1i}$ and $\tilde{\beta}_{2i,k} \triangleq \beta_{2i,k} - \bar{\beta}_{2i}$. Letting $e_{i,k+1} \triangleq x_{k+1} - \hat{x}_{i,k+1}$ be the filtering error of node i, and combining with (8.1)–(8.4), one has the following filtering error dynamics:

$$\begin{aligned} e_{i,k+1} = &\left(A_k - L_{i,k}C_{i,k} - \bar{\beta}_{1i}\Delta L_{i,k}C_{i,k} - \sum_{j\in\mathcal{N}_i}(K_{ij,k} + \bar{\beta}_{2i}\Delta K_{ij,k}) \right) e_{i,k} + \tilde{f}(e_{i,k}) \\ &+ \tilde{g}(e_{i,k-\tau_k}) - (L_{i,k}D_{i,k} + \beta_{1i,k}\Delta L_{i,k}D_{i,k})v_{i,k} - \tilde{\beta}_{1i,k}\Delta L_{i,k}C_{i,k}e_{i,k} \\ &+ B_k w_k + \tilde{\beta}_{2i,k}\sum_{j\in\mathcal{N}_i}\Delta K_{ij,k}(e_{j,k} - e_{i,k}) + \sum_{j\in\mathcal{N}_i}(K_{ij,k} + \bar{\beta}_{2i}\Delta K_{ij,k})e_{j,k} \end{aligned} \tag{8.8}$$

where

$$\begin{aligned} \tilde{f}(e_{i,k}) &\triangleq f(x_k) - f(\hat{x}_{i,k}), \\ \tilde{g}(e_{i,k-\tau_k}) &\triangleq g(x_{k-\tau_k}) - g(\hat{x}_{i,k-\tau_k}). \end{aligned} \tag{8.9}$$

Correspondingly, from (8.3), one obtains the following constraints:

$$\begin{aligned} &\left[\tilde{f}(e_{i,k}) - \Phi_{1i}e_{i,k}\right]^T \left[\tilde{f}(e_{i,k}) - \Phi_{2i}e_{i,k}\right] \leq 0, \\ &\left[\tilde{g}(e_{i,k-\tau_k}) - \Psi_{1i}e_{i,k-\tau_k}\right]^T \left[\tilde{g}(e_{i,k-\tau_k}) - \Psi_{2i}e_{i,k-\tau_k}\right] \leq 0. \end{aligned} \tag{8.10}$$

Before proceeding further, a desirable performance index with respect to the dynamics (8.8) is presented.

Definition 8.1 For given a sequence of weighted matrices $R_{i,k}$ $(k \geq 1)$, $T_{i,s}$ $(s = -\tau_M, \ldots, -1, 0)$, and scalars θ_i $(0 < \theta_i < 1)$, (8.4) is said to be an exponentially bounded state filter in the mean square if there exist constants $\kappa_i > 0$ and $d_{i,t} > 0$, such that the filtering error system (8.8) satisfies

$$\sum_{i=1}^{N}\mathbb{E}\left\{\|e_{i,k}\|^2_{R_{i,k}}\right\} \leq \sum_{i=1}^{N}\theta_i^k \kappa_i + \sum_{t=0}^{k-1}\sum_{i=1}^{N}\theta_i^{k-t} d_{i,t}. \tag{8.11}$$

The objective of this chapter is to find the filter gains $L_{i,k}$ and $K_{ij,k}$ such that the filtering error $e_{i,k}$ satisfies the inequality constraint (8.11) and the upper bound $d_{i,t}$ in (8.11) is minimized.

8.2 Distributed Resilient Filtering over Sensor Networks

In this section, the distributed resilient filtering problem is dealt with for the discrete-time nonlinear system with time-varying delays (8.1). As mentioned in the preceding chapters, the vector dissipativity theory plays a key role to achieve the local algorithm design for each node. Nevertheless, it is worth mentioning that the corresponding criterion with respect to the time-delayed systems has not yet been presented in the existing literature. According to our investigations, the main challenge is how to characterize the impact of time-delays in the vector dissipation inequality. In the traditional vector dissipativity theory, the dissipation matrix is utilized to formulate the complicated couplings among different nodes at fixed time instants. As is well known, the couplings among nodes belong to the scope of space, while time-delays fall into the category of time-horizon. In order to put these two essentially different kinds of information into a unified framework, a more general vector dissipation definition is proposed for the filtering error dynamics (8.8).

Denote

$$\begin{aligned} S_{i,k} \triangleq\ & \mathbb{E}\left\{v_{i,k}^T(L_{i,k}D_{i,k} + \beta_{1i,k}\Delta L_{i,k}D_{i,k})^T P_{1i,k+1}(L_{i,k}D_{i,k} + \beta_{1i,k}\Delta L_{i,k}D_{i,k})v_{i,k}\right\} \\ & + \mathbb{E}\left\{w_k^T B_k^T P_{1i,k+1} B_k w_k\right\} \end{aligned} \tag{8.12}$$

where $P_{1i,k+1}$ is defined later in Lemma 8.1.

Definition 8.2 The class of discrete time-varying system dynamics (8.8) is strictly vector dissipative-like with respect to a vector-type supply rate function $\mathbf{S}_k \triangleq [S_{1,k} \ \cdots \ S_{N,k}]^T$, if there exist q_0 vector-type storage functions $\mathbf{V}_{s,k} = [V_{s1,k} \ \cdots \ V_{sN,k}]^T$ ($V_{si,k} \geq 0, i = 1, 2, \ldots N, s = 1, 2, \ldots, q_0$), and q_0 sequences of nonsingular column substochastic matrices $W_{s,k} \in \mathbb{R}^{N\times N}$ such that the following vector dissipation inequality is satisfied for $k \geq 0$

$$\sum_{s=1}^{q_0}\mathbf{V}_{s,k+1} \ll \sum_{s=1}^{q_0} W_{s,k}\mathbf{V}_{s,k} + \mathbf{S}_k. \tag{8.13}$$

Moreover, the scalar form of (8.13) can be given as:

$$\sum_{s=1}^{q_0} V_{si,k+1} < \sum_{s=1}^{q_0} [W_{s,k}\mathbf{V}_{s,k}]_i + S_{i,k} \tag{8.14}$$

where $[W_{s,k}\mathbf{V}_{s,k}]_i$ is the ith element of the vector $W_{s,k}\mathbf{V}_{s,k}$.

Let $\mathfrak{I}_{q_i}$ be an interval-valued function dependent on the out-degree q_i:

$$\mathfrak{I}_{q_i} = \begin{cases} (0, \frac{1+q_i}{1+2q_i}), & \text{if } q_i \neq 0; \\ (0, 1], & \text{if } q_i = 0. \end{cases} \tag{8.15}$$

Lemma 8.1 *Let scalars $\alpha_i \in \mathfrak{I}_{q_i}$, positive definite matrices $R_{i,k}$ and $\{T_{i,s}\}_{-\tau_M \le s \le 0}$, and filter gains $L_{i,k}$ and $K_{ij,k}$ ($i \in \mathscr{V}$, $j \in \mathscr{N}_i$) be given. The filtering error dynamics (8.8) is strictly vector dissipative-like if there exist two sets of scalars $\lambda_{i,k}$ and $\rho_{i,k}$, two sets of positive definite matrices $P_{1i,k}$ and $P_{2i,k}$ with the initial conditions $P_{1i,0} \le T_{i,0}$ and $\{P_{2i,s} \le (\tau_M - \tau_m)T_{i,s}\}_{-\tau_M \le s \le -1}$ such that the following conditions hold*

$$\begin{bmatrix} \Xi_{i,k} & * \\ \mathscr{G}_{i,k} & -P_{1i,k+1}^{-1} \end{bmatrix} < 0, \tag{8.16}$$

$$P_{1i,k} \ge R_{i,k} \tag{8.17}$$

where

$$\Xi_{i,k} = \begin{bmatrix} \Xi_{i,k}^{11} & * & * & * & * \\ \Xi_{i,k}^{21} & \Xi_{i,k}^{22} & * & * & * \\ \Xi_{i,k}^{31} & 0 & \Xi_{i,k}^{33} & * & * \\ \Xi_{i,k}^{41} & 0 & 0 & \Xi_{i,k}^{44} & * \\ 0 & 0 & 0 & 0 & \Xi_{i,k}^{55} \end{bmatrix},$$

$$\Xi_{i,k}^{11} = \hat{\beta}_{1i}^2 C_{i,k}^T \Delta L_{i,k}^T P_{1i,k+1} \Delta L_{i,k} C_{i,k} + \hat{\beta}_{2i}^2 \sum_{j\in\mathscr{N}_i} \Delta K_{ij,k}^T P_{1i,k+1} \sum_{j\in\mathscr{N}_i} \Delta K_{ij,k}$$
$$- (1-\alpha_i)P_{1i,k} - \bar{\Psi}_{i,k} - \bar{\Phi}_{i,k}, \quad \Xi_{i,k}^{33} = -\lambda_{i,k} I,$$

$$\Xi_{i,k}^{21} = -\hat{\beta}_{2i}^2 \left[\Delta K_{ij_1,k}^T P_{1i,k+1} \sum_{j\in\mathscr{N}_i} \Delta K_{ij,k} \quad \cdots \quad \Delta K_{ij_{p_i},k}^T P_{1i,k+1} \sum_{j\in\mathscr{N}_i} \Delta K_{ij,k} \right]^T,$$

$$\Xi_{i,k}^{22} = \hat{\beta}_{2i}^2 \left[\Delta K_{ij_1,k} \quad \cdots \quad \Delta K_{ij_{p_i},k} \right]^T P_{1i,k+1} \left[\Delta K_{ij_1,k} \quad \cdots \quad \Delta K_{ij_{p_i},k} \right]$$
$$- \operatorname{diag}\left\{ \frac{\alpha_{ji_1}}{1+q_{ji_1}} P_{1j_{i_1},k}, \ldots, \frac{\alpha_{ji_{p_i}}}{1+q_{ji_{p_i}}} P_{1j_{i_{p_i}},k} \right\},$$

$$\mathscr{K}_{\mathscr{N}_i,k} = \left[K_{ij_1,k} + \bar{\beta}_{2i}\Delta K_{ij_1,k} \quad \cdots \quad K_{ij_{p_i},k} + \bar{\beta}_{2i}\Delta K_{ij_{p_i},k} \right],$$

$$\Xi_{i,k}^{31} = \frac{\lambda_{i,k}}{2}(\Phi_{1i} + \Phi_{2i}), \quad \Xi_{i,k}^{41} = \frac{\rho_{i,k}}{2}(\Psi_{1i} + \Psi_{2i}), \quad \bar{\Psi}_{i,k} = \frac{\rho_{i,k}}{2}(\Psi_{1i}^T \Psi_{2i} + \Psi_{2i}^T \Psi_{1i}),$$

$$\Xi_{i,k}^{44} = (1+\tau_M-\tau_m)P_{2i,k}-\rho_{i,k}I,\quad \mu_i=\frac{\alpha_i}{1+q_i},\quad \bar{\Phi}_{i,k}=\frac{\lambda_{i,k}}{2}(\Phi_{1i}^T\Phi_{2i}+\Phi_{2i}^T\Phi_{1i}),$$
$$\Xi_{i,k}^{55} = -(1-\mu_i)^{\tau_M}P_{2i,k-\tau_k},\quad \hat{\beta}_{1i}=\sqrt{\bar{\beta}_{1i}(1-\bar{\beta}_{1i})},\quad \hat{\beta}_{2i}=\sqrt{\bar{\beta}_{2i}(1-\bar{\beta}_{2i})},$$
$$\mathscr{G}_{i,k}=\begin{bmatrix} A_k-L_{i,k}C_{i,k}-\bar{\beta}_{1i}\Delta L_{i,k}C_{i,k}-\sum_{j\in\mathscr{N}_i}(K_{ij,k}+\bar{\beta}_{2i}\Delta K_{ij,k}) & \mathscr{K}_{\mathscr{N}_i,k} & I & 0 & I\end{bmatrix}.$$

Proof Choose the following Lyapunov–Krasovskii functional $V_{i,k}$ for node i to deal with the distributed filtering problem for the filtering error dynamics (8.8):

$$V_{i,k}=V_{1i,k}+V_{2i,k}+V_{3i,k} \tag{8.18}$$

where

$$\begin{aligned}
V_{1i,k} &= e_{i,k}^T P_{1i,k}e_{i,k},\\
V_{2i,k} &= \sum_{s=k-\tau_k}^{k-1}(1-\mu_i)^{k-s-1}\tilde{g}^T(e_{i,s})P_{2i,s}\tilde{g}(e_{i,s}),\\
V_{3i,k} &= \sum_{p=k+1-\tau_M}^{k-\tau_m}\sum_{s=p}^{k-1}(1-\mu_i)^{k-s-1}\tilde{g}^T(e_{i,s})P_{2i,s}\tilde{g}(e_{i,s}).
\end{aligned}$$

Meanwhile, the difference of the Lyapunov–Krasovskii functional is described as follows:

$$\mathbb{E}\{\Delta V_{i,k}\}=\mathbb{E}\{V_{i,k+1}\}-V_{i,k}=\sum_{t=1}^{3}\mathbb{E}\{\Delta V_{ti,k}\}. \tag{8.19}$$

Then, along the state trajectory of system dynamics (8.8), we have

$$\begin{aligned}
&\mathbb{E}\{\Delta V_{1i,k}\}\\
&=\mathbb{E}\{V_{1i,k+1}\}-V_{1i,k}\\
&=-2\hat{\beta}_{2i}^2\sum_{j\in\mathscr{N}_i}e_{j,k}^T\Delta K_{ij,k}^T P_{1i,k+1}\sum_{j\in\mathscr{N}_i}\Delta K_{ij,k}e_{i,k}\\
&\quad+\hat{\beta}_{2i}^2\sum_{j\in\mathscr{N}_i}e_{i,k}^T\Delta K_{ij,k}^T P_{1i,k+1}\sum_{j\in\mathscr{N}_i}\Delta K_{ij,k}e_{i,k}\\
&\quad+\hat{\beta}_{1i}^2 e_{i,k}^T C_{i,k}^T\Delta L_{i,k}^T P_{1i,k+1}\Delta L_{i,k}C_{i,k}e_{i,k}+S_{i,k}-e_{i,k}^T P_{1i,k}e_{i,k}\\
&\quad+\hat{\beta}_{2i}^2\sum_{j\in\mathscr{N}_i}e_{j,k}^T\Delta K_{ij,k}^T P_{1i,k+1}\sum_{j\in\mathscr{N}_i}\Delta K_{ij,k}e_{j,k}+\varpi_{i,k}^T\mathscr{G}_{i,k}^T P_{1i,k+1}\mathscr{G}_{i,k}\varpi_{i,k}
\end{aligned} \tag{8.20}$$

where

$$e^T_{\mathscr{N}_i,k} \triangleq \begin{bmatrix} e^T_{j_{i_1},k} & e^T_{j_{i_2},k} & \cdots & e^T_{j_{i_{p_i}},k} \end{bmatrix}^T,$$

$$\varpi_{i,k} \triangleq \begin{bmatrix} e^T_{i,k} & e^T_{\mathscr{N}_i,k} & \tilde{f}^T(e_{i,k}) & \tilde{g}^T(e_{i,k}) & \tilde{g}^T(e_{i,k-\tau_k}) \end{bmatrix}^T.$$

Similarly, one calculates

$$\begin{aligned}
&\mathbb{E}\{\Delta V_{2i,k}\} \\
=&\mathbb{E}\{V_{2i,k+1}\} - V_{2i,k} \\
=&\sum_{s=k+1-\tau_{k+1}}^{k} (1-\mu_i)^{k-s}\tilde{g}^T(e_{i,s})P_{2i,s}\tilde{g}(e_{i,s}) - \sum_{s=k-\tau_k}^{k-1} (1-\mu_i)^{k-s-1}\tilde{g}^T(e_{i,s})P_{2i,s}\tilde{g}(e_{i,s}) \\
\le&\tilde{g}^T(e_{i,k})P_{2i,k}\tilde{g}(e_{i,k}) + \sum_{s=k+1-\tau_k}^{k-1} (1-\mu_i)^{k-s}\tilde{g}^T(e_{i,s})P_{2i,s}\tilde{g}(e_{i,s}) \\
&- \sum_{s=k-\tau_k}^{k-1} (1-\mu_i)^{k-s-1}\tilde{g}^T(e_{i,s})P_{2i,s}\tilde{g}(e_{i,s}) + \sum_{s=k+1-\tau_M}^{k-\tau_m} (1-\mu_i)^{k-s}\tilde{g}^T(e_{i,s})P_{2i,s}\tilde{g}(e_{i,s}) \\
=&\tilde{g}^T(e_{i,k})P_{2i,k}\tilde{g}(e_{i,k}) - \mu_i V_{2i,k} - (1-\mu_i)^{\tau_k}\tilde{g}^T(e_{i,k-\tau_k})P_{2i,k-\tau_k}\tilde{g}(e_{i,k-\tau_k}) \\
&+ \sum_{s=k+1-\tau_M}^{k-\tau_m} (1-\mu_i)^{k-s}\tilde{g}^T(e_{i,s})P_{2i,s}\tilde{g}(e_{i,s}),
\end{aligned} \tag{8.21}$$

which implies that

$$\begin{aligned}
&\mathbb{E}\{\Delta V_{2i,k}\} + \mu_i V_{2i,k} \\
\le&\tilde{g}^T(e_{i,k})P_{2i,k}\tilde{g}(e_{i,k}) - (1-\mu_i)^{\tau_M}\tilde{g}^T(e_{i,k-\tau_k})P_{2i,k-\tau_k}\tilde{g}(e_{i,k-\tau_k}) \\
&+ \sum_{s=k+1-\tau_M}^{k-\tau_m} (1-\mu_i)^{k-s}\tilde{g}^T(e_{i,s})P_{2i,s}\tilde{g}(e_{i,s}).
\end{aligned} \tag{8.22}$$

Furthermore, it follows that

$$\begin{aligned}
&\mathbb{E}\{\Delta V_{3i,k}\} \\
=&\mathbb{E}\{V_{3i,k+1}\} - V_{3i,k} \\
=&\sum_{p=k+2-\tau_M}^{k-\tau_m+1}\sum_{s=p}^{k}(1-\mu_i)^{k-s}\tilde{g}^T(e_{i,s})P_{2i,s}\tilde{g}(e_{i,s}) \\
&- \sum_{p=k-\tau_M+1}^{k-\tau_m}\sum_{s=p}^{k-1}(1-\mu_i)^{k-1-s}\tilde{g}^T(e_{i,s})P_{2i,s}\tilde{g}(e_{i,s}) \\
=&(\tau_M-\tau_m)\tilde{g}^T(e_{i,k})P_{2i,k}\tilde{g}(e_{i,k}) + \sum_{p=k+2-\tau_M}^{k-\tau_m+1}\sum_{s=p}^{k-1}(1-\mu_i)^{k-s}\tilde{g}^T(e_{i,s})P_{2i,s}\tilde{g}(e_{i,s}) \\
&- \sum_{p=k-\tau_M+1}^{k-\tau_m}\sum_{s=p}^{k-1}(1-\mu_i)^{k-1-s}\tilde{g}^T(e_{i,s})P_{2i,s}\tilde{g}(e_{i,s})
\end{aligned}$$

$$
\begin{aligned}
&= (\tau_M - \tau_m)\tilde{g}^T(e_{i,k})P_{2i,k}\tilde{g}(e_{i,k}) + \sum_{p=k+1-\tau_M}^{k-\tau_m}\sum_{s=p+1}^{k-1}(1-\mu_i)^{k-s}\tilde{g}^T(e_{i,s})P_{2i,s}\tilde{g}(e_{i,s}) \\
&\quad - \sum_{p=k-\tau_M+1}^{k-\tau_m}\sum_{s=p}^{k-1}(1-\mu_i)^{k-1-s}\tilde{g}^T(e_{i,s})P_{2i,s}\tilde{g}(e_{i,s}) \\
&= (\tau_M - \tau_m)\tilde{g}^T(e_{i,k})P_{2i,k}\tilde{g}(e_{i,k}) + \sum_{p=k+1-\tau_M}^{k-\tau_m}\sum_{s=p}^{k-1}(1-\mu_i)^{k-s}\tilde{g}^T(e_{i,s})P_{2i,s}\tilde{g}(e_{i,s}) \\
&\quad - \sum_{p=k-\tau_M+1}^{k-\tau_m}\sum_{s=p}^{k-1}(1-\mu_i)^{k-1-s}\tilde{g}^T(e_{i,s})P_{2i,s}\tilde{g}(e_{i,s}) \\
&\quad - \sum_{s=k+1-\tau_M}^{k-\tau_m+1}(1-\mu_i)^{k-s}\tilde{g}^T(e_{i,s})P_{2i,s}\tilde{g}(e_{i,s}) \\
&= -\mu_i V_{3i,k} + (\tau_M - \tau_m)\tilde{g}^T(e_{i,k})P_{2i,k}\tilde{g}(e_{i,k}) - \sum_{s=k+1-\tau_M}^{k-\tau_m+1}(1-\mu_i)^{k-s}\tilde{g}^T(e_{i,s})P_{2i,s}\tilde{g}(e_{i,s}),
\end{aligned}
\tag{8.23}
$$

which means that

$$
\begin{aligned}
&\mathbb{E}\{\Delta V_{3i,k}\} + \mu_i V_{3i,k} \\
&= (\tau_M - \tau_m)\tilde{g}^T(e_{i,k})P_{2i,k}\tilde{g}(e_{i,k}) - \sum_{s=k+1-\tau_M}^{k-\tau_m+1}(1-\mu_i)^{k-s}\tilde{g}^T(e_{i,s})P_{2i,s}\tilde{g}(e_{i,s}).
\end{aligned}
\tag{8.24}
$$

Then, taking (8.20)–(8.24) into consideration, one has

$$
\begin{aligned}
&\mathbb{E}\{\Delta V_{i,k}\} + \mu_i V_{2i,k} + \mu_i V_{3i,k} \\
&< \varpi_{i,k}^T \mathscr{G}_{i,k}^T P_{1i,k+1}\mathscr{G}_{i,k}\varpi_{i,k} + (1+\tau_M-\tau_m)\tilde{g}^T(e_{i,k})P_{2i,k}\tilde{g}(e_{i,k}) \\
&\quad - (1-\mu_i)^{\tau_M}\tilde{g}^T(e_{i,k-\tau_k})P_{2i,k-\tau_k}\tilde{g}(e_{i,k-\tau_k}) - e_{i,k}^T P_{1i,k}e_{i,k} \\
&\quad + \hat{\beta}_{1i}^2 e_{i,k}^T C_{i,k}^T \Delta L_{i,k}^T P_{1i,k+1}\Delta L_{i,k}C_{i,k}e_{i,k} + S_{i,k} \\
&\quad + \hat{\beta}_{2i}^2 \sum_{j\in\mathcal{N}_i} e_{j,k}^T \Delta K_{ij,k}^T P_{1i,k+1}\sum_{j\in\mathcal{N}_i}\Delta K_{ij,k}e_{j,k} \\
&\quad + \hat{\beta}_{2i}^2 \sum_{j\in\mathcal{N}_i} e_{i,k}^T \Delta K_{ij,k}^T P_{1i,k+1}\sum_{j\in\mathcal{N}_i}\Delta K_{ij,k}e_{i,k} \\
&\quad - 2\hat{\beta}_{2i}^2 \sum_{j\in\mathcal{N}_i} e_{j,k}^T \Delta K_{ij,k}^T P_{1i,k+1}\sum_{j\in\mathcal{N}_i}\Delta K_{ij,k}e_{i,k}.
\end{aligned}
\tag{8.25}
$$

By introducing two positive scalars $\lambda_{i,k}$ and $\rho_{i,k}$, the above inequality can be further enlarged by adding the sector-bounded constraints (8.10):

$$
\begin{aligned}
&\mathbb{E}\{\Delta V_{i,k}\} + \mu_i V_{2i,k} + \mu_i V_{3i,k} \\
<&\varpi_{i,k}^T \mathcal{G}_{i,k}^T P_{1i,k+1} \mathcal{G}_{i,k} \varpi_{i,k} + (1 + \tau_M - \tau_m) \tilde{g}^T(e_{i,k}) P_{2i,k} \tilde{g}(e_{i,k}) \\
&- (1 - \mu_i)^{\tau_M} \tilde{g}^T(e_{i,k-\tau_k}) P_{2i,k-\tau_k} \tilde{g}(e_{i,k-\tau_k}) - e_{i,k}^T P_{1i,k} e_{i,k} \\
&+ \hat{\beta}_{1i}^2 e_{i,k}^T C_{i,k}^T \Delta L_{i,k}^T P_{1i,k+1} \Delta L_{i,k} C_{i,k} e_{i,k} + S_{i,k} \\
&+ \hat{\beta}_{2i}^2 \sum_{j \in \mathcal{N}_i} e_{j,k}^T \Delta K_{ij,k}^T P_{1i,k+1} \sum_{j \in \mathcal{N}_i} \Delta K_{ij,k} e_{j,k} \\
&+ \hat{\beta}_{2i}^2 \sum_{j \in \mathcal{N}_i} e_{i,k}^T \Delta K_{ij,k}^T P_{1i,k+1} \sum_{j \in \mathcal{N}_i} \Delta K_{ij,k} e_{i,k} \\
&- 2\hat{\beta}_{2i}^2 \sum_{j \in \mathcal{N}_i} e_{j,k}^T \Delta K_{ij,k}^T P_{1i,k+1} \sum_{j \in \mathcal{N}_i} \Delta K_{ij,k} e_{i,k} \\
&- \lambda_{i,k} \left[\tilde{f}(e_{i,k}) - \Phi_{1i} e_{i,k} \right]^T \left[\tilde{f}(e_{i,k}) - \Phi_{2i} e_{i,k} \right] \\
&- \rho_{i,k} \left[\tilde{g}(e_{i,k}) - \Psi_{1i} e_{i,k} \right]^T \left[\tilde{g}(e_{i,k}) - \Psi_{2i} e_{i,k} \right].
\end{aligned} \tag{8.26}
$$

From (8.26) and the definition of $S_{i,k}$, one can obtain that

$$
\begin{aligned}
&\mathbb{E}\{V_{1i,k+1} + V_{2i,k+1} + V_{3i,k+1}\} - \sum_{j \in \mathcal{N}_i} \frac{\alpha_j}{1 + q_j} V_{1j,k} - S_{i,k} \\
&- (1 - \alpha_i) V_{1i,k} - (1 - \mu_i) V_{2i,k} - (1 - \mu_i) V_{3i,k} \\
<& \varpi_{i,k}^T \mathcal{G}_{i,k}^T P_{1i,k+1} \mathcal{G}_{i,k} \varpi_{i,k} + (1 + \tau_M - \tau_m) \tilde{g}^T(e_{i,k}) P_{2i,k} \tilde{g}(e_{i,k}) \\
&- (1 - \mu_i)^{\tau_M} \tilde{g}^T(e_{i,k-\tau_k}) P_{2i,k-\tau_k} \tilde{g}(e_{i,k-\tau_k}) - (1 - \alpha_i) e_{i,k}^T P_{1i,k} e_{i,k} \\
&+ \hat{\beta}_{1i}^2 e_{i,k}^T C_{i,k}^T \Delta L_{i,k}^T P_{1i,k+1} \Delta L_{i,k} C_{i,k} e_{i,k} - \sum_{j \in \mathcal{N}_i} \frac{\alpha_j}{1 + q_j} e_{j,k}^T P_{1j,k+1} e_{j,k} \\
&+ \hat{\beta}_{2i}^2 \sum_{j \in \mathcal{N}_i} e_{j,k}^T \Delta K_{ij,k}^T P_{1i,k+1} \sum_{j \in \mathcal{N}_i} \Delta K_{ij,k} e_{j,k} \\
&+ \hat{\beta}_{2i}^2 \sum_{j \in \mathcal{N}_i} e_{i,k}^T \Delta K_{ij,k}^T P_{1i,k+1} \sum_{j \in \mathcal{N}_i} \Delta K_{ij,k} e_{i,k} \\
&- 2\hat{\beta}_{2i}^2 \sum_{j \in \mathcal{N}_i} e_{j,k}^T \Delta K_{ij,k}^T P_{1i,k+1} \sum_{j \in \mathcal{N}_i} \Delta K_{ij,k} e_{i,k} \\
&- \lambda_{i,k} \left[\tilde{f}(e_{i,k}) - \Phi_{1i} e_{i,k} \right]^T \left[\tilde{f}(e_{i,k}) - \Phi_{2i} e_{i,k} \right] \\
&- \rho_{i,k} \left[\tilde{g}(e_{i,k}) - \Psi_{1i} e_{i,k} \right]^T \left[\tilde{g}(e_{i,k}) - \Psi_{2i} e_{i,k} \right] \\
\triangleq& \varpi_{i,k}^T \left(\Xi_{i,k} + \mathcal{G}_{i,k}^T P_{1i,k+1} \mathcal{G}_{i,k} \right) \varpi_{i,k}
\end{aligned} \tag{8.27}
$$

where $\Xi_{i,k}$ is defined in (8.18).

According to (8.16) and the Schur Complement Lemma, one can know that

$$\varpi_{i,k}^T \left(\Xi_{i,k} + \mathscr{G}_{i,k}^T P_{1i,k+1} \mathscr{G}_{i,k}\right) \varpi_{i,k} < 0. \tag{8.28}$$

Thus, it follows from (8.27) and (8.28) that

$$\begin{aligned}
&\mathbb{E}\{V_{1i,k+1} + V_{2i,k+1} + V_{3i,k+1}\} \\
<&(1-\alpha_i)V_{1i,k} + (1-\mu_i)V_{2i,k} + (1-\mu_i)V_{3i,k} + S_{i,k} + \sum_{j\in\mathscr{N}_i} \frac{\alpha_j}{1+q_j} V_{1j,k}.
\end{aligned}$$

Then, one constructs matrix W_1 by taking the row vector $\left[\frac{a_{i1}\alpha_1}{1+q_1} \ \cdots \ 1-\alpha_i \ \cdots \ \frac{a_{iN}\alpha_N}{1+q_N}\right]$ as its ith row. Since $\alpha_i \in \mathfrak{I}_{q_i}$, W_1 is strictly diagonally dominant, and thus it is nonsingular. Moreover, one can obtain $\mathbf{1}^T W_1 = [1-\mu_1 \ 1-\mu_2 \ \cdots \ 1-\mu_N] \ll \mathbf{1}^T$. As a result, W_1 is the dissipation matrix. Meanwhile, denote $W_2 = W_3 = \text{diag}_N\{1-\mu_i\}$. Clearly, both W_2 and W_3 are also the dissipation matrices. Hence, it follows from the matrices W_1, W_2 and W_3 that

$$\mathbb{E}\left\{V_{1i,k+1} + V_{2i,k+1} + V_{3i,k+1}\right\} < \left[W_1\mathbf{V}_{1,k}\right]_i + \left[W_2\mathbf{V}_{2,k}\right]_i + \left[W_3\mathbf{V}_{3,k}\right]_i + S_{i,k}$$

where $\mathbf{V}_{t,k} = \left[V_{t1,k} \ \ V_{t2,k} \ \ \cdots \ \ V_{tN,k}\right]^T$ $(t = 1, 2, 3)$. The above inequalities can be rewritten as a vector form:

$$\mathbb{E}\left\{\mathbf{V}_{1,k+1} + \mathbf{V}_{2,k+1} + \mathbf{V}_{3,k+1}\right\} \ll \mathbb{E}\{W_1\mathbf{V}_{1,k} + W_2\mathbf{V}_{2,k} + W_3\mathbf{V}_{3,k}\} + \mathbb{E}\{\mathbf{S}_k\} \tag{8.29}$$

where $\mathbf{S}_k = \left[S_{1,k} \ \ S_{2,k} \ \ \cdots \ \ S_{N,k}\right]^T$. Finally, recalling Definition 8.2, the proof of Lemma 8.1 is therefore complete.

Theorem 8.1 *Let scalars $\alpha_i \in \mathfrak{I}_{q_i}$, positive definite matrices $R_{i,k}$ and $\{T_{i,s}\}_{-\tau_M\le s\le 0}$, and the filter gains $L_{i,k}$ and $K_{ij,k}$ $(i \in \mathscr{V}, j \in \mathscr{N}_i)$ be given. Equation (8.4) is an exponentially bounded state filter of the plant (8.1) in the mean square, if there exist two sequences of positive definite matrices $P_{1i,k}$ and $P_{2i,k}$ with the initial conditions $P_{1i,0} \le T_{i,0}$ and $\left\{P_{2i,s} \le (\tau_M - \tau_m)T_{i,s}\right\}_{-\tau_M\le s\le -1}$, and two sequences of constants $\lambda_{i,k}$ and $\rho_{i,k}$ such that (8.16)–(8.17) hold.*

Proof First and foremost, recalling the expressions of W_1, W_2 and W_3 and multiplying by $\mathbf{1}^T$ on both sides of the vector dissipation inequality (8.29), one has the following result:

$$\begin{aligned}
&\mathbf{1}^T\mathbb{E}\left\{\mathbf{V}_{1,k+1} + \mathbf{V}_{2,k+1} + \mathbf{V}_{3,k+1}\right\} \\
<&\mathbf{1}^T W_1\mathbb{E}\left\{\mathbf{V}_{1,k}\right\} + \mathbf{1}^T W_2\mathbb{E}\left\{\mathbf{V}_{2,k}\right\} + \mathbf{1}^T W_3\mathbb{E}\left\{\mathbf{V}_{3,k}\right\} + \mathbf{1}^T\mathbb{E}\left\{\mathbf{S}_k\right\},
\end{aligned}$$

which means that

$$\sum_{i=1}^{N}\mathbb{E}\left\{V_{1i,k+1}+V_{2i,k+1}+V_{3i,k+1}\right\}<\sum_{i=1}^{N}(1-\mu_i)\mathbb{E}\left\{V_{1i,k}+V_{2i,k}+V_{3i,k}\right\}+\sum_{i=1}^{N}\mathbb{E}\left\{S_{i,k}\right\}. \tag{8.30}$$

By means of the denotation $V_{i,k}=V_{1i,k}+V_{2i,k}+V_{3i,k}$, we further have

$$\sum_{i=1}^{N}\mathbb{E}\left\{V_{i,k+1}\right\}<\sum_{i=1}^{N}(1-\mu_i)\mathbb{E}\left\{V_{i,k}\right\}+\sum_{i=1}^{N}\mathbb{E}\left\{S_{i,k}\right\},$$

which implies that

$$\begin{aligned}\sum_{i=1}^{N}\mathbb{E}\left\{V_{i,k+1}\right\}&<\sum_{i=1}^{N}(1-\mu_i)^{k+1}\mathbb{E}\left\{V_{i,0}\right\}+\sum_{t=0}^{k}\sum_{i=1}^{N}(1-\mu_i)^{k-t}\mathbb{E}\left\{S_{i,t}\right\}\\&\triangleq\sum_{i=1}^{N}\theta_i^{k+1}\mathbb{E}\left\{V_{i,0}\right\}+\sum_{t=0}^{k}\sum_{i=1}^{N}\theta_i^{k-t}\mathbb{E}\left\{S_{i,t}\right\}.\end{aligned} \tag{8.31}$$

Considering $P_{1i,k+1}\geq R_{i,k+1}$, one obtains

$$\begin{aligned}\sum_{i=1}^{N}\mathbb{E}\left\{V_{i,k+1}\right\}&\geq\sum_{i=1}^{N}\mathbb{E}\left\{e_{i,k+1}^{T}P_{1i,k+1}e_{i,k+1}\right\}\\&\geq\sum_{i=1}^{N}\mathbb{E}\left\{e_{i,k+1}^{T}R_{i,k+1}e_{i,k+1}\right\}\\&=\sum_{i=1}^{N}\mathbb{E}\|e_{i,k+1}\|_{R_{i,k+1}}^{2}.\end{aligned} \tag{8.32}$$

As such, from (8.31) and (8.32), one can know

$$\sum_{i=1}^{N}\mathbb{E}\left\{\|e_{i,k+1}\|_{R_{i,k+1}}^{2}\right\}<\sum_{i=1}^{N}\theta_i^{k+1}\mathbb{E}\left\{V_{i,0}\right\}+\sum_{t=0}^{k}\sum_{i=1}^{N}\theta_i^{k-t}\mathbb{E}\left\{S_{i,t}\right\}. \tag{8.33}$$

Finally, according to the expression of $V_{i,k}$ and the initial conditions $P_{1i,0}\leq T_{i,0}$ and $P_{2i,s}\leq(\tau_M-\tau_m)T_{2i,s}$, one then obtains

$$
\begin{aligned}
V_{i,0} =& e_{i,0}^T P_{1i,0} e_{i,0} + \sum_{s=-\tau_k}^{-1} (1-\mu_i)^{-s-1} \tilde{g}^T(e_{i,s}) P_{2i,s} \tilde{g}(e_{i,s}) \\
&+ \sum_{p=1-\tau_M}^{-\tau_m} \sum_{s=p}^{-1} (1-\mu_i)^{-s-1} \tilde{g}^T(e_{i,s}) P_{2i,s} \tilde{g}(e_{i,s}) \\
<& e_{i,0}^T T_{i,0} e_{i,0} + (\tau_M - \tau_m) \sum_{s=-\tau_M}^{-1} \tilde{g}^T(e_{i,s}) P_{2i,s} \tilde{g}(e_{i,s}) \\
\le& e_{i,0}^T T_{i,0} e_{i,0} + \sum_{s=-\tau_M}^{-1} \tilde{g}^T(e_{i,s}) T_{i,s} \tilde{g}(e_{i,s}).
\end{aligned} \tag{8.34}
$$

Therefore, according to Definition 8.1, the proof is complete.

Up to now, the desirable local conditions are constructed for each node by means of the extended vector dissipativity in Lemma 8.1. Moreover, these local conditions can guarantee that the filtering error in (8.8) is exponentially bounded in the mean square.

8.3 Performance Analysis and Distributed Optimization

As is well known, the filtering accuracy depends on the upper bound in the performance index. Naturally, the smaller this upper bound is, the better the quality of this filtering is. In this section, we consider the performance optimization problem of the proposed algorithm. More specifically, we will minimize the upper bound in the performance index (8.11).

According to (8.33), the upper bound to be optimized is given as follows:

$$
J_k \triangleq \sum_{i=1}^{N} \sum_{t=0}^{k} \theta_i^{k-t} \mathbb{E}\left\{ S_{i,t} \right\}. \tag{8.35}
$$

Denote $J_k^* \triangleq \min J_k$ as the optimal upper bound of the performance index at time instant k. Then, from the recursive nature of the algorithm proposed in Theorem 8.1 and the forward dynamic programming, one further has

$$
\begin{aligned}
J_k^* &= \min\left(\sum_{i=1}^{N}\sum_{t=0}^{k}\theta_i^{k-t}\mathbb{E}\{S_{i,t}\}\right) \\
&= \sum_{i=1}^{N}\min\left(\sum_{t=0}^{k}\theta_i^{k-t}\mathbb{E}\{S_{i,t}\}\right) \\
&= \sum_{i=1}^{N}\min\left(\sum_{t=0}^{k-1}\theta_i^{k-1-t}\mathbb{E}\{S_{i,t}\}\right) + \sum_{i=1}^{N}\min\mathbb{E}\left\{S_{i,k}\right\} \\
&= J_{k-1}^* + \sum_{i=1}^{N}\min\mathbb{E}\left\{S_{i,k}\right\}.
\end{aligned}
\tag{8.36}
$$

In view of (8.36), J_{k-1}^* is a constant at time instant k and thus could be neglected when solving (8.36). As a result, only $\mathbb{E}\left\{S_{i,k}\right\}$ needs to be minimized at time instant k. Recalling the expression of $\mathbb{E}\left\{S_{i,k}\right\}$, one has

$$
\begin{aligned}
&\mathbb{E}\left\{S_{i,k}\right\} \\
&= \mathbb{E}\left\{v_{i,k}^T D_{i,k}^T (L_{i,k} + \beta_{1i,k}\Delta L_{i,k})^T P_{1i,k+1}(L_{i,k} + \beta_{1i,k}\Delta L_{i,k}) D_{i,k} v_{i,k}\right\} \\
&\quad + \mathbb{E}\left\{w_k^T B_k^T P_{1i,k+1} B_k w_k\right\} \\
&= \mathbf{1}_{n_v}^T((D_{i,k}^T(L_{i,k}^T P_{1i,k+1} L_{i,k} + \bar{\beta}_{1i}\Delta L_{i,k}^T P_{1i,k+1}\Delta L_{i,k}) D_{i,k}) \circ R_i)\mathbf{1}_{n_v} \\
&\quad + \mathbf{1}_{n_w}^T((B_k^T P_{1i,k+1} B_k) \circ Q)\mathbf{1}_{n_w} \\
&= \mathbf{1}_{n_w}^T((B_k^T P_{1i,k+1} B_k) \circ Q)\mathbf{1}_{n_w} + \sum_{t=1}^{n_v}\sigma_{it}^2 \mathbf{1}_{n_v}^T F_t^T D_{i,k}^T L_{i,k}^T P_{1i,k+1} L_{i,k} D_{i,k} F_t \mathbf{1}_{n_v} \\
&\quad + \sum_{t=1}^{n_v}\bar{\beta}_{1i}\sigma_{it}^2 \mathbf{1}_{n_v}^T F_t^T D_{i,k}^T \Delta L_{i,k}^T P_{1i,k+1}\Delta L_{i,k} D_{i,k} F_t \mathbf{1}_{n_v}
\end{aligned}
\tag{8.37}
$$

where

$$
F_t \triangleq \mathrm{diag}\{\underbrace{0,\ldots,0}_{t-1}, 1, \underbrace{0,\cdots,0}_{n_v-t}\}.
$$

Then, based on the above arguments together with the existence of $\Delta L_{i,k}$, the following robust optimization problem can be proposed to obtain the minimal upper bound in the performance index. That is,

$$
\begin{aligned}
&\min\max_{\Delta L_{i,k}}\mathbb{E}\left\{S_{i,k}\right\} \\
&\text{s.t.}\quad (8.16), (8.17).
\end{aligned}
\tag{8.38}
$$

It is noted that the objective function and the constraints are all uncertain and nonlinear in this optimization problem (8.38). To make this problem tractable, the first step is to introduce an auxiliary variable $\rho_{i,k}$, such that

$$\mathbb{E}\left\{S_{i,k}\right\} \leq \rho_{i,k}, \tag{8.39}$$

for any feasible $\Delta L_{i,k}$.

As a result, the robust optimization problem (8.38) can be reformulated as:

$$\begin{aligned} &\min \rho_{i,k} \\ &\text{s.t.} \quad (8.16), (8.17), (8.39). \end{aligned} \tag{8.40}$$

The second step is to change the inequality constraints (8.16), (8.17), and (8.39) into linear matrix inequalities. Meanwhile, the filter gains can be calculated. For this purpose, a useful lemma will be firstly given in the following.

Lemma 8.2 (S-procedure [2]) *Let $\mathscr{U}$, $\mathscr{V}_k$, $\mathscr{W}$ and $\mathscr{Q}$ be real matrices of appropriate dimensions with $\mathscr{Q}$ satisfying $\mathscr{Q} = \mathscr{Q}^T$. Then, for all $\mathscr{V}_k \mathscr{V}_k^T \leq I$,*

$$\mathscr{Q} + \mathscr{U}\mathscr{V}_k\mathscr{W} + \mathscr{W}^T\mathscr{V}_k^T\mathscr{U}^T < 0$$

holds if and only if there exists a positive scalar ψ such that

$$\mathscr{Q} + \psi^{-1}\mathscr{U}\mathscr{U}^T + \psi\mathscr{W}^T\mathscr{W} < 0.$$

In the following theorem, let $\rho_{i,k}$ not be temporarily optimized.

Theorem 8.2 *Let scalars $\alpha_i \in \mathfrak{I}_{q_i}$, positive definite matrices $R_{i,k}$ and $\left\{T_{i,s}\right\}_{-\tau_M \leq s \leq 0}$ $(i \in \mathscr{V})$ be given. Equation (8.4) is an exponentially bounded state filter of the plant (8.1) in the mean square, if there exist two sets of positive definite matrices $P_{1i,k}$ and $P_{2i,k}$ with the initial conditions $P_{1i,0} \leq T_{i,0}$ and $\left\{P_{2i,s} \leq (\tau_M - \tau_m)T_{i,s}\right\}_{-\tau_M \leq s \leq -1}$, and positive scalars $\varsigma_{ij,k} (j \in \mathscr{N}_i)$, $\rho_{i,k}$, $\lambda_{i,k}$, $\rho_{i,k}$, $\gamma_{1i,k}$, $\varepsilon_{1i,k}$ and $\varepsilon_{2i,k}$ such that the following RLMIs hold for $k \geq 0$*

$$\begin{bmatrix} \overline{\Xi}_{i,k} & * \\ \bar{\$}_{i,k} & \Omega_{i,k} \end{bmatrix} < 0, \tag{8.41}$$

$$R_{i,k} - P_{1i,k} \leq 0, \tag{8.42}$$

$$\begin{bmatrix} \tilde{\Pi}_{i,k}^{11} & * & * & * \\ \tilde{\Pi}_{i,k}^{21} & \tilde{\Pi}_{i,k}^{22} & * & * \\ 0 & 0 & \tilde{\Pi}_{i,k}^{33} & * \\ 0 & 0 & \tilde{\Pi}_{i,k}^{43} & \tilde{\Pi}_{i,k}^{44} \end{bmatrix} < 0 \tag{8.43}$$

where

$$\bar{\Xi}_{i,k}=\begin{bmatrix}\bar{\Xi}^{11}_{i,k} & * & * & * & * & * & * & * \\ 0 & \bar{\Xi}^{22}_{i,k} & * & * & * & * & * & * \\ \Xi^{31}_{i,k} & 0 & \Xi^{33}_{i,k} & * & * & * & * & * \\ \Xi^{41}_{i,k} & 0 & 0 & \Xi^{44}_{i,k} & * & * & * & * \\ 0 & 0 & 0 & 0 & \Xi^{55}_{i,k} & * & * & * \\ \bar{\Xi}^{61}_{i,k} & \bar{\Xi}^{62}_{i,k} & I & 0 & I & \bar{\Xi}^{66}_{i,k} & * & * \\ 0 & 0 & 0 & 0 & 0 & 0 & \bar{\Xi}^{77}_{i,k} & * \\ 0 & 0 & 0 & 0 & 0 & 0 & 0 & \bar{\Xi}^{88}_{i,k}\end{bmatrix},$$

$$\bar{\$}_{2i,k}=\begin{bmatrix}-\bar{\beta}_{1i}M^T_{1i,k}P_{1i,k+1} & \hat{\beta}_{1i}M^T_{1i,k}P_{1i,k+1} & 0 \\ \bar{\beta}_{2i}M^T_{2i,k}P_{1i,k+1} & 0 & \hat{\beta}_{2i}M^T_{2i,k}P_{1i,k+1} \\ -\bar{\beta}_{2i}M^T_{2i,k}P_{1i,k+1} & 0 & -\hat{\beta}_{2i}M^T_{2i,k}P_{1i,k+1} \\ \vdots & \vdots & \vdots \\ -\bar{\beta}_{2i}M^T_{2i,k}P_{1i,k+1} & 0 & -\hat{\beta}_{2i}M^T_{2i,k}P_{1i,k+1}\end{bmatrix},$$

$$\Omega_{i,k}=\operatorname{diag}\left\{-\varepsilon_{1i,k}I,-\frac{1}{p_i}\varepsilon_{2i,k}I,-\varsigma_{ij_1,k}I,\ldots,-\varsigma_{ij_{p_i},k}I\right\},$$

$$\bar{\Xi}^{11}_{i,k}=-(1-\alpha_i)P_{1i,k}-\bar{\Psi}_{i,k}-\bar{\Phi}_{i,k}+\varepsilon_{1i,k}C^T_{i,k}N^T_{1i,k}N_{1i,k}C_{i,k}+\sum_{j\in\mathcal{N}_i}\varsigma_{ij,k}N^T_{2i,k}N_{2i,k},$$

$$\bar{\Xi}^{22}_{i,k}=\operatorname{diag}\left\{-\frac{\alpha_{j_{i_1}}}{1+q_{j_{i_1}}}P_{1j_{i_1},k}+\varepsilon_{2i,k}N^T_{2i,k}N_{2i,k},\ldots,-\frac{\alpha_{j_{ip_i}}}{1+q_{j_{ip_i}}}P_{1j_{ip_i},k}+\varepsilon_{2i,k}N^T_{2i,k}N_{2i,k}\right\},$$

$$\bar{\Xi}^{61}_{i,k}=P_{1i,k+1}A_k-E_{i,k}C_{i,k}-\sum_{j\in\mathcal{N}_i}F_{ij,k},\quad \bar{\Xi}^{62}_{i,k}=\begin{bmatrix}F_{ij_1,k} & \cdots & F_{ij_{p_i},k}\end{bmatrix},$$

$$\bar{\Xi}^{66}_{i,k}=-P_{1i,k+1},\quad \bar{\Xi}^{77}_{i,k}=-P_{1i,k+1},\quad \bar{\Xi}^{88}_{i,k}=-P_{1i,k+1},$$

$$\tilde{\Pi}^{11}_{i,k}=\mathbf{1}^T_{n_w}((B^T_kP_{1i,k+1}B_k)\circ Q)\mathbf{1}_{n_w}-\rho_{i,k}+\gamma_{1i,k}\sum_{t=1}^{n_v}\mathbf{1}^T_{n_v}F^T_tD^T_{i,k}N^T_{1i,k}N_{1i,k}D_{i,k}F_t\mathbf{1}_{n_v},$$

$$\tilde{\Pi}^{21}_{i,k}=\left[(E_{i,k}D_{i,k}F_1\mathbf{1}_{n_v})^T\ \cdots\ (E_{i,k}D_{i,k}F_{n_v}\mathbf{1}_{n_v})^T\right]^T,\quad \tilde{\Pi}^{22}_{i,k}=-\operatorname{diag}_{n_v}\left\{\frac{1}{\sigma^2_{it}}P_{1i,k+1}\right\},$$

$$\tilde{\Pi}^{33}_{i,k}=-\operatorname{diag}_{n_v}\left\{\frac{1}{\bar{\beta}_{1i}\sigma^2_{it}}P_{1i,k+1}\right\},\quad \tilde{\Pi}^{43}_{i,k}=\operatorname{diag}_{n_v}\left\{M^T_{1i,k}P_{i,k+1}\right\},$$

$$\tilde{\Pi}^{44}_{i,k}=-\operatorname{diag}_{n_v}\{\gamma_{1i,k}I\},\quad \bar{\$}_{i,k}=\begin{bmatrix}0_{(2+p_i)\times 5} & \bar{\$}_{2i,k}\end{bmatrix}.$$

Moreover, the filter gains can be calculated as

$$L_{i,k}=P^{-1}_{1i,k+1}E_{i,k},\quad K_{ij,k}=P^{-1}_{1i,k+1}F_{ij,k}. \tag{8.44}$$

Proof According to the Schur Complement Lemma, (8.16) holds if and only if the following inequality holds:

$$\begin{bmatrix} \widehat{\Xi}^{11}_{i,k} & * & * & * & * & * & * & * \\ 0 & \widehat{\Xi}^{22}_{i,k} & * & * & * & * & * & * \\ \Xi^{31}_{i,k} & 0 & \Xi^{33}_{i,k} & * & * & * & * & * \\ \Xi^{41}_{i,k} & 0 & 0 & \Xi^{44}_{i,k} & * & * & * & * \\ 0 & 0 & 0 & 0 & \Xi^{55}_{i,k} & * & * & * \\ \widehat{\Xi}^{61}_{i,k} & \widehat{\Xi}^{62}_{i,k} & I & 0 & I & \widehat{\Xi}^{66}_{i,k} & * & * \\ \widehat{\Xi}^{71}_{i,k} & 0 & 0 & 0 & 0 & 0 & \widehat{\Xi}^{77}_{i,k} & * \\ \widehat{\Xi}^{81}_{i,k} & \widehat{\Xi}^{82}_{i,k} & 0 & 0 & 0 & 0 & 0 & \widehat{\Xi}^{88}_{i,k} \end{bmatrix} < 0 \tag{8.45}$$

where

$$\begin{aligned}
\widehat{\Xi}^{11}_{i,k} &= -(1-\alpha_i)P_{1i,k} - \bar{\Psi}_{i,k} - \bar{\Phi}_{i,k}, \quad \widehat{\Xi}^{88}_{i,k} = -P^{-1}_{1i,k+1}, \\
\widehat{\Xi}^{22}_{i,k} &= \operatorname{diag}\left\{ -\frac{\alpha_{j_{i_1}}}{1+q_{j_{i_1}}} P_{1 j_{i_1},k}, \ldots, -\frac{\alpha_{j_{i p_i}}}{1+q_{j_{i p_i}}} P_{1 j_{i p_i},k} \right\}, \\
\widehat{\Xi}^{62}_{i,k} &= \begin{bmatrix} K_{i j_1,k} + \bar{\beta}_{2i}\Delta K_{i j_1,k} & \cdots & K_{i j_{p_i},k} + \bar{\beta}_{2i}\Delta K_{i j_{p_i},k} \end{bmatrix}, \\
\widehat{\Xi}^{66}_{i,k} &= -P^{-1}_{1i,k+1}, \quad \widehat{\Xi}^{71}_{i,k} = \hat{\beta}_{1i}\Delta L_{i,k}C_{i,k}, \quad \widehat{\Xi}^{77}_{i,k} = -P^{-1}_{1i,k+1}, \\
\widehat{\Xi}^{61}_{i,k} &= A_k - L_{i,k}C_{i,k} - \sum_{j\in\mathcal{N}_i}(K_{ij,k} + \bar{\beta}_{2i}\Delta K_{ij,k}) - \bar{\beta}_{1i}\Delta L_{i,k}C_{i,k}, \\
\widehat{\Xi}^{81}_{i,k} &= -\hat{\beta}_{2i}\sum_{j\in\mathcal{N}_i}\Delta K_{ij,k}, \quad \widehat{\Xi}^{82}_{i,k} = \begin{bmatrix} \hat{\beta}_{2i}\Delta K_{i j_1,k} & \cdots & \hat{\beta}_{2i}\Delta K_{i j_{p_i},k} \end{bmatrix}.
\end{aligned}$$

According to (8.5), (8.45) can be rewritten as

$$\begin{aligned}
&\widetilde{\Xi}_{i,k} + \mathscr{M}_{1i,k}F_{1i,k}\mathscr{N}_{1i,k} + \mathscr{N}^T_{1i,k}F^T_{1i,k}\mathscr{M}^T_{1i,k} + \mathscr{M}_{2i,k}F_{2i,k}\mathscr{N}_{2i,k} + \mathscr{N}^T_{2i,k}F^T_{2i,k}\mathscr{M}^T_{2i,k} \\
&+ \sum_{j\in\mathcal{N}_i}\left(\mathscr{M}_{3i,k}F_{2ij,k}\mathscr{N}_{3i,k} + \mathscr{N}^T_{3i,k}F^T_{2ij,k}\mathscr{M}^T_{3i,k}\right) < 0
\end{aligned} \tag{8.46}$$

where

$$\widetilde{\Xi}_{i,k} = \begin{bmatrix} \widehat{\Xi}^{11}_{i,k} & * & * & * & * & * & * & * \\ 0 & \widehat{\Xi}^{22}_{i,k} & * & * & * & * & * & * \\ \Xi^{31}_{i,k} & 0 & \Xi^{33}_{i,k} & * & * & * & * & * \\ \Xi^{41}_{i,k} & 0 & 0 & \Xi^{44}_{i,k} & * & * & * & * \\ 0 & 0 & 0 & 0 & \Xi^{55}_{i,k} & * & * & * \\ \widetilde{\Xi}^{61}_{i,k} & \widetilde{\Xi}^{62}_{i,k} & I & 0 & I & \widehat{\Xi}^{66}_{i,k} & * & * \\ 0 & 0 & 0 & 0 & 0 & 0 & \widehat{\Xi}^{77}_{i,k} & * \\ 0 & 0 & 0 & 0 & 0 & 0 & 0 & \widehat{\Xi}^{88}_{i,k} \end{bmatrix},$$

$$\mathscr{M}_{1i,k} = \begin{bmatrix} 0\,0\,0\,0\,0 & -\bar{\beta}_{1i}M^T_{1i,k} & \hat{\beta}_{1i}M^T_{1i,k} & 0 \end{bmatrix}^T,$$

$$\mathscr{M}_{2i,k} = \begin{bmatrix} 0\,0\,0\,0\,0\; \bar{\beta}_{2i}\Upsilon_{1i,k}^T\; 0\; \hat{\beta}_{2i}\Upsilon_{1i,k}^T \end{bmatrix}^T,$$
$$\mathscr{N}_{1i,k} = [N_{1i,k}C_{i,k}\; 0\,0\,0\,0\,0\,0\,0], \quad \Upsilon_{1i,k} = \mathrm{Vec}_{p_i}[M_{2i,k}],$$
$$\mathscr{M}_{3i,k} = \begin{bmatrix} 0\,0\,0\,0\,0\; -\bar{\beta}_{2i}M_{2i,k}^T\; 0\; -\hat{\beta}_{2i}M_{2i,k}^T \end{bmatrix}^T,$$
$$\mathscr{N}_{2i,k} = \begin{bmatrix} 0\; \Upsilon_{2i,k}\; 0\,0\,0\,0\,0\,0 \end{bmatrix}, \quad \Upsilon_{2i,k} = \mathrm{diag}_{p_i}\left\{N_{2i,k}\right\},$$
$$F_{2i,k} = \mathrm{diag}\left\{F_{2ij_1,k}, \ldots, F_{2ij_{p_i},k}\right\}, \quad \widetilde{\Xi}_{i,k}^{61} = A_k - L_{i,k}C_{i,k} - \sum_{j \in \mathscr{N}_i} K_{ij,k},$$
$$\mathscr{N}_{3i,k} = \begin{bmatrix} N_{2i,k}\; 0\,0\,0\,0\,0\,0\,0 \end{bmatrix}, \quad \widetilde{\Xi}_{i,k}^{62} = \begin{bmatrix} K_{ij_1,k} & \cdots & K_{ij_{p_i},k} \end{bmatrix}.$$

According to Lemma 8.2, (8.46) holds if and only if there exist positive scalars $\varepsilon_{1i,k}$, $\varepsilon_{2i,k}$ and $\varsigma_{ij,k}$ ($j \in \mathscr{N}_i$) such that the following inequality holds

$$\begin{aligned} &\widetilde{\Xi}_{i,k} + \varepsilon_{1i,k}^{-1}\mathscr{M}_{1i,k}\mathscr{M}_{1i,k}^T + \varepsilon_{1i,k}\mathscr{N}_{1i,k}^T\mathscr{N}_{1i,k} + \varepsilon_{2i,k}^{-1}\mathscr{M}_{2i,k}\mathscr{M}_{2i,k}^T + \varepsilon_{2i,k}\mathscr{N}_{2i,k}^T\mathscr{N}_{2i,k} \\ &+ \sum_{j \in \mathscr{N}_i}\left(\varsigma_{ij,k}^{-1}\mathscr{M}_{3i,k}\mathscr{M}_{3i,k}^T + \varsigma_{ij,k}\mathscr{N}_{3i,k}^T\mathscr{N}_{3i,k}\right) < 0. \end{aligned} \tag{8.47}$$

That is,

$$\begin{bmatrix} \overline{\Xi}_{i,k}^{11} & * & * & * & * & * & * & * \\ 0 & \overline{\Xi}_{i,k}^{22} & * & * & * & * & * & * \\ \Xi_{i,k}^{31} & 0 & \Xi_{i,k}^{33} & * & * & * & * & * \\ \Xi_{i,k}^{41} & 0 & 0 & \Xi_{i,k}^{44} & * & * & * & * \\ 0 & 0 & 0 & 0 & \Xi_{i,k}^{55} & * & * & * \\ \widetilde{\Xi}_{i,k}^{61} & \widetilde{\Xi}_{i,k}^{62} & I & 0 & I & \widetilde{\Xi}_{i,k}^{66} & * & * \\ 0 & 0 & 0 & 0 & 0 & \widetilde{\Xi}_{i,k}^{76} & \widetilde{\Xi}_{i,k}^{77} & * \\ 0 & 0 & 0 & 0 & 0 & \widetilde{\Xi}_{i,k}^{86} & 0 & \widetilde{\Xi}_{i,k}^{88} \end{bmatrix} < 0 \tag{8.48}$$

where

$$\widetilde{\Xi}_{i,k}^{66} = -P_{1i,k+1}^{-1} + \varepsilon_{1i,k}^{-1}\bar{\beta}_{1i}^2 M_{1i,k}M_{1i,k}^T + \varepsilon_{2i,k}^{-1}\bar{\beta}_{2i}^2 p_i M_{2i,k}M_{2i,k}^T + \sum_{j \in \mathscr{N}_i}\varsigma_{ij,k}^{-1}\bar{\beta}_{2i}^2 M_{2i,k}M_{2i,k}^T,$$
$$\widetilde{\Xi}_{i,k}^{76} = -\varepsilon_{1i,k}^{-1}\bar{\beta}_{1i}\hat{\beta}_{1i}M_{1i,k}M_{1i,k}^T, \quad \widetilde{\Xi}_{i,k}^{77} = -P_{1i,k+1}^{-1} + \varepsilon_{1i,k}^{-1}\hat{\beta}_{1i}^2 M_{1i,k}M_{1i,k}^T,$$
$$\widetilde{\Xi}_{i,k}^{86} = \sum_{j \in \mathscr{N}_i}\varsigma_{ij,k}^{-1}\bar{\beta}_{2i}\hat{\beta}_{2i}M_{2i,k}M_{2i,k}^T + \varepsilon_{2i,k}^{-1}\bar{\beta}_{2i}\hat{\beta}_{2i}p_i M_{2i,k}M_{2i,k}^T,$$
$$\widetilde{\Xi}_{i,k}^{88} = \widehat{\Xi}_{i,k}^{88} + \varepsilon_{2i,k}^{-1}\hat{\beta}_{2i}^2 p_i M_{2i,k}M_{2i,k}^T + \sum_{j \in \mathscr{N}_i}\varsigma_{ij,k}^{-1}\hat{\beta}_{2i}^2 M_{2i,k}M_{2i,k}^T.$$

Then, by resorting to the Schur Complement Lemma, (8.48) holds if and only if

$$\begin{bmatrix} \widehat{\Xi}_{i,k} & * \\ \$_{i,k} & \Omega_{i,k} \end{bmatrix} < 0 \tag{8.49}$$

where

$$\widehat{\Xi}_{i,k} = \begin{bmatrix} \overline{\Xi}_{i,k}^{11} & * & * & * & * & * & * & * \\ 0 & \overline{\Xi}_{i,k}^{22} & * & * & * & * & * & * \\ \Xi_{i,k}^{31} & 0 & \Xi_{i,k}^{33} & * & * & * & * & * \\ \Xi_{i,k}^{41} & 0 & 0 & \Xi_{i,k}^{44} & * & * & * & * \\ 0 & 0 & 0 & 0 & \Xi_{i,k}^{55} & * & * & * \\ \widetilde{\Xi}_{i,k}^{61} & \widetilde{\Xi}_{i,k}^{62} & I & 0 & I & \widehat{\Xi}_{i,k}^{66} & * & * \\ 0 & 0 & 0 & 0 & 0 & 0 & \widehat{\Xi}_{i,k}^{77} & * \\ 0 & 0 & 0 & 0 & 0 & 0 & 0 & \widehat{\Xi}_{i,k}^{88} \end{bmatrix},$$

$$\$_{i,k} = \begin{bmatrix} 0_{(2+p_i)\times 5} & \$_{2i,k} \end{bmatrix},$$

$$\$_{2i,k} = \begin{bmatrix} -\bar{\beta}_{1i} M_{1i,k}^T & \hat{\beta}_{1i} M_{1i,k}^T & 0 \\ \bar{\beta}_{2i} M_{2i,k}^T & 0 & \hat{\beta}_{2i} M_{2i,k}^T \\ -\bar{\beta}_{2i} M_{2i,k}^T & 0 & -\hat{\beta}_{2i} M_{2i,k}^T \\ \vdots & \vdots & \vdots \\ -\bar{\beta}_{2i} M_{2i,k}^T & 0 & -\hat{\beta}_{2i} M_{2i,k}^T \end{bmatrix}.$$

Then, performing the congruent transformation

$$\text{diag}\left\{ I, I, I, I, I, P_{1i,k+1}, P_{1i,k+1}, P_{1i,k+1}, I, I, \underbrace{I, \ \ldots, \ I}_{p_i} \right\}$$

to (8.49), one obtains (8.41).

Next, (8.39) can be further changed into

$$\mathbb{E}\{S_{i,k}\} - \rho_{i,k} < 0, \tag{8.50}$$

which implies that

$$\Pi_{i,k} = \begin{bmatrix} \Pi_{i,k}^{11} & * & * \\ \Pi_{i,k}^{21} & \Pi_{i,k}^{22} & * \\ \Pi_{i,k}^{31} & 0 & \Pi_{i,k}^{33} \end{bmatrix} < 0 \tag{8.51}$$

where

$$\Pi_{i,k}^{11} = \mathbf{1}_{n_w}^T((B_k^T P_{1i,k+1} B_k) \circ Q)\mathbf{1}_{n_w} - \rho_{i,k}$$
$$\Pi_{i,k}^{21} = \left[(L_{i,k} D_{i,k} F_1 \mathbf{1}_{n_v})^T \ \cdots \ (L_{i,k} D_{i,k} F_{n_v} \mathbf{1}_{n_v})^T\right]^T,$$
$$\Pi_{i,k}^{31} = \left[(\Delta L_{i,k} D_{i,k} F_1 \mathbf{1}_{n_v})^T \ \cdots \ (\Delta L_{i,k} D_{i,k} F_{n_v} \mathbf{1}_{n_v})^T\right]^T,$$
$$\Pi_{i,k}^{22} = -\operatorname{diag}_{n_v}\left\{\frac{1}{\sigma_{it}^2} P_{1i,k+1}^{-1}\right\}, \quad \Pi_{i,k}^{33} = -\operatorname{diag}_{n_v}\left\{\frac{1}{\bar{\beta}_{1i}\sigma_{it}^2} P_{1i,k+1}^{-1}\right\}.$$

In consideration of $\Delta L_{i,k} = M_{1i,k} F_{1i,k} N_{1i,k}$, one has

$$\Pi_{i,k} = \Pi_{i,k}^0 + \mathscr{M}_{4i,k} F_{1i,k} \mathscr{N}_{4i,k} + \mathscr{N}_{4i,k}^T F_{1i,k}^T \mathscr{M}_{4i,k}^T < 0 \tag{8.52}$$

where

$$\Pi_{i,k}^0 = \begin{bmatrix} \Pi_{i,k}^{11} & * & * \\ \Pi_{i,k}^{21} & \Pi_{i,k}^{22} & * \\ 0 & 0 & \Pi_{i,k}^{33} \end{bmatrix},$$
$$\Upsilon_{4i,k} = \left[(N_{1i,k} D_{i,k} F_1 \mathbf{1}_{n_v})^T \ \cdots \ (N_{1i,k} D_{i,k} F_{n_v} \mathbf{1}_{n_v})^T\right]^T,$$
$$\mathscr{M}_{4i,k} = \left[0 \ 0 \ \Upsilon_{3i,k}\right]^T, \quad \Upsilon_{3i,k} = \operatorname{diag}_{n_v}\left\{M_{1i,k}\right\}, \quad \mathscr{N}_{4i,k} = \left[\Upsilon_{4i,k} \ 0 \ 0\right]. \tag{8.53}$$

According to Lemma 8.2, (8.52) holds if and only if there exists a positive real scalar $\gamma_{1i,k}$ such that the following inequality holds:

$$\Pi_{i,k}^0 + \gamma_{1i,k}^{-1} \mathscr{M}_{4i,k} \mathscr{M}_{4i,k}^T + \gamma_{1i,k} \mathscr{N}_{4i,k}^T \mathscr{N}_{4i,k} < 0, \tag{8.54}$$

which implies that

$$\begin{bmatrix} \tilde{\Pi}_{i,k}^{11} & * & * \\ \Pi_{i,k}^{21} & \Pi_{i,k}^{22} & * \\ 0 & 0 & \Pi_{i,k}^{33} + \gamma_{1i,k}^{-1} \Upsilon_{3i,k} \Upsilon_{3i,k}^T \end{bmatrix} < 0. \tag{8.55}$$

By utilizing the Schur Complement Lemma, one further has

$$\begin{bmatrix} \tilde{\Pi}_{i,k}^{11} & * & * & * \\ \Pi_{i,k}^{21} & \Pi_{i,k}^{22} & * & * \\ 0 & 0 & \Pi_{i,k}^{33} & * \\ 0 & 0 & \Pi_{i,k}^{43} & \tilde{\Pi}_{i,k}^{44} \end{bmatrix} < 0 \tag{8.56}$$

where

$$\Pi_{i,k}^{43} = \operatorname{diag}_{n_v}\left\{M_{1i,k}^T\right\}.$$

Next, perform the congruent transformation

$$\text{diag}\left\{I, \underbrace{P_{1i,k+1}, \ldots, P_{1i,k+1}}_{n_v}, \underbrace{P_{1i,k+1}, \ldots, P_{1i,k+1}}_{n_v}, \underbrace{I, \ldots, I}_{n_v}\right\}$$

to (8.56), then we can obtain (8.43). Therefore, the proof is complete.

According to our observations, the optimization of the upper bound sometimes will cause the unacceptable level of filtering errors, i.e., over-optimization. In order to address this problem, an effective method is to give an additional constraint associated with $\rho_{i,k}$. A commonly used constraint is of the form:

$$\rho_{i,k} \leq \varpi_i \tag{8.57}$$

where ϖ_i is a given positive constant.

As a result, by means of Theorem 8.2, the robust semidefinite optimization problem (8.38) can be finally reformulated as the following semidefinite optimization problem:

$$\begin{aligned} &\min \rho_{i,k} \\ &\text{s.t.} \quad (8.41), (8.42), (8.43), \textit{and}(8.57). \end{aligned} \tag{8.58}$$

Moreover, the filter gains can be calculated from (8.44).

In the end, according to Theorem 8.2 and the optimization problem (8.58), the algorithm executed for node i is summarized in Table 8.1.

Remark 8.1 Actually, the conventional Kalman filtering to address the Gaussian noises has relatively invariant structure including five equations. When encountering some complicated plant and network-induced phenomena, it is difficult for the Kalman filtering method to deal with them, while H_∞ filteringH_∞ filtering method has shown powerful abilities to deal with these challenges from the existing ref-

Table 8.1 Distributed resilient filtering with stochastic perturbations

Filtering Algorithm Executed on Node i	
Step 1	Set the initial states of x_k and $\hat{x}_{i,k}$ $(k = -\tau_M, -\tau_M + 1, \ldots, 0)$, in-degree p_i and out-degree q_i, scalars α_i, weighted matrix $R_{i,k}$ and the lower bound Z_i; determine initial matrices $P_{1i,0}$ and $P_{2i,s}(-\tau_M \leq s \leq -1)$ and set $k = 0$
Step 2	Compute the optimization problem (8.58) to obtain the positive definite matrices $P_{1i,k+1}$ and $P_{2i,k}$, matrices $F_{i,k}$, $E_{ij,k}$ and scalars $\varsigma_{ij,k}(j \in \mathcal{N}_i)$, $\rho_{i,k}$, $\lambda_{i,k}$, $\rho_{i,k}$, $\gamma_{1i,k}$, $\varepsilon_{1i,k}$ and $\varepsilon_{2i,k}$ at time instant k by solving the RLMIs (8.41), (8.42), (8.43), and (8.57)
Step 3	Calculate the parameter matrices $L_{i,k}$, $K_{ij,k}$ of the distributed filter of node i from (8.44), and compute $\hat{x}_{i,k+1}$ according to (8.4)
Step 4	If $k < n$, then return to Step 2, else proceed to Step 5
Step 5	Stop

erences [3–11]. Based on these observations, H_∞ filteringH_∞ filtering method is employed to deal with the stochastic noises in this chapter.

Remark 8.2 By means of the generalized vector dissipativity, a design scheme of resilient filter has been developed in Theorem 8.2 to reduce the impact from stochastic perturbations. In comparison with other ones, the proposed method exposes the following prominent advantages: (1) the desired gain via the solutions of recursive linear matrix inequalities achieves the exponential boundedness, whose bound is further optimized, while the approaches in [7, 12] can guarantee a suboptimal bound of error covariance under the framework of Kalman filtering, and cannot deal with the time-varying time-delayed case; (2) the computational burden is only dependent on the scale of neighboring sensor but not the scale of whole sensor networks and therefore the proposed method is really calculated in a distributed way while the approaches in [6, 13–15] are derived based on the augmented error dynamics and the corresponding condition cannot be simplified; and (3) a novel dissipation matrix with a diagonal structure is creatively proposed to formulate the impact from time-delays although the similar local performance analysis [1] is adopted.

Remark 8.3 This chapter investigates the distributed resilient filtering problem for a class of discrete time-varying nonlinear systems with time-varying delays and stochastic perturbations within the framework of local performance analysis. The more general vector dissipation definition is proposed including multiple energy functions and multiple dissipation matrices to cope with the time-varying delays system. A local optimization problem subject to a set of RLMIs is established for each node in SNs such that the filtering error system satisfies the desirable performance index, where the upper bound is optimized for each node with respect to the corresponding decision variables. Compared with the existing works, our result exhibits the following distinguished features: (1) a new exponential boundedness definition is proposed for the time-varying system; (2) both the stochastic perturbations and the time-varying delays are addressed via the local design method inspired by the extended vector dissipation definition; (3) the upper bound of the performance index is locally optimized for each node subject to linear matrix inequalities.

8.4 An Illustrative Example

In this section, a simulation example is illustrated to verify the effectiveness of the distributed resilient filtering algorithm developed in this chapter.

Consider an SN with 5 nodes, whose topology is represented by a digraph $\mathscr{G} = (\mathscr{V}, \mathscr{E}, \mathcal{A})$ with the set of nodes $\mathscr{V} = \{1, 2, 3, 4, 5\}$ and the set of edges $\mathscr{E} = \{(2, 1), (2, 4), (3, 2), (4, 3), (4, 5), (4, 1)\}$, as shown in Fig. 8.1. Obviously, the in-degree and out-degree of every node can be calculated as $p_1 = p_3 = p_4 = 1$, $p_2 = p_5 = 2$ and $q_1 = 2, q_2 = q_3 = 1, q_4 = 3, q_5 = 0$.

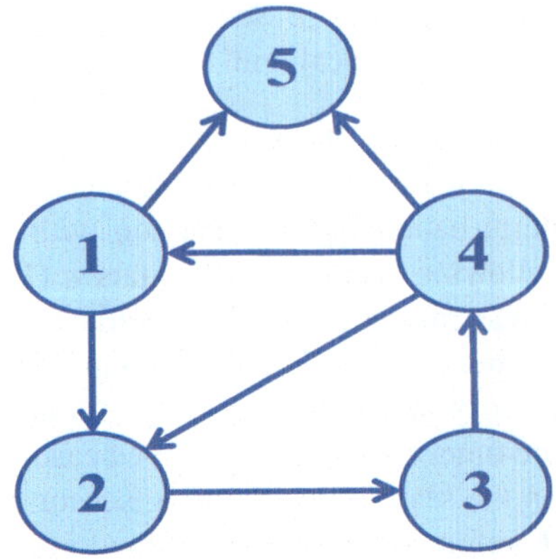

Fig. 8.1 Topology of sensor network

Similar to the application in [16], consider a maneuvering target tracking system where the state vector consists of its position and the velocity. The maneuvering target is tracked through a monitoring system furnished with a set of distributed radars (i.e., sensor nodes) communicating through networks. The dynamics of the maneuvering target could be subject to state-dependent nonlinear disturbances due to sudden environment changes, abnormal power outputs, and so forth. Meanwhile, the desirable filter designed for each node could be subject to the gain perturbations during its implementation.

Selecting the sampling period as T, one has the discrete-time maneuvering target with nonlinearities and time-varying delays, where the corresponding parameters are given as follows:

$$A_k = \begin{bmatrix} 1 & T \\ 0 & 1 \end{bmatrix}, \quad B_k = \begin{bmatrix} \frac{T^2}{2} \\ T \end{bmatrix}.$$

The state is $x_k = [x_k^{(1)} \ x_k^{(2)}]^T$, where $x_k^{(1)}$ and $x_k^{(2)}$ are the position and the velocity at time instant k, respectively. Suppose that the target is not moving too fast, and we take $T = 0.15$ s in the simulation.

The other parameters are set as:

$$\begin{aligned} C_{i,k} &= \begin{bmatrix} 0.8 & 1.2\sin(k) \end{bmatrix}, \quad D_{i,k} = 0.01, \quad f(x_k) = g(x_k), \\ f(x_k) &= \begin{bmatrix} -0.1x_k^{(1)} + 0.2\sin(x_k^{(1)}) + 0.1x_k^{(2)} \\ \tanh(-0.2x_k^{(2)}) \end{bmatrix}, \\ \Delta K_{ij,k} &= \begin{bmatrix} 0.04\sin(k) & 0 \\ 0 & 0 \end{bmatrix}, \quad \Delta L_{i,k} = \begin{bmatrix} 0.04\sin(k) \\ 0 \end{bmatrix}, \end{aligned}$$

$$F_k = \sin(k), \quad N_{1i,k} = 0.2, \quad \tau_k = 2 + \frac{1+(-1)^k}{2},$$

$$M_{1i,k} = M_{2i,k} = \begin{bmatrix} 0.2 \\ 0 \end{bmatrix}, \quad N_{2i,k} = \begin{bmatrix} 0.2 \\ 0 \end{bmatrix}^T.$$

With the parameters set above, it can be easily verified that

$$\Phi_{1i} = \Psi_{1i} = \begin{bmatrix} -0.1 & 0.1 \\ 0 & -0.2 \end{bmatrix}, \quad \Phi_{2i} = \Psi_{2i} = \begin{bmatrix} 0.1 & 0.1 \\ 0 & 0 \end{bmatrix}.$$

The time horizon $n-1$, scalar α_i, probabilities $\bar{\beta}_{1i}$ and $\bar{\beta}_{2i}$, weighted matrix $R_{i,k}$, and initial condition $P_{1i,0}$ and $P_{2i,s}$ $(s = 0, -1, -2, -3)$ are, respectively, given by $n = 51$, $\alpha_i = 0.5$, $\bar{\beta}_{1i} = 0.1$, $\bar{\beta}_{2i} = 0.1$, $R_{i,k} = I$, $P_{1i,0} = 50I$ and $P_{2i,s} = 50I$.

By using the MATLAB software with the YALMIP toolbox, all the desired filter parameters are recursively calculated from the optimization problem (8.58) and Theorem 8.2. Furthermore, the initial states of the plant and all nodes are set as follows: $x_0 = [0.3 \;\; -0.5]^T$, $\hat{x}_{1,0} = [0.4 \;\; -0.6]^T$, $\hat{x}_{2,0} = [0.1 \;\; -0.3]^T$, $\hat{x}_{3,0} = [0.8 \;\; -0.5]^T$, $\hat{x}_{4,0} = [-0.1 \;\; -0.5]^T$ and $\hat{x}_{5,0} = [0.3 \;\; -0.2]^T$. For the two cases of pre and post optimizations, the corresponding simulation results are presented in Figs. 8.2, 8.3, 8.4 and

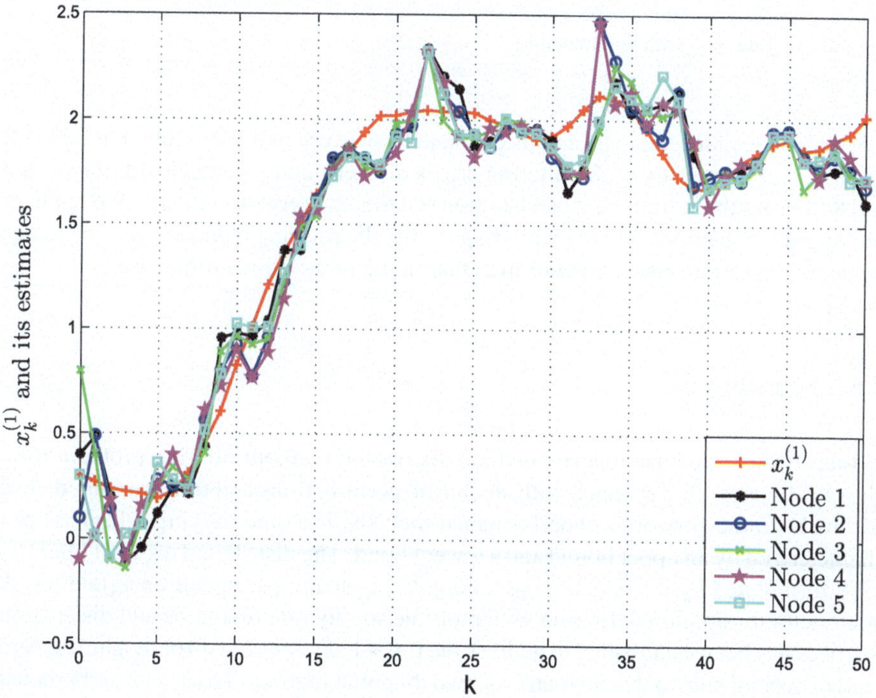

Fig. 8.2 $x_k^{(1)}$ and $\hat{x}_{i,k}^{(1)}$ without optimization

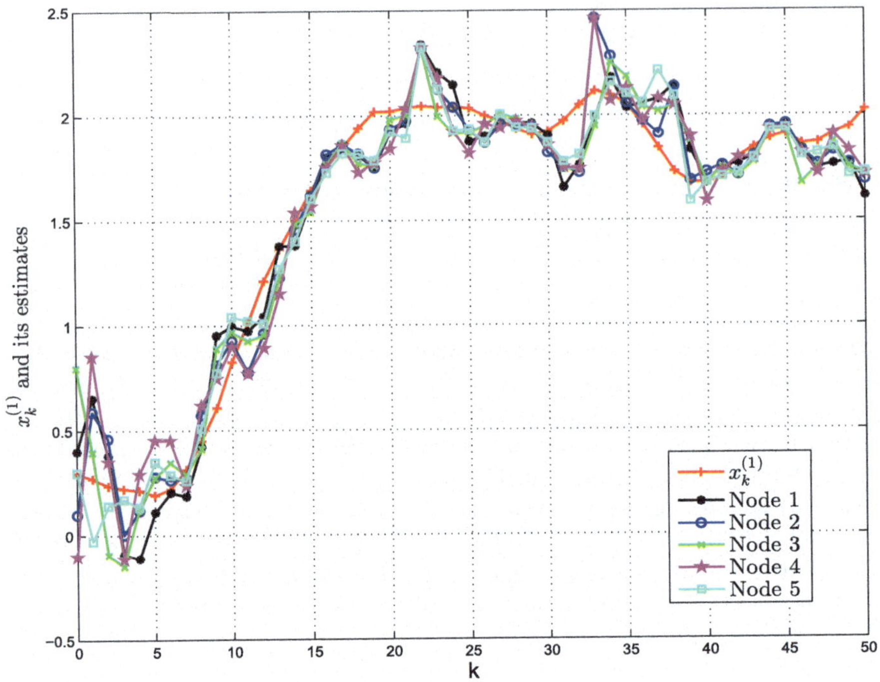

Fig. 8.3 $x_k^{(1)}$ and $\hat{x}_{i,k}^{(1)}$ with optimization

8.5, where the plant state $x_k^{(1)}$ and its estimate from every node are plotted in Figs. 8.2 and 8.3, and the norms of the filtering errors of every node are depicted in Figs. 8.4 and 8.5. From these figures, it can be observed that the proposed distributed resilient filtering scheme performs very well. Especially, the filtering behaviors among nodes in the optimization case are better than that in the non-optimization case.

8.5 Summary

In this chapter, we have investigated the distributed resilient filtering problem for a class of discrete-time systems with stochastic perturbations and time-varying delays within the framework of local performance analysis. The time-varying delay has been characterized by an upper bound and a lower bound. The distributed resilient filter has been formulated by introducing the randomly occurring parameter uncertainties. A new vector dissipation definition with multiple supply rate functions and dissipation matrices has been established to address the time-delay filtering error dynamics. As a kind of special substochastic matrices, two diagonal matrices have been constructed

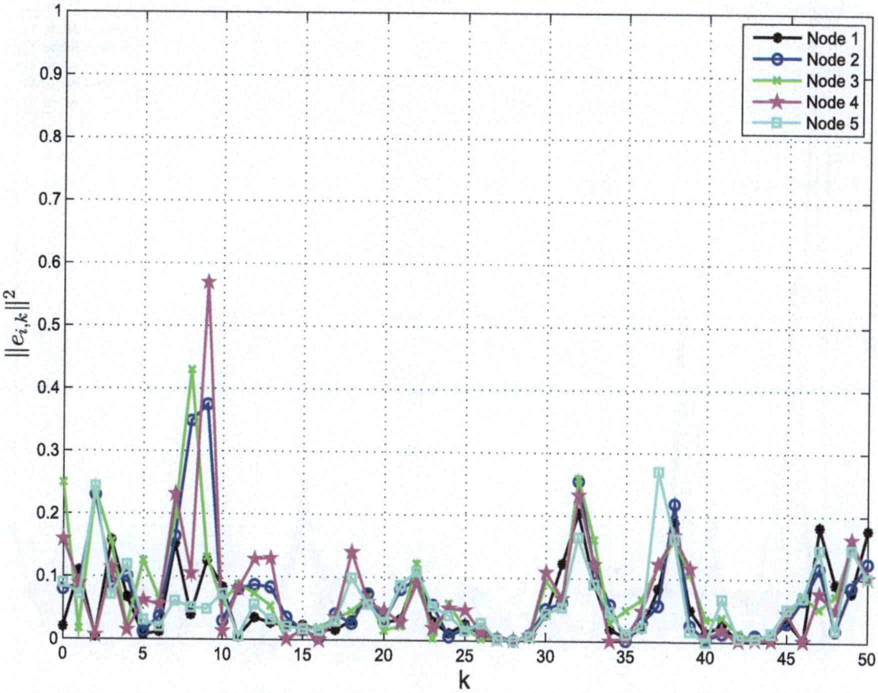

Fig. 8.4 $\|e_{i,k}\|^2$ without optimization

to formulate the impact of time-delay information within the framework of vector dissipativity theory. Through skillfully constructing three dissipation matrices, the desirable vector dissipation has been established for a class of time-delay nonlinear systems by constructing a set of sufficient conditions. Next, the distributed filtering algorithm has been designed to guarantee the desirable exponential boundness. Then, the performance analysis and the optimization problem have been conducted on each node, respectively. Finally, a simulation example has been presented to verify the effectiveness of the developed distributed filtering algorithm.

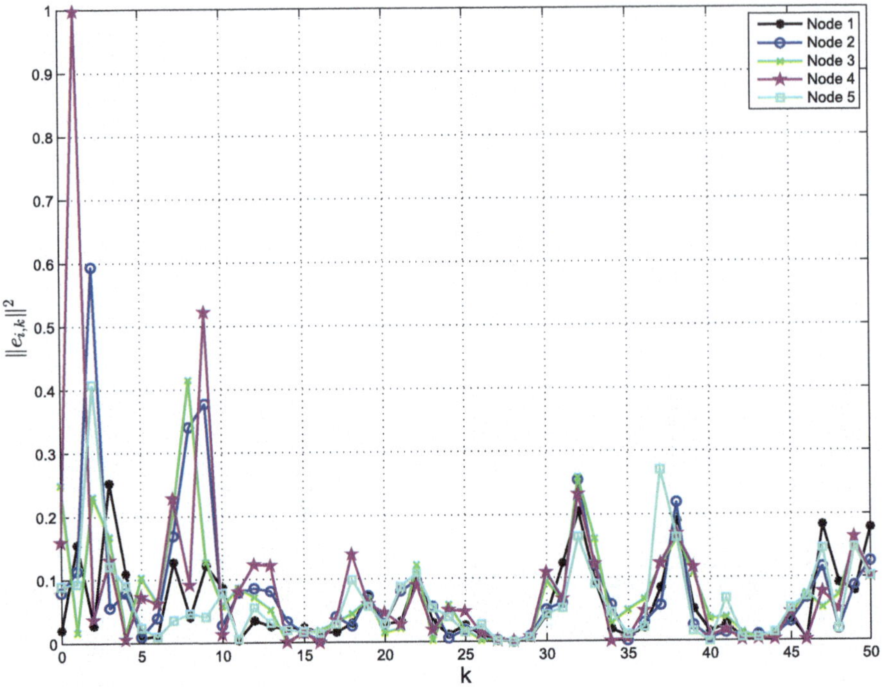

Fig. 8.5 $\|e_{i,k}\|^2$ with optimization

References

1. Han, F., Wei, G., Ding, D., Song, Y.: Local condition-based consensus filtering with stochastic nonlinearities and multiple missing measurements. IEEE Trans. Autom. Control **62**(9), 4784–4790 (2017)
2. Boyd, S., Ghaoui, L., Feron, E., Balakrishnan, V.: Linear Matrix Inequalities in System and Control Theory. SIAM, Philadelphia (1994)
3. Liu, Y., Wang, Z., Liang, J., Liu, X.: Synchronization and state estimation for discrete-time complex networks with distributed delays. IEEE Trans. Syst. Man Cybern. Part B: Cybern. **38**(5), 1314–1325 (2008)
4. Dong, H., Wang, Z., Gao, H.: Distributed H_∞ filtering for a class of Markovian jump nonlinear time-delay systems over lossy sensor networks. IEEE Trans. Ind. Electron. **60**(10), 4665–4672 (2013)
5. Dong, H., Wang, Z., Lam, J., Gao, H.: Distributed filtering in sensor networks with randomly occurring saturations and successive packet dropouts. Int. J. Robust Nonlinear Control **24**(12), 1743–1759 (2014)
6. Liang, J., Wang, Z., Liu, X.: Distributed state estimation for uncertain Markov-type sensor networks with mode-dependent distributed delays. Int. J. Robust Nonlinear Control **22**(3), 331–346 (2012)
7. Wang, F., Wang, Z., Liang, J., Liu, X.: Resilient filtering for linear time-varying repetitive processes under uniform quantizations and Round-Robin protocols. IEEE Trans. Circuits Syst.-I: Regul. Pap. **65**(9), 2992–3004 (2018)

8. Wang, D., Wang, Z., Shen, B., Alsaadi, E.: Security-guaranteed filtering for discrete-time stochastic delayed systems with randomly occurring sensor saturations and deception attacks. Int. J. Robust Nonlinear Control **27**(7), 1194–1208 (2017)
9. Yan, H., Yang, Q., Zhang, H., Yang, F., Zhan, X.: Distributed H_∞ state estimation for a class of filtering networks with time-varying switching topologies and packet losses. IEEE Trans. Syst. Man Cybern.: Syst. **48**(12), 2047–2057 (2018)
10. Yang, G., Che, W.: Non-fragile H_∞ filter design for linear continuous-time systems. Automatica **44**(11), 2849–2856 (2008)
11. Zhu, Y., Zhang, L., Zheng, W.: Distributed H_∞ filtering for a class of discrete-time Markov jump Lur'e systems with redundant channels. IEEE Trans. Ind. Electron. **63**(3), 1876–1885 (2016)
12. Chen, W., Ding, D., Dong, H., Wei, G.: Distributed resilient filtering for power systems subject to denial-of-service attacks. IEEE Trans. Syst. Man Cybern.: Syst. **49**(8), 1688–1697 (2019)
13. Liu, Q., Wang, Z., He, X., Ghinea, G., Alsaadi, E.: A resilient approach to distributed filter design for time-varying systems under stochastic nonlinearities and sensor degradation. IEEE Trans. Signal Proc. **65**(5), 1300–1309 (2017)
14. Zhang, D., Cai, W., Xie, L., Wang, Q.: Nonfragile distributed filtering for T-S fuzzy systems in sensor networks. IEEE Trans. Fuzzy Syst. **23**(5), 1883–1890 (2015)
15. Zhang, D., Xu, Z., Karimi, H., Wang, Q.: Distributed filtering for switched linear systems with sensor networks in presence of packet dropouts and quantization. IEEE Trans. Circuits Syst.-Part I **64**(10), 2783–2796 (2017)
16. Zhang, W., Feng, G., Yu, L.: Multi-rate distributed fusion estimation for sensor networks with packet losses. Automatica **48**, 2016–2028 (2012)

Chapter 9
Finite-Horizon H_∞-Consensus Control for Multi-agent Systems with Random Parameters

The local performance analysis method has been employed to design the distributed filtering schemes in Chaps. 3–8. The typical feature of these distributed schemes is local calculation, which implies that every node implements its own scheme via the available information. Such a design idea efficiently solves the heavy computational burden resulted from the traditional globally augmented method. In the rest part of this book, this method will be applied in multi-agent systems and complex networks.

This chapter is concerned with the distributed H_∞-consensus control problem over the finite horizon for a class of discrete time-varying multi-agent systems with random parameters. First of all, the statistical property of a random matrix is described by the information matrix, which is employed to calculate the weighted covariance matrix of the random matrix. Next, by allowing every agent to track the average value of the neighboring agents, a novel local H_∞-consensus performance constraint is presented to cater to the local performance analysis. Then, by means of the proposed definition of the stochastic vector dissipativity-like over the finite horizon, a set of sufficient conditions is obtained for every agent such that the controlled outputs of the closed-loop multi-agent systems satisfy the proposed H_∞-consensus performance constraint. As a result, the proposed consensus control algorithm can be executed on each agent in an indeed distributed manner. Finally, a simulation example is employed to verify the effectiveness of the proposed algorithm.

9.1 Problem Formulation

First of all, the topology of multi-agent systems is also represented by a digraph, which is the same as the previous chapters. A directed tree is a digraph, where every agent, except the root, has exactly one parent. A spanning tree of a digraph is a

F. Han et al., *Distributed Filtering, Control and Synchronization*,
Studies in Systems, Decision and Control 428,
https://doi.org/10.1007/978-3-030-97075-8_9

directed tree formed by graph edges that connect all the agents of the graph. A graph has a spanning tree if a subset of the edges forms a spanning tree. Throughout this chapter, it is assumed that the digraph $\mathscr{G}$ has a spanning tree and $p_i > 0$ for any $i \in \mathscr{V}$.

Next, consider a multi-agent system consisting N agents, in which the dynamics of agent i is described by the following discrete time-varying stochastic system defined on $k \in \mathscr{K} \triangleq \{0, 1, \ldots, n-1\}$:

$$\begin{cases} x_{i,k+1} = A_{i,k}x_{i,k} + B_{i,k}u_{i,k} + E_{i,k}w_{i,k}, \\ y_{i,k} = \gamma_{i,k}C_{i,k}x_{i,k} + D_{i,k}\vartheta_{i,k}, \\ z_{i,k} = L_{i,k}x_{i,k}, \quad i = 1, 2, \ldots, N \end{cases} \tag{9.1}$$

where $x_{i,k} \in \mathbb{R}^{n_x}$, $y_{i,k} \in \mathbb{R}^{n_y}$, $u_{i,k} \in \mathbb{R}^{n_u}$, and $z_{i,k} \in \mathbb{R}^{n_z}$ are the state, measurement output, control input, and control output of agent i, respectively. $w_{i,k} \in \mathbb{R}^{n_w}$ and $\vartheta_{i,k} \in \mathbb{R}^{n_\vartheta}$ are the external disturbances belonging to $l_2[0, n-1]$. $B_{i,k} \in \mathbb{R}^{n_x \times n_u}$, $D_{i,k} \in \mathbb{R}^{n_y \times n_\vartheta}$, and $E_{i,k} \in \mathbb{R}^{n_x \times n_w}$ are all known deterministic time-varying matrices, wherein $B_{i,k}$ is of full-column rank.

$A_{i,k}$ and $C_{i,k}$ are the time-varying random parameter matrices and mutually independent. Without loss of generality, assume the random matrices $A_{i,k}$ and $C_{i,k}$ have the following structures:

$$A_{i,k} = \bar{A}_{i,k} + \tilde{A}_{i,k}, \quad C_{i,k} = \bar{C}_{i,k} + \tilde{C}_{i,k}. \tag{9.2}$$

Also, their statistical properties are given as follows:

$$\begin{aligned} &\mathbb{E}\{A_{i,k}\} = \bar{A}_{i,k}, \quad \mathbb{E}\left\{\tilde{A}_{i,k}\right\} = \mathbb{E}\{[\tilde{a}^{st}_{i,k}]\} = 0, \\ &\mathbb{E}\{C_{i,k}\} = \bar{C}_{i,k}, \quad \mathbb{E}\left\{\tilde{C}_{i,k}\right\} = \mathbb{E}\{[\tilde{c}^{st}_{i,k}]\} = 0 \end{aligned}$$

where $\tilde{a}^{st}_{i,k}$ and $\tilde{c}^{st}_{i,k}$ are the (s,t)th element of matrix $\mathbb{E}\{\tilde{A}_{i,k}\}$ and $\mathbb{E}\{\tilde{C}_{i,k}\}$, respectively. Furthermore, for simplicity, denote the matrices composed of all covariance elements in $\tilde{A}_{i,k}$ and $\tilde{C}_{i,k}$ as the information matrices $\mathcal{A}_{i,k} \in \mathbb{R}^{n_x n_x \times n_x n_x}$ and $\mathcal{C}_{i,k} \in \mathbb{R}^{n_x n_y \times n_x n_y}$:

$$\begin{aligned} &\mathcal{A}_{i,k} = \left[\mathcal{A}^{sl}_{i,k}\right], \quad \mathcal{A}^{sl}_{i,k} = \mathbb{E}\left\{\left[\tilde{a}^{st}_{i,k}\tilde{a}^{lm}_{i,k}\right]\right\} \triangleq \left[\hat{a}^{stlm}_{i,k}\right], \\ &\mathcal{C}_{i,k} = \left[\mathcal{C}^{sl}_{i,k}\right], \quad \mathcal{C}^{sl}_{i,k} = \mathbb{E}\left\{\left[\tilde{c}^{st}_{i,k}\tilde{c}^{lm}_{i,k}\right]\right\} \triangleq \left[\hat{c}^{stlm}_{i,k}\right] \end{aligned}$$

where $\mathcal{A}^{sl}_{i,k}$ and $\mathcal{C}^{sl}_{i,k}$ respectively are the (s,l)th block matrices of $\mathcal{A}_{i,k}$ and $\mathcal{C}_{i,k}$, $\hat{a}^{stlm}_{i,k}$ and $\hat{c}^{stlm}_{i,k}$ respectively are the (t,m)th entries of the matrix blocks $\mathcal{A}^{sl}_{i,k}$ and $\mathcal{C}^{sl}_{i,k}$.

$\gamma_{i,k} \in \mathbb{R}$ is a Bernoulli distributed white sequence taking values of 1 or 0 with $\text{Prob}\{\gamma_{i,k} = 1\} = \bar{\gamma}_i$ and $\text{Prob}\{\gamma_{i,k} = 0\} = 1 - \bar{\gamma}_i$, which is employed to describe random nature of the missing measurements. Here, $\bar{\gamma}_i \in [0, 1]$ is a known constant. In addition, it is assumed that all random variables $\gamma_{i,k}$ are independent in k and

i. Moreover, random matrices $\tilde{A}_{i,k}$, $\tilde{C}_{i,k}$ and random variable $\gamma_{i,k}$ are also mutually independent.

In this chapter, the following distributed controller for agent i is proposed:

$$u_{i,k} = K_{i,k} \sum_{j \in \mathcal{N}_i} (y_{j,k} - y_{i,k}). \tag{9.3}$$

Letting $\xi_{i,k} \triangleq [w_{i,k}^T \ \vartheta_{i,k}^T \ \vartheta_{j_{i_1},k}^T \ \cdots \ \vartheta_{j_{i p_i},k}^T]^T$, the closed-loop system of agent i is described as follows:

$$\begin{cases} x_{i,k+1} = \left(\bar{A}_{i,k} - \bar{\gamma}_i p_i B_{i,k} K_{i,k} \bar{C}_{i,k}\right) x_{i,k} - \left(\gamma_{i,k} - \bar{\gamma}_i\right) p_i B_{i,k} K_{i,k} \bar{C}_{i,k} x_{i,k} \\ \quad + B_{i,k} K_{i,k} \sum\limits_{j \in \mathcal{N}_i} (\gamma_{j,k} - \bar{\gamma}_j) \bar{C}_{j,k} x_{j,k} + B_{i,k} K_{i,k} \sum\limits_{j \in \mathcal{N}_i} \bar{\gamma}_j \bar{C}_{j,k} x_{j,k} \\ \quad - \gamma_{i,k} p_i B_{i,k} K_{i,k} \tilde{C}_{i,k} x_{i,k} + B_{i,k} K_{i,k} \sum\limits_{j \in \mathcal{N}_i} \gamma_{j,k} \tilde{C}_{j,k} x_{j,k} \\ \quad + \tilde{A}_{i,k} x_{i,k} + F_{i,k} \xi_{i,k}, \\ z_{i,k} = L_{i,k} x_{i,k} \end{cases} \tag{9.4}$$

where

$$F_{i,k} \triangleq \left[E_{i,k} \quad -p_i B_{i,k} K_{i,k} D_{i,k} \quad B_{i,k} K_{i,k} D_{j_{i_1},k} \quad \cdots \quad B_{i,k} K_{i,k} D_{j_{i p_i},k} \right].$$

To begin with, introduce a local average value of the neighbors of agent i, i.e., $z_{i,k}^* \triangleq \frac{1}{p_i} \sum_{j \in \mathcal{N}_i} z_{j,k}$, which is the consensus aim tracked by agent i. The following proposition indicates that if every agent tracks its local average value under certain topology conditions, then it implies the global consensus, i.e., $z_{i,k} = z_{j,k}, \forall i, j \in \mathcal{V}$.

Lemma 9.1 ([1]) *Given a matrix $H = [h_{ij}] \in \mathbb{R}^{N \times N}$, where $h_{ii} < 0$, $h_{ij} \geq 0$, $\forall i \neq j$, and $\sum_{j=1}^{N} h_{ij} = 0$ for each i, then $r(H) = N - 1$ if and only if the digraph associated with H has a spanning tree.*

Proposition 9.1 *For a digraph $\mathcal{G}$ which has a spanning tree and in-degree $p_i > 0$ ($i \in \mathcal{V}$), if $z_{i,k} = z_{i,k}^*$ for each agent i, then $z_{1,k} = z_{2,k} = \cdots = z_{N,k}$.*

Proof From $z_{i,k}^* = \frac{1}{p_i} \sum_{j \in \mathcal{N}_i} z_{j,k}$, one has

$$\begin{cases} z_{1,k} = \frac{1}{p_1} \sum_{j \in \mathcal{N}_1} z_{j,k} \\ \quad \vdots \\ z_{N,k} = \frac{1}{p_N} \sum_{j \in \mathcal{N}_N} z_{j,k} \end{cases} \Leftrightarrow \mathbb{F}_N z_k = 0 \tag{9.5}$$

where

$$\mathbb{F}_N \triangleq \begin{bmatrix} -1 & \frac{a_{12}}{p_1} & \cdots & \frac{a_{1N-1}}{p_1} & \frac{a_{1N}}{p_1} \\ \frac{a_{21}}{p_2} & -1 & \cdots & \frac{a_{2N-1}}{p_2} & \frac{a_{2N}}{p_2} \\ \vdots & \vdots & \ddots & \vdots & \vdots \\ \frac{a_{N-11}}{p_{N-1}} & \frac{a_{N-12}}{p_{N-1}} & \cdots & -1 & \frac{a_{N-1N}}{p_{N-1}} \\ \frac{a_{N1}}{p_N} & \frac{a_{N2}}{p_N} & \cdots & \frac{a_{NN-1}}{p_N} & -1 \end{bmatrix}, \quad z_k = \begin{bmatrix} z_{1,k} \\ z_{2,k} \\ \vdots \\ z_{N-1,k} \\ z_{N,k} \end{bmatrix}.$$

It is easy to verify that $\mathbb{F}_N$ satisfies the conditions of Lemma 9.1. From Lemma 9.1, it follows obviously that $r(\mathbb{F}_N) = N - 1$. Then, the equation group (9.5) has a nonzero solution with one free variable. In addition, it is obviously verified that $z_{1,k} = z_{2,k} = \cdots = z_{N,k}$ is a solution to the equation group (9.5). As a result, $z_{1,k} = z_{2,k} = \cdots = z_{N,k}$ is the nonzero solution to the equation group (9.5). Therefore, the proof is complete.

Now, via the local average value $z^*_{i,k}$, a novel H_∞-consensus performance constraint over the finite horizon is presented.

Definition 9.1 Given a prescribed disturbance attention level $\gamma > 0$, a set of positive definite matrices S_i, $\{Q_{i,k}\}_{k\in\mathscr{K}}$, and $\{T_{i,k}\}_{k\in\mathscr{K}}$. The controlled output $z_{i,k}$ (9.4) is said to satisfy the H_∞-consensus performance constraint over the finite horizon $\mathscr{K}$ if the following inequality holds

$$\mathbb{E}\left\{\sum_{k\in\mathscr{K}} \Phi_{\mathscr{G}}(z_k)\right\} \le \frac{\gamma^2}{N}\sum_{i\in\mathscr{V}} \mathbb{E}\left\{x^T_{i,0}S_i x_{i,0} + \sum_{k\in\mathscr{K}} \|\xi_{i,k}\|^2_{T_{i,k}}\right\} \tag{9.6}$$

where $\Phi_{\mathscr{G}}(z_k) = \frac{1}{N}\sum_{i\in\mathscr{V}} \|z_{i,k} - z^*_{i,k}\|^2_{Q_{i,k}}$ and $T_{i,k} = \mathrm{diag}\left\{T^1_{i,k}, T^2_{i,k}, T^{j_1}_{i,k}, \ldots, T^{j_{p_i}}_{i,k}\right\}$.

Remark 9.1 As is well known, the consensus of all agents implies $z_{i,k} = z_{j,k}$, for all $i, j \in \mathscr{V}$. For the sake of analysis, some specific consensus descriptions have been proposed in [2–16]. The static average consensus has been first presented in [2], where every agent tracks the average of the initial values of all agents, i.e., $z^\diamond_{i,k} = \frac{1}{N}\sum_{j\in\mathscr{V}} z_{j,0}$. Subsequently, the dynamically average consensus has been proposed, where the consensus aim is the state average of all agents at each time in [3], i.e., $z^\star_{i,k} = \frac{1}{N}\sum_{j\in\mathscr{V}} z_{j,k}$, or, the average of all their reference inputs in [12–16], i.e., $r_{i,k} = \frac{1}{N}\sum_{j\in\mathscr{V}} r_{j,k}$, where $r_{j,k}$ is measured by agent j at time k. As a matter of fact, all these existing definitions of consensus imply that every agent can know the information of all the other agents, which violates the basic local communication constraint. To resolve this problem, we put forward a novel consensus in Definition 9.1, where every agent needs to track the average value of the controlled output of its neighboring agents. Therefore, it is no doubt that this consensus is conducive to the following local performance analysis.

In this chapter, the objective is to design a distributed controller of the form (9.3) for the system (9.1). In other words, we are going to design the controller gains $K_{i,k}$ such that the controlled output of the closed-loop multi-agent system $z_{i,k}$ (9.4) satisfies the H_∞-consensus performance constraint (9.6) over the finite-horizon $\mathscr{K}$.

9.2 Analysis of Distributed H_∞-Consensus Control Based on Local Disagreement Function

In this section, by means of the vector dissipativity theory, a set of locally coupled sufficient conditions is obtained to guarantee the controlled output of the multi-agent dynamics satisfies the H_∞-consensus constraint with a prescribed attenuation level γ. Notably, the traditional dissipativity conception cannot be directly applied to the following local performance analysis due to the introduction of random parameters and missing measurements. On this account, the stochastic vector dissipativity-like over the finite horizon for the closed-loop multi-agent dynamics (9.4) is proposed.

From the H_∞-consensus constraint (9.6), denote

$$S_i(z_k, \xi_{i,k}) \triangleq \gamma^2 \|\xi_{i,k}\|^2_{T_{i,k}} - \|z_{i,k} - z^*_{i,k}\|^2_{Q_{i,k}}. \tag{9.7}$$

Definition 9.2 The discrete time-varying closed-loop multi-agent system (9.4) is stochastic vector dissipative-like over the finite horizon concerning a vector of supply rate functions $\mathbf{S}(z_k, \xi_k) \triangleq \begin{bmatrix} S_1(z_k, \xi_{1,k}) & \cdots & S_N(z_k, \xi_{N,k}) \end{bmatrix}^T$, if there exists a vector of nonnegative definite storage functions $\mathbf{V}(x_k) \triangleq \begin{bmatrix} V_1(x_{1,k}) & \cdots & V_N(x_{N,k}) \end{bmatrix}^T$, and a sequence of nonsingular column substochastic dissipation matrices $W_k \in \mathbb{R}^{N\times N}$ such that the following vector dissipation inequality is satisfied for any $k \in \mathscr{K}$:

$$\mathbb{E}\{\mathbf{V}(x_{k+1})\} \leq\leq W_k \mathbb{E}\{\mathbf{V}(x_k)\} + \mathbb{E}\{\mathbf{S}(z_k, \xi_k)\}. \tag{9.8}$$

Remark 9.2 Notably, the differences between new and old definitions concerning the vector dissipativity lie in the following aspects: (1) different vector dissipation inequalities. This new conception presents a vector dissipation inequality in statistic sense for addressing the discrete time-varying stochastic system. (2) different dissipation matrix. The novel time-varying dissipation matrix is formulated by a sequence of nonsingular column substochastic matrices. Nevertheless, the existing dissipation matrix is semistable [17] or Hurwitz [18].

For a random vector x_k, its covariance matrix is defined as $\mathbb{E}\{x_k x_k^T\}$. Inspired by this definition, we also define the covariance matrix for the random matrix $\tilde{A}_{i,k}$ and $\tilde{C}_{i,k}$, i.e., $\mathbb{E}\{\tilde{A}^T_{i,k}\tilde{A}_{i,k}\}$ and $\mathbb{E}\{\tilde{C}^T_{i,k}\tilde{C}_{i,k}\}$. If deterministic weighted matrices $W_k > 0$ and $R_k > 0$ are considered, then weighted covariance matrices (WCMs) associated with $\tilde{A}_{i,k}$ and $\tilde{C}_{i,k}$ can be defined as $\mathbb{E}\{\tilde{A}^T_{i,k} W_k \tilde{A}_{i,k}\}$ and $\mathbb{E}\{\tilde{C}^T_{i,k} R_k \tilde{C}_{i,k}\}$, where $W_k \in$

$\mathbb{R}^{n_x \times n_x}$ and $R_k \in \mathbb{R}^{n_y \times n_y}$. These two WCMs will be encountered in the following local performance analysis, and thus an urgent work is how to calculate $\mathbb{E}\{\tilde{A}_{i,k}^T W_k \tilde{A}_{i,k}\}$ and $\mathbb{E}\{\tilde{C}_{i,k}^T R_k \tilde{C}_{i,k}\}$. Now, two formulas for the WCMs are presented via the Kronecker product and the Hadamard product. Actually, they are simple generalizations of Proposition 4.1 in Chap. 4.

Proposition 9.2 *Given two deterministic weighted positive definite matrices $W_k > 0$ and $R_k > 0$, one has*

$$\mathbb{E}\{\tilde{A}_{i,k}^T W_k \tilde{A}_{i,k}\} = \mathbf{I}_{n_x^2 \times n_x}^T (W_k \otimes \mathbf{1}_{n_x \times n_x} \circ \mathcal{A}_{i,k}) \mathbf{I}_{n_x^2 \times n_x}, \tag{9.9}$$

$$\mathbb{E}\{\tilde{C}_{i,k}^T R_k \tilde{C}_{i,k}\} = \mathbf{I}_{n_x n_y \times n_x}^T (R_k \otimes \mathbf{1}_{n_x \times n_x} \circ \mathcal{C}_{i,k}) \mathbf{I}_{n_x n_y \times n_x} \tag{9.10}$$

where $\mathbf{I}_{n_x^2 \times n_x} \triangleq \mathbf{1}_{n_x \times 1} \otimes I_{n_x}$ and $\mathbf{I}_{n_x n_y \times n_x} \triangleq \mathbf{1}_{n_y \times 1} \otimes I_{n_x}$.

In what follows, via Proposition 9.2, the stochastic vector dissipativity-like for the closed-loop system (9.4) will be established. Let $\mathcal{I}_{q_i}$ be an interval-valued function with respect to out-degree q_i with the form

$$\mathcal{I}_{q_i} = \begin{cases} \left(0, \frac{1+q_i}{2q_i}\right), & \text{if } q_i \neq 0; \\ (0, 1], & \text{if } q_i = 0, \end{cases} \tag{9.11}$$

and denote

$$W_k(\alpha_{1,k}, \ldots, \alpha_{N,k}) \triangleq \begin{bmatrix} \frac{1+q_1(1-\alpha_{1,k})}{1+q_1} & a_{12}\frac{\alpha_{2,k}}{1+q_2} & \cdots & a_{1N}\frac{\alpha_{N,k}}{1+q_N} \\ a_{21}\frac{\alpha_{1,k}}{1+q_1} & \frac{1+q_2(1-\alpha_{2,k})}{1+q_2} & \cdots & a_{2N}\frac{\alpha_{N,k}}{1+q_N} \\ \vdots & \vdots & \ddots & \vdots \\ a_{N1}\frac{\alpha_{1,k}}{1+q_1} & a_{N2}\frac{\alpha_{2,k}}{1+q_2} & \cdots & \frac{1+q_N(1-\alpha_{N,k})}{1+q_N} \end{bmatrix} \tag{9.12}$$

where $\{\alpha_{i,k}\}_{k \in \mathcal{K}} \in \mathcal{I}_{q_i} (i \in \mathcal{V})$. For simplicity, $W_k(\alpha_{1,k}, \ldots, \alpha_{N,k})$ is further denoted as W_k. In addition, the sum of the ith column other than the diagonal element of W_k is $\frac{q_i \alpha_{i,k}}{1+q_i}$, and the corresponding diagonal element is $\frac{1+q_i(1-\alpha_{i,k})}{1+q_i}$. Obviously, $\frac{q_i \alpha_{i,k}}{1+q_i} < \frac{1+q_i(1-\alpha_{i,k})}{1+q_i}$ if and only if $\alpha_{i,k} \in \mathcal{I}_{q_i}$. Therefore, W_k is a strictly diagonally dominant matrix.

For convenience, let $\sigma_i = \sqrt{\bar{\gamma}_i(1-\bar{\gamma}_i)}$.

Lemma 9.2 *Consider any agent i and its neighboring agents $j \in \mathcal{N}_i$. Given the disturbance attenuation level $\gamma > 0$, the scalar sequence $\{\alpha_{i,k}\}_{k \in \mathcal{K}} \in \mathcal{I}_{q_i}$ and $\{\alpha_{j,k}\}_{k \in \mathcal{K}} \in \mathcal{I}_{q_j}$, positive definite weight matrices $\{Q_{i,k}\}_{k \in \mathcal{K}}$, $\{T_{i,k}\}_{k \in \mathcal{K}}$ and S_i and the controller gains $\{K_{i,k}\}_{k \in \mathcal{K}}$. The closed-loop multi-agent system (9.4) is stochastic vector dissipative-like over the finite horizon $\mathcal{K}$ with respect to the vector supply rate function $\mathbf{S}(z_k, \xi_k) = [S_1(z_k, \xi_{1,k}) \;\; \cdots \;\; S_N(z_k, \xi_{N,k})]^T$, the vector storage function $\mathbf{V}(x_k) = [V_1(x_{1,k}) \;\; \cdots \;\; V_N(x_{N,k})]^T$ with $V_i(x_{i,k}) \triangleq x_{i,k}^T P_{i,k} x_{i,k}$, and the dissipation* *matrix sequence* $\{W_k\}_{k\in\mathscr{K}}$*, if there exist three sequences of positive definite matrices* $\{P_{i,k}\}_{k\in\mathscr{K}\cup\{n\}}$*,* $\{P_{j,k}\}_{k\in\mathscr{K}\cup\{n\}}$*, and* $\{\mathscr{B}_{i,k+1}\}_{k\in\mathscr{K}}$ *such that the following conditions hold*

$$\begin{bmatrix} \Psi_{i,k}^{11} & * & * & * & * \\ \Psi_{i,k}^{21} & \Psi_{i,k}^{22} & * & * & * \\ \Psi_{i,k}^{31} & \Psi_{i,k}^{32} & \Psi_{i,k}^{33} & * & * \\ \Psi_{i,k}^{41} & 0 & 0 & -P_{i,k+1}^{-1} & * \\ 0 & \Psi_{i,k}^{52} & 0 & 0 & \Psi_{i,k}^{55} \end{bmatrix} \leq 0, \tag{9.13}$$

$$K_{i,k}^T B_{i,k}^T P_{i,k+1} B_{i,k} K_{i,k} \leq \mathscr{B}_{i,k+1} \tag{9.14}$$

where

$$\begin{aligned}\Psi_{i,k}^{11} =& -\frac{1+q_i(1-\alpha_{i,k})}{1+q_i}P_{i,k} + \bar{\gamma}_i p_i^2 \mathbf{I}_{n_y n_x \times n_x}^T(\mathscr{B}_{i,k+1}\otimes \mathbf{1}_{n_x\times n_x}\circ \mathfrak{C}_{i,k})\mathbf{I}_{n_y n_x\times n_x} \\ &+ \mathbf{I}_{n_x^2\times n_x}^T(P_{i,k+1}\otimes \mathbf{1}_{n_x\times n_x}\circ \mathcal{A}_{i,k})\mathbf{I}_{n_x^2\times n_x} + L_{i,k}^T Q_{i,k} L_{i,k},\end{aligned}$$

$$\Psi_{i,k}^{21} = \left[-\frac{1}{p_i}L_{i,k}^T Q_{i,k} L_{j_{i_1},k} \ \cdots \ -\frac{1}{p_i}L_{i,k}^T Q_{i,k} L_{j_{ip_i},k}\right]^T, \quad \Psi_{i,k}^{41} = \sigma_i p_i B_{i,k} K_{i,k} \bar{C}_{i,k},$$

$$\begin{aligned}\Psi_{i,k}^{22} = \mathrm{diag}\Bigg\{&-\frac{\alpha_{j_{i_1},k}}{1+q_{j_{i_1}}}P_{j_{i_1},k} + \bar{\gamma}_{j_{i_1}}\mathbf{I}_{n_y n_x\times n_x}^T(\mathscr{B}_{i,k+1}\otimes \mathbf{1}_{n_x\times n_x}\circ \mathfrak{C}_{j_{i_1},k})\mathbf{I}_{n_y n_x\times n_x}, \ldots, \\ &-\frac{\alpha_{j_{ip_i},k}}{1+q_{j_{ip_i}}}P_{j_{ip_i},k} + \bar{\gamma}_{j_{ip_i}}\mathbf{I}_{n_y n_x\times n_x}^T(\mathscr{B}_{i,k+1}\otimes \mathbf{1}_{n_x\times n_x}\circ \mathfrak{C}_{j_{ip_i},k})\mathbf{I}_{n_y n_x\times n_x}\Bigg\} + \Upsilon_{i,k},\end{aligned}$$

$$\Upsilon_{i,k} = \frac{1}{p_i^2}\left[L_{j_{i_1},k} \ \cdots \ L_{j_{ip_i},k}\right]^T Q_{i,k}\left[L_{j_{i_1},k} \ \cdots \ L_{j_{ip_i},k}\right], \quad \Psi_{i,k}^{31} = \bar{A}_{i,k} - \bar{\gamma}_i p_i B_{i,k} K_{i,k} \bar{C}_{i,k},$$

$$\Psi_{i,k}^{32} = \left[\bar{\gamma}_{j_{i_1}} B_{i,k} K_{i,k} \bar{C}_{j_{i_1},k} \ \cdots \ \bar{\gamma}_{j_{ip_i}} B_{i,k} K_{i,k} \bar{C}_{j_{ip_i},k}\right], \quad \Psi_{i,k}^{33} = -P_{i,k+1}^{-1} + \frac{1}{\gamma^2}F_{i,k} T_{i,k}^{-1} F_{i,k}^T,$$

$$\Psi_{i,k}^{52} = \mathrm{diag}\left\{\sigma_{j_{i_1}} B_{i,k} K_{i,k} \bar{C}_{j_{i_1},k}, \ldots, \sigma_{j_{ip_i}} B_{i,k} K_{i,k} \bar{C}_{j_{ip_i},k}\right\}, \quad \Psi_{i,k}^{55} = \mathrm{diag}_{p_i}\left\{-P_{i,k+1}^{-1}\right\}.$$

Proof First of all, calculating $V_i(x_{i,k+1})$ along the trajectory of system (9.4) and taking the conditional mathematical expectation, it obtains

$$
\begin{aligned}
&\mathbb{E}\left\{V_i(x_{i,k+1})|x_{i,k}, x_{j,k}, j \in \mathcal{N}_i\right\} \\
&= \mathbb{E}\left\{x_{i,k+1}^T P_{i,k+1} x_{i,k+1} | x_{i,k}, x_{j,k}, j \in \mathcal{N}_i\right\} \\
&= x_{i,k}^T \mathbb{E}\left\{\tilde{A}_{i,k}^T P_{i,k+1} \tilde{A}_{i,k}\right\} x_{i,k} \\
&\quad + \xi_{i,k}^T (F_{i,k}^T P_{i,k+1} F_{i,k} - \gamma^2 T_{i,k}) \xi_{i,k} \\
&\quad + 2\xi_{i,k}^T F_{i,k}^T P_{i,k+1} B_{i,k} K_{i,k} \sum_{j \in \mathcal{N}_i} \bar{\gamma}_j \bar{C}_{j,k} x_{j,k} \\
&\quad + \sigma_i^2 p_i^2 x_{i,k}^T \bar{C}_{i,k}^T K_{i,k}^T B_{i,k}^T P_{i,k+1} B_{i,k} K_{i,k} \bar{C}_{i,k} x_{i,k} \\
&\quad + 2\xi_{i,k}^T F_{i,k}^T P_{i,k+1} (\bar{A}_{i,k} - \bar{\gamma}_i p_i B_{i,k} K_{i,k} \bar{C}_{i,k}) x_{i,k} \\
&\quad + \bar{\gamma}_i p_i^2 x_{i,k}^T \mathbb{E}\{\tilde{C}_{i,k}^T K_{i,k}^T B_{i,k}^T P_{i,k+1} B_{i,k} K_{i,k} \tilde{C}_{i,k}\} x_{i,k} \\
&\quad + \sum_{j \in \mathcal{N}_i} \bar{\gamma}_j x_{j,k}^T \mathbb{E}\{\tilde{C}_{j,k}^T K_{i,k}^T B_{i,k}^T P_{i,k+1} B_{i,k} K_{i,k} \tilde{C}_{j,k}\} x_{j,k} \\
&\quad + \sum_{j \in \mathcal{N}_i} (\bar{\gamma}_j x_{j,k}^T \bar{C}_{j,k}^T K_{i,k}^T B_{i,k}^T) P_{i,k+1} B_{i,k} K_{i,k} \sum_{j \in \mathcal{N}_i} \bar{\gamma}_j \bar{C}_{j,k} x_{j,k} \\
&\quad + 2 \sum_{j \in \mathcal{N}_i} (\bar{\gamma}_j x_{j,k}^T \bar{C}_{j,k}^T K_{i,k}^T B_{i,k}^T) P_{i,k+1} (\bar{A}_{i,k} - \bar{\gamma}_i p_i B_{i,k} K_{i,k} \bar{C}_{i,k}) x_{i,k} \\
&\quad + \gamma^2 \xi_{i,k}^T T_{i,k} \xi_{i,k} + \sum_{j \in \mathcal{N}_i} \sigma_j^2 x_{j,k}^T \bar{C}_{j,k}^T K_{i,k}^T B_{i,k}^T P_{i,k+1} B_{i,k} K_{i,k} \bar{C}_{j,k} x_{j,k} \\
&\quad + x_{i,k}^T (\bar{A}_{i,k} - \bar{\gamma}_i p_i B_{i,k} K_{i,k} \bar{C}_{i,k})^T P_{i,k+1} (\bar{A}_{i,k} - \bar{\gamma}_i p_i B_{i,k} K_{i,k} \bar{C}_{i,k}) x_{i,k}. \qquad (9.15)
\end{aligned}
$$

From Proposition 9.2 and (9.14), it follows that

$$
x_{i,k}^T \mathbb{E}\{\tilde{A}_{i,k}^T P_{i,k+1} \tilde{A}_{i,k}\} x_{i,k} = x_{i,k}^T \mathbf{I}_{n_x^2 \times n_x}^T (P_{i,k+1} \otimes \mathbf{1}_{n_x \times n_x} \circ \mathcal{A}_{i,k}) \mathbf{I}_{n_x^2 \times n_x} x_{i,k}. \qquad (9.16)
$$

$$
\begin{aligned}
&x_{i,k}^T \mathbb{E}\{\tilde{C}_{i,k}^T K_{i,k}^T B_{i,k}^T P_{i,k+1} B_{i,k} K_{i,k} \tilde{C}_{i,k}\} x_{i,k} \\
&= x_{i,k}^T \mathbf{I}_{n_y n_x \times n_x}^T ((K_{i,k}^T B_{i,k}^T P_{i,k+1} B_{i,k} K_{i,k}) \otimes \mathbf{1}_{n_x \times n_x} \circ \mathcal{C}_{i,k}) \mathbf{I}_{n_y n_x \times n_x} x_{i,k} \qquad (9.17) \\
&\le x_{i,k}^T \mathbf{I}_{n_y n_x \times n_x}^T (\mathcal{B}_{i,k+1} \otimes \mathbf{1}_{n_x \times n_x} \circ \mathcal{C}_{i,k}) \mathbf{I}_{n_y n_x \times n_x} x_{i,k}.
\end{aligned}
$$

$$
\begin{aligned}
&x_{j,k}^T \mathbb{E}\{\tilde{C}_{j,k}^T K_{i,k}^T B_{i,k}^T P_{i,k+1} B_{i,k} K_{i,k} \tilde{C}_{j,k}\} x_{j,k} \\
&= x_{j,k}^T \mathbf{I}_{n_y n_x \times n_x}^T ((K_{i,k}^T B_{i,k}^T P_{i,k+1} B_{i,k} K_{i,k}) \otimes \mathbf{1}_{n_x \times n_x} \circ \mathcal{C}_{j,k}) \mathbf{I}_{n_y n_x \times n_x} x_{j,k} \qquad (9.18) \\
&\le x_{j,k}^T \mathbf{I}_{n_y n_x \times n_x}^T (\mathcal{B}_{i,k+1} \otimes \mathbf{1}_{n_x \times n_x} \circ \mathcal{C}_{j,k}) \mathbf{I}_{n_y n_x \times n_x} x_{j,k}.
\end{aligned}
$$

Now, one needs to deal with three terms with respect to $\xi_{i,k}$ in (9.15) other than $\gamma^2 \xi_{i,k}^T T_{i,k} \xi_{i,k}$. Above all, from $\Psi_{i,k}^{33} = -P_{i,k+1}^{-1} + \frac{1}{\gamma^2} F_{i,k} T_{i,k}^{-1} F_{i,k}^T < 0$, it follows that $\Lambda_{i,k} \triangleq \gamma^2 T_{i,k} - F_{i,k}^T P_{i,k+1} F_{i,k} > 0$. Next, through the completing-square technique, one has $(\xi_{i,k} - \xi_{i,k}^*)^T \Lambda_{i,k} (\xi_{i,k} - \xi_{i,k}^*) \ge 0$, where

$$
\xi_{i,k}^* \triangleq \Lambda_{i,k}^{-1} F_{i,k}^T P_{i,k+1} (\bar{A}_{i,k} - \bar{\gamma}_i p_i B_{i,k} K_{i,k} \bar{C}_{i,k}) x_{i,k} + \Lambda_{i,k}^{-1} F_{i,k}^T P_{i,k+1} B_{i,k} K_{i,k} \sum_{j \in \mathcal{N}_i} \bar{\gamma}_j \bar{C}_{j,k} x_{j,k}.
$$

Actually, $\xi^*_{i,k}$ is called the worst disturbance. Then, $(\xi_{i,k} - \xi^*_{i,k})^T \Lambda_{i,k}(\xi_{i,k} - \xi^*_{i,k}) \geq 0$ can become

$$\begin{aligned}
&\xi^T_{i,k}(F^T_{i,k}P_{i,k+1}F_{i,k} - \gamma^2 T_{i,k})\xi_{i,k} \\
&+ 2\xi^T_{i,k}F^T_{i,k}P_{i,k+1}B_{i,k}K_{i,k}\sum_{j\in\mathscr{N}_i}\bar{\gamma}_j\bar{C}_{j,k}x_{j,k} \\
&+ 2\xi^T_{i,k}F^T_{i,k}P_{i,k+1}(\bar{A}_{i,k} - \bar{\gamma}_i p_i B_{i,k}K_{i,k}\bar{C}_{i,k})x_{i,k} \\
\leq& \sum_{j\in\mathscr{N}_i}\bar{\gamma}_j x^T_{j,k}\bar{C}^T_{j,k}K^T_{i,k}B^T_{i,k}P_{i,k+1}F_{i,k}\Lambda^{-1}_{i,k}F^T_{i,k}P_{i,k+1}B_{i,k}K_{i,k}\sum_{j\in\mathscr{N}_i}\bar{\gamma}_j\bar{C}_{j,k}x_{j,k} \\
&+ 2\sum_{j\in\mathscr{N}_i}\bar{\gamma}_j x^T_{j,k}\bar{C}^T_{j,k}K^T_{i,k}B^T_{i,k}P_{i,k+1}F_{i,k}\Lambda^{-1}_{i,k}F^T_{i,k}P_{i,k+1}(\bar{A}_{i,k} - \bar{\gamma}_i p_i B_{i,k}K_{i,k}\bar{C}_{i,k})x_{i,k} \\
&+ x^T_{i,k}(\bar{A}_{i,k} - \bar{\gamma}_i p_i B_{i,k}K_{i,k}\bar{C}_{i,k})^T P_{i,k+1}F_{i,k}\Lambda^{-1}_{i,k}F^T_{i,k}P_{i,k+1}(\bar{A}_{i,k} - \bar{\gamma}_i p_i B_{i,k}K_{i,k}\bar{C}_{i,k})x_{i,k}.
\end{aligned} \tag{9.19}$$

Then, from (9.16)–(9.19), (9.15) becomes

$$\begin{aligned}
&\mathbb{E}\left\{V_i(x_{i,k+1})|x_{i,k}, x_{j,k}, j \in \mathscr{N}_i\right\} \\
\leq& x^T_{i,k}\mathbf{I}^T_{n_x^2\times n_x}(P_{i,k+1}\otimes \mathbf{1}_{n_x\times n_x} \circ \mathcal{A}_{i,k})\mathbf{I}_{n_x^2\times n_x}x_{i,k} \\
&+ \sigma_i^2 p_i^2 x^T_{i,k}\bar{C}^T_{i,k}K^T_{i,k}B^T_{i,k}P_{i,k+1}B_{i,k}K_{i,k}\bar{C}_{i,k}x_{i,k} \\
&+ \bar{\gamma}_i p_i^2 x^T_{i,k}\mathbf{I}^T_{n_y n_x\times n_x}(\mathscr{B}_{i,k+1}\otimes \mathbf{1}_{n_x\times n_x} \circ \mathfrak{C}_{i,k})\mathbf{I}_{n_y n_x\times n_x}x_{i,k} \\
&+ \sum_{j\in\mathscr{N}_i}\bar{\gamma}_j x^T_{j,k}\mathbf{I}^T_{n_y n_x\times n_x}(\mathscr{B}_{i,k+1}\otimes \mathbf{1}_{n_x\times n_x} \circ \mathfrak{C}_{j,k})\mathbf{I}_{n_y n_x\times n_x}x_{j,k} \\
&+ \sum_{j\in\mathscr{N}_i}\bar{\gamma}_j x^T_{j,k}\bar{C}^T_{j,k}K^T_{i,k}B^T_{i,k}\bar{\Gamma}_{i,k}B_{i,k}K_{i,k}\sum_{j\in\mathscr{N}_i}\bar{\gamma}_j\bar{C}_{j,k}x_{j,k} \\
&+ 2\sum_{j\in\mathscr{N}_i}\bar{\gamma}_j x^T_{j,k}\bar{C}^T_{j,k}K^T_{i,k}B^T_{i,k}\bar{\Gamma}_{i,k}(\bar{A}_{i,k} - \bar{\gamma}_i p_i B_{i,k}K_{i,k}\bar{C}_{i,k})x_{i,k} \\
&+ x^T_{i,k}(\bar{A}_{i,k} - \bar{\gamma}_i p_i B_{i,k}K_{i,k}\bar{C}_{i,k})^T\bar{\Gamma}_{i,k}(\bar{A}_{i,k} - \bar{\gamma}_i p_i B_{i,k}K_{i,k}\bar{C}_{i,k})x_{i,k} \\
&+ \gamma^2\xi^T_{i,k}T_{i,k}\xi_{i,k} + \sum_{j\in\mathscr{N}_i}\bar{\gamma}_j(1-\bar{\gamma}_j)x^T_{j,k}\bar{C}^T_{j,k}K^T_{i,k}B^T_{i,k}P_{i,k+1}B_{i,k}K_{i,k}\bar{C}_{j,k}x_{j,k}.
\end{aligned} \tag{9.20}$$

where

$$P_{i,k+1}F_{i,k}\Lambda^{-1}_{i,k}F^T_{i,k}P_{i,k+1} + P_{i,k+1} \triangleq \bar{\Gamma}_{i,k}.$$

Via the matrix inverse formula, one has

$$P_{i,k+1}F_{i,k}\Lambda^{-1}_{i,k}F^T_{i,k}P_{i,k+1} + P_{i,k+1} = \left(P^{-1}_{i,k+1} - \frac{1}{\gamma^2}F_{i,k}T^{-1}_{i,k}F^T_{i,k}\right)^{-1} \triangleq \Pi^{-1}_{i,k}.$$

Then, (9.20) can be simplified to

$$\begin{aligned}
&\mathbb{E}\{V_i(x_{i,k+1})|x_{i,k}, x_{j,k}, j \in \mathcal{N}_i\} \\
&\leq x_{i,k}^T \mathbf{I}_{n_x^2 \times n_x}^T (P_{i,k+1} \otimes \mathbf{1}_{n_x \times n_x} \circ \mathcal{A}_{i,k}) \mathbf{I}_{n_x^2 \times n_x} x_{i,k} \\
&\quad + \bar{\gamma}_i p_i^2 x_{i,k}^T \mathbf{I}_{n_y n_x \times n_x}^T (\mathscr{B}_{i,k+1} \otimes \mathbf{1}_{n_x \times n_x} \circ \mathcal{C}_{i,k}) \mathbf{I}_{n_y n_x \times n_x} x_{i,k} \\
&\quad + \sum_{j \in \mathcal{N}_i} \bar{\gamma}_j x_{j,k}^T \mathbf{I}_{n_y n_x \times n_x}^T (\mathscr{B}_{i,k+1} \otimes \mathbf{1}_{n_x \times n_x} \circ \mathcal{C}_{j,k}) \mathbf{I}_{n_y n_x \times n_x} x_{j,k} \\
&\quad + \sum_{j \in \mathcal{N}_i} \bar{\gamma}_j x_{j,k}^T \bar{C}_{j,k}^T K_{i,k}^T B_{i,k}^T \Pi_{i,k}^{-1} B_{i,k} K_{i,k} \sum_{j \in \mathcal{N}_i} \bar{\gamma}_j \bar{C}_{j,k} x_{j,k} \\
&\quad + 2 \sum_{j \in \mathcal{N}_i} \bar{\gamma}_j x_{j,k}^T \bar{C}_{j,k}^T K_{i,k}^T B_{i,k}^T \Pi_{i,k}^{-1} (\bar{A}_{i,k} - \bar{\gamma}_i p_i B_{i,k} K_{i,k} \bar{C}_{i,k}) x_{i,k} \\
&\quad + x_{i,k}^T (\bar{A}_{i,k} - \bar{\gamma}_i p_i B_{i,k} K_{i,k} \bar{C}_{i,k})^T \Pi_{i,k}^{-1} (\bar{A}_{i,k} - \bar{\gamma}_i p_i B_{i,k} K_{i,k} \bar{C}_{i,k}) x_{i,k} \\
&\quad + \sigma_i^2 p_i^2 x_{i,k}^T \bar{C}_{i,k}^T K_{i,k}^T B_{i,k}^T P_{i,k+1} B_{i,k} K_{i,k} \bar{C}_{i,k} x_{i,k} \\
&\quad + \gamma^2 \xi_{i,k}^T T_{i,k} \xi_{i,k} + \sum_{j \in \mathcal{N}_i} \sigma_j^2 x_{j,k}^T \bar{C}_{j,k}^T K_{i,k}^T B_{i,k}^T P_{i,k+1} B_{i,k} K_{i,k} \bar{C}_{j,k} x_{j,k} \\
&\triangleq \Gamma_{i,k}.
\end{aligned} \tag{9.21}$$

Next, in view of the Schur Complement Lemma, it follows from (9.13) that

$$\begin{aligned}
&\Gamma_{i,k} - S_i(z_k, \xi_{i,k}) - \sum_{j \in \mathcal{N}_i} \frac{\alpha_{j,k}}{1+q_j} x_{j,k}^T P_{j,k} x_{j,k} - \frac{1+q_i(1-\alpha_{i,k})}{1+q_i} x_{i,k}^T P_{i,k} x_{i,k} \\
&= \eta_{i,k}^T \Bigg(\begin{bmatrix} \Psi_{i,k}^{11} & * \\ \Psi_{i,k}^{21} & \Psi_{i,k}^{22} \end{bmatrix} - \begin{bmatrix} (\Psi_{i,k}^{31})^T \\ (\Psi_{i,k}^{32})^T \end{bmatrix} (\Psi_{i,k}^{33})^{-1} \begin{bmatrix} (\Psi_{i,k}^{31})^T \\ (\Psi_{i,k}^{32})^T \end{bmatrix}^T \\
&\quad + \begin{bmatrix} \Psi_{i,k}^{41} & 0 \end{bmatrix}^T P_{i,k+1} \begin{bmatrix} \Psi_{i,k}^{41} & 0 \end{bmatrix} - \begin{bmatrix} 0 & \Psi_{i,k}^{52} \end{bmatrix}^T (\Psi_{i,k}^{55})^{-1} \begin{bmatrix} 0 & \Psi_{i,k}^{52} \end{bmatrix} \Bigg) \eta_{i,k} \\
&\leq 0
\end{aligned} \tag{9.22}$$

where $\eta_{i,k} \triangleq \begin{bmatrix} x_{i,k}^T & x_{\mathcal{N}_i,k}^T \end{bmatrix}^T$ and $x_{\mathcal{N}_i,k} \triangleq \begin{bmatrix} x_{j_{i_1},k}^T & \cdots & x_{j_{i p_i},k}^T \end{bmatrix}^T$.

According to (9.22), we then have

$$\begin{aligned}
&\mathbb{E}\{V_i(x_{i,k+1})|x_{i,k}, x_{j,k}, j \in \mathcal{N}_i\} \\
&\leq S_i(z_k, \xi_{i,k}) + \sum_{j \in \mathcal{N}_i} \frac{\alpha_{j,k}}{1+q_j} x_{j,k}^T P_{j,k} x_{j,k} + \frac{1+q_i(1-\alpha_{i,k})}{1+q_i} x_{i,k}^T P_{i,k} x_{i,k} \\
&= S_i(z_k, \xi_{i,k}) + \begin{bmatrix} \frac{a_{i1}\alpha_{1,k}}{1+q_1} & \cdots & \frac{1+q_i(1-\alpha_{i,k})}{1+q_i} & \cdots & \frac{a_{iN}\alpha_{N,k}}{1+q_N} \end{bmatrix} \mathbf{V}(x_k) \\
&\triangleq S_i(z_k, \xi_{i,k}) + W_k^i \mathbf{V}(x_k)
\end{aligned} \tag{9.23}$$

where $W_k^i \triangleq \begin{bmatrix} \frac{a_{i1}\alpha_{1,k}}{1+q_1} & \cdots & \frac{1+q_i(1-\alpha_{i,k})}{1+q_i} & \cdots & \frac{a_{iN}\alpha_{N,k}}{1+q_N} \end{bmatrix}$ is the ith row of matrix W_k. It is not difficult to see W_k is a column substochastic matrix due to $\mathbf{1}^T W_k = \mathbf{1}^T$. And, it is also a nonnegative column strictly diagonally dominant matrix. As a result, W_k is the desired dissipation matrix. Clearly, one has

$$\mathbb{E}\left\{V_i(x_{i,k+1})|x_{i,k}, x_{j,k}, j \in \mathcal{N}_i\right\} \leq S_i(z_k, \xi_{i,k}) + [W_k \mathbf{V}(x_k)]_i \tag{9.24}$$

where $[W_k V(x_k)]_i$ is the ith element of vector $W_k \mathbf{V}(x_k)$. Furthermore, an augmented form of the above inequality is obtained as follows:

$$\mathbb{E}\left\{\mathbf{V}(x_{k+1})\right\} \leq\leq W_k \mathbb{E}\left\{\mathbf{V}(x_k)\right\} + \mathbb{E}\left\{\mathbf{S}(z_k, \xi_k)\right\}. \tag{9.25}$$

Finally, from Definition 9.2, the proof is complete.

So far, a set of locally coupled analysis conditions has been obtained for each agent to guarantee the vector dissipativity-like of the closed-loop multi-agent systems. Next, the following theorem is to obtain the desirable H_∞-consensus performance constraint (9.6) by utilizing the obtained vector dissipation inequality (9.25).

Theorem 9.1 *Consider any agent i and its neighboring agents $j \in \mathcal{N}_i$. Given the disturbance attenuation level $\gamma > 0$, the scalar sequences $\{\alpha_{i,k}\}_{k\in\mathcal{K}} \in \mathfrak{I}_{q_i}$, $\{\alpha_{j,k}\}_{k\in\mathcal{K}} \in \mathfrak{I}_{q_j}$, positive definite weighted matrices $\{Q_{i,k}\}_{k\in\mathcal{K}}$, $\{T_{i,k}\}_{k\in\mathcal{K}}$, S_i and the controller gains $\{K_{i,k}\}_{k\in\mathcal{K}}$, the closed-loop multi-agent system (9.4) satisfies the H_∞-consensus performance constraint (9.6), if there exist positive definite matrix sequences $\{P_{i,k}\}_{k\in\mathcal{K}\cup\{n\}}$, $\{P_{j,k}\}_{k\in\mathcal{K}\cup\{n\}}$ with the initial conditions $P_{i,0} \leq \gamma^2 S_i$ and $P_{j,0} \leq \gamma^2 S_j$, and $\{\mathcal{B}_{i,k+1}\}_{k\in\mathcal{K}} > 0$ for any $k \in \mathcal{K}$ such that (9.13) and (9.14) hold.*

Proof First of all, pre-multiplying on both sides of the vector dissipation inequality (9.25) by $\mathbf{1}^T$ yields

$$\mathbf{1}^T \mathbb{E}\{\mathbf{V}(x_{k+1})\} \leq \mathbf{1}^T W_k \mathbb{E}\{\mathbf{V}(x_k)\} + \mathbf{1}^T \mathbb{E}\{\mathbf{S}(z_k, \xi_k)\}.$$

Denote $v(x_k) \triangleq \mathbf{1}^T \mathbb{E}\{\mathbf{V}(x_k)\}$. Then, from the definition of the dissipation matrix W_k and (9.15), it is obvious that

$$v(x_{k+1}) - v(x_k) + N\mathbb{E}\{\Phi_{\mathcal{G}}(z_k)\} \leq \sum_{i=1}^{N} \gamma^2 \xi_{i,k}^T T_{i,k} \xi_{i,k},$$

which implies

$$v(x_n) - v(x_0) + \sum_{k=0}^{n-1} N\mathbb{E}\{\Phi_{\mathcal{G}}(z_k)\} \leq \sum_{k=0}^{n-1} \sum_{i=1}^{N} \gamma^2 \xi_{i,k}^T T_{i,k} \xi_{i,k}.$$

That is,

$$\sum_{k=0}^{n-1} \mathbb{E}\{\Phi_{\mathscr{G}}(z_k)\} \leq \frac{1}{N}\sum_{i=1}^{N}\left(x_{i,0}^T P_{i,0} x_{i,0} + \gamma^2 \sum_{k=0}^{n-1} \|\xi_{i,k}\|^2_{T_{i,k}}\right).$$

Finally, according to the initial condition $P_{i,0} \leq \gamma^2 S_i$, the H_∞-consensus performance constraint (9.6) is satisfied. From Definition 9.1, the proof is complete.

Now, the local performance analysis of the closed-loop system (9.4) has been completed. Namely, with the aid of the vector dissipativity-like over the finite horizon, a set of desirable local design conditions (9.13)–(9.14) for each agent are obtained such that all agent dynamics satisfy the H_∞-consensus performance constraint. In what follows, the aim is to design the controller gains $K_{i,k}$.

Theorem 9.2 *Consider any agent i and its neighboring agents $j \in \mathscr{N}_i$. Given the disturbance attenuation level $\gamma > 0$, the initial weighted matrix sequences $\{Q_{i,k}\}_{k\in\mathscr{K}}$, $\{T_{i,k}\}_{k\in\mathscr{K}}$, and $S_i > 0$, and the scalar sequences $\{\alpha_{i,k}\}_{k\in\mathscr{K}} \in \mathfrak{I}_{q_i}$, $\{\alpha_{j,k}\}_{k\in\mathscr{K}} \in \mathfrak{I}_{q_j}$, a sequence of positive definite matrices $\{P_{i,k}\}_{k\in\mathscr{K}\cup\{n\}}$ with the initial condition $P_{i,0} \leq \gamma^2 S_i$. The finite-horizon distributed H_∞-consensus control problem is solvable if there exists the matrix sequence $\{\bar{K}_{i,k}\}_{k\in\mathscr{K}}$, positive definite matrix sequences $\{\mathscr{B}_{i,k+1}\}_{k\in\mathscr{K}}$ and $\{\tilde{P}_{i,k}\}_{k\in\mathscr{K}\cup\{n\}}$ such that the following conditions hold,*

$$\begin{bmatrix} \Psi_{i,k}^{11} & * & * & * & * & * \\ \Psi_{i,k}^{21} & \Psi_{i,k}^{22} & * & * & * & * \\ \bar{\Psi}_{i,k}^{31} & \bar{\Psi}_{i,k}^{32} & -P_{i,k+1} & * & * & * \\ \bar{\Psi}_{i,k}^{41} & 0 & 0 & -P_{i,k+1} & * & * \\ 0 & \bar{\Psi}_{i,k}^{52} & 0 & 0 & \bar{\Psi}_{i,k}^{55} & * \\ 0 & 0 & \Psi_{i,k}^{63} & 0 & 0 & -\gamma^2 T_{i,k} \end{bmatrix} \leq 0, \tag{9.26}$$

$$\begin{bmatrix} \mathscr{B}_{i,k+1} & * \\ B_{i,k}\bar{K}_{i,k} & P_{i,k+1} \end{bmatrix} \geq 0, \tag{9.27}$$

$$P_{i,k+1}B_{i,k} = B_{i,k}\tilde{P}_{i,k+1} \tag{9.28}$$

where

$$\begin{aligned}
\bar{\Psi}_{i,k}^{31} &= P_{i,k+1}\bar{A}_{i,k} - \bar{\gamma}_i p_i B_{i,k}\bar{K}_{i,k}\bar{C}_{i,k}, \\
\bar{\Psi}_{i,k}^{32} &= \left[\bar{\gamma}_{j_{i_1}} B_{i,k}\bar{K}_{i,k}\bar{C}_{j_{i_1},k} \quad \cdots \quad \bar{\gamma}_{j_{ip_i}} B_{i,k}\bar{K}_{i,k}\bar{C}_{j_{ip_i},k}\right], \\
\bar{\Psi}_{i,k}^{41} &= \sigma_i p_i B_{i,k}\bar{K}_{i,k}\bar{C}_{i,k}, \quad \bar{\Psi}_{i,k}^{55} = \mathrm{diag}_{p_i}\left\{-P_{i,k+1}\right\}, \\
\bar{\Psi}_{i,k}^{52} &= \mathrm{diag}\left\{\sigma_{j_{i_1}} B_{i,k}\bar{K}_{i,k}\bar{C}_{j_{i_1},k}, \ldots, \sigma_{j_{ip_i}} B_{i,k}\bar{K}_{i,k}\bar{C}_{j_{ip_i},k}\right\}, \\
\Psi_{i,k}^{63} &= \left[P_{i,k+1}E_{i,k} \quad -p_i B_{i,k}\bar{K}_{i,k}D_{i,k} \quad B_{i,k}\bar{K}_{i,k}D_{j_{i_1},k} \quad \cdots \quad B_{i,k}\bar{K}_{i,k}D_{j_{ip_i},k}\right]^T.
\end{aligned}$$

Moreover, the distributed controller gains $K_{i,k} = \tilde{P}_{i,k+1}^{-1}\bar{K}_{i,k}$.

Proof In terms of Lemma 9.2 and Theorem 9.1, performing respectively congruence transformation to (9.13) and (9.14) by $\text{diag}\{I, I, P_{i,k+1}, P_{i,k+1}, P_{i,k+1}, I\}$ and $\text{diag}\{I, P_{i,k+1}\}$, one has

$$\begin{bmatrix} \Psi_{i,k}^{11} & * & * & * & * & * \\ \Psi_{i,k}^{21} & \Psi_{i,k}^{22} & * & * & * & * \\ P_{i,k+1}\Psi_{i,k}^{31} & P_{i,k+1}\Psi_{i,k}^{32} & -P_{i,k+1} & * & * & * \\ P_{i,k+1}\Psi_{i,k}^{41} & 0 & 0 & -P_{i,k+1} & * & * \\ 0 & P_{i,k+1}\Psi_{i,k}^{52} & 0 & 0 & \bar{\Psi}_{i,k}^{55} & * \\ 0 & 0 & F_{i,k}^T P_{i,k+1} & 0 & 0 & -\gamma^2 T_{i,k} \end{bmatrix} \le 0,$$

$$\begin{bmatrix} \mathscr{B}_{i,k+1} & * \\ P_{i,k+1}B_{i,k}K_{i,k} & P_{i,k+1} \end{bmatrix} \ge 0.$$

From (9.28), one has $P_{i,k+1}B_{i,k}K_{i,k} = B_{i,k}\tilde{P}_{i,k+1}K_{i,k}$ and denote $\bar{K}_{i,k} = \tilde{P}_{i,k+1}K_{i,k}$, $\bar{\Psi}_{i,k}^{31} = P_{i,k+1}\Psi_{i,k}^{31}$, $\bar{\Psi}_{i,k}^{32} = P_{i,k+1}\Psi_{i,k}^{32}$, $\bar{\Psi}_{i,k}^{52} = P_{i,k+1}\Psi_{i,k}^{52}$, $\Psi_{i,k}^{63} = F_{i,k}^T P_{i,k+1}$, then we can get (9.26) and (9.27). And thus the proof of this theorem is complete by resorting to Theorem 9.1.

Now, we are in a position to deal with the equality constraint (9.28). That is, we need to establish the relationship between $P_{i,k+1}$ and $\tilde{P}_{i,k+1}$. For this purpose, a lemma is presented in term of the singular value decomposition theory.

Lemma 9.3 *For the matrix $B_{i,k} \in \mathbb{R}^{n_x \times n_u}$ with full-column rank ($n_x > n_u$), there always exist two orthogonal matrices $U_{i,k} \in \mathbb{R}^{n_x \times n_x}$ and $V_{i,k} \in \mathbb{R}^{n_u \times n_u}$ such that*

$$B_{i,k} = U_{i,k}\begin{bmatrix} \Sigma_{i,k} \\ 0 \end{bmatrix} V_{i,k} = \begin{bmatrix} U_{i,k}^1 & U_{i,k}^2 \end{bmatrix}\begin{bmatrix} \Sigma_{i,k} \\ 0 \end{bmatrix} V_{i,k} \tag{9.29}$$

where $U_{i,k}^1 \in \mathbb{R}^{n_x \times n_u}$, $U_{i,k}^2 \in \mathbb{R}^{n_x \times (n_x - n_u)}$, $\Sigma_{i,k} = \text{diag}\{\sigma_{i,k}^1, \ldots, \sigma_{i,k}^{n_u}\}$, and $\sigma_{i,k}^j (j = 1, 2, \ldots, n_u)$ are nonzero singular values of $B_{i,k}$. Thus, if matrix $P_{i,k+1}$ has the following structure

$$\begin{aligned} P_{i,k+1} &= \begin{bmatrix} U_{i,k}^1 & U_{i,k}^2 \end{bmatrix} \text{diag}\left\{P_{i,k+1}^1, P_{i,k+1}^2\right\} \begin{bmatrix} U_{i,k}^1 & U_{i,k}^2 \end{bmatrix}^T \\ &= U_{i,k}^1 P_{i,k+1}^1 (U_{i,k}^1)^T + U_{i,k}^2 P_{i,k+1}^2 (U_{i,k}^2)^T \end{aligned} \tag{9.30}$$

where $P_{i,k+1}^1 \in \mathbb{R}^{n_u \times n_u} > 0$ and $P_{i,k+1}^2 \in \mathbb{R}^{(n_x - n_u) \times (n_x - n_u)} > 0$, then there exists a nonsingular matrix $\tilde{P}_{i,k+1} \in \mathbb{R}^{n_u \times n_u}$ such that $B_{i,k}\tilde{P}_{i,k+1} = P_{i,k+1}B_{i,k}$. Moreover, $\tilde{P}_{i,k+1} = V_{i,k}^T \Sigma_{i,k}^{-1} P_{i,k+1}^1 \Sigma_{i,k} V_{i,k}$.

Proof From (9.29), one has

$$B_{i,k} = U_{i,k}\begin{bmatrix} \Sigma_{i,k} \\ 0 \end{bmatrix} V_{i,k} = U_{i,k}^1 \Sigma_{i,k} V_{i,k}.$$

Then, according to (9.30), one obtains

$$\begin{aligned}P_{i,k+1}B_{i,k} &= U_{i,k}\begin{bmatrix} P^1_{i,k} & 0 \\ 0 & P^2_{i,k}\end{bmatrix}U^T_{i,k}U_{i,k}\begin{bmatrix}\Sigma_{i,k} \\ 0\end{bmatrix}V_{i,k} \\ &= \begin{bmatrix} U^1_{i,k} & U^2_{i,k}\end{bmatrix}\begin{bmatrix}P^1_{i,k}\Sigma_{i,k} \\ 0\end{bmatrix}V_{i,k} \\ &= U^1_{i,k}P^1_{i,k}\Sigma_{i,k}V_{i,k} \\ &= U^1_{i,k}\Sigma_{i,k}V_{i,k}V^T_{i,k}\Sigma^{-1}_{i,k}P^1_{i,k}\Sigma_{i,k}V_{i,k} \\ &= B_{i,k}V^T_{i,k}\Sigma^{-1}_{i,k}P^1_{i,k}\Sigma_{i,k}V_{i,k}.\end{aligned}$$

Thus, one acquires that $\tilde{P}_{i,k+1} = V^T_{i,k}\Sigma^{-1}_{i,k}P^1_{i,k+1}\Sigma_{i,k}V_{i,k}$. Obviously, $\tilde{P}_{i,k+1}$ is nonsingular. Therefore, the proof is complete.

From Theorem 9.2 and Lemma 9.3, the following result can be readily obtained, and its proof is therefore omitted.

Theorem 9.3 *Consider agent i and its neighboring agents $j \in \mathcal{N}_i$. Given the disturbance attenuation level $\gamma > 0$, the initial weighted matrices $\{Q_{i,k}\}_{k\in\mathcal{K}}$, $\{T_{i,k}\}_{k\in\mathcal{K}}$, S_{1i}, $S_{2i} > 0$, and the scalar sequences $\{\alpha_{i,k}\}_{k\in\mathcal{K}} \in \mathfrak{I}_{q_i}$, $\{\alpha_{j,k}\}_{k\in\mathcal{K}} \in \mathfrak{I}_{q_j}$, the finite-horizon distributed H_∞-consensus control problem is solvable if there exist the matrix sequences $\{\bar{K}_{i,k}\}_{k\in\mathcal{K}}$, several sets of positive definite matrix sequences $\{\mathscr{B}_{i,k+1}\}_{k\in\mathcal{K}}$, $\{P^1_{i,k}\}_{k\in\mathcal{K}\cup\{n\}}$ and $\{P^2_{i,k}\}_{k\in\mathcal{K}\cup\{n\}}$ with initial conditions $P^1_{i,0} \le \gamma^2 S_{1i}$ and $P^2_{i,0} \le \gamma^2 S_{2i}$ such that* (9.26) *and* (9.27) *hold. Moreover, the distributed controller gains $K_{i,k} = V_{i,k}\sum_{i,k}(P^1_{i,k+1})^{-1}\sum^{-1}_{i,k}V^T_{i,k}\bar{K}_{i,k}$, where $P_{i,k+1} = (U^1_{i,k})^T P^1_{i,k+1}U^1_{i,k} + (U^2_{i,k})^T P^2_{i,k+1}U^2_{i,k}$, $U^1_{i,k}$, $\sum_{i,k}$, $U^2_{i,k}$ and $V_{i,k}$ come from* (9.29), *and $S_i = \mathrm{diag}\{S_{1i}, S_{2i}\}$.*

Notably, since the Kronecker product and the Hadamard product are essentially linear operations, the Yalmip toolbox in [19] can be employed to solve $\mathscr{B}_{i,k+1}$ in inequality (9.26). Finally, two special cases are both considered. First and foremost, if multi-agent system is time-invariant and the concern focus is its steady-state property over an infinite horizon, then an LMIs-based asymptotical consensus condition can be easily deduced from RLMIs (9.26)–(9.27) as long as the constant variables replace the time-varying variables, respectively. Second, a special form of the random matrix is considered. The corresponding result can be easily available along the same line of Theorem 9.3. Accordingly, here we only list this result and its proof is also omitted.

Corollary 9.1 *Let random matrices $\tilde{A}_{i,k}$ and $\tilde{C}_{i,k}$ be of form $\tilde{A}_{i,k} = \mu_A\hat{A}_{i,k}$, $\tilde{C}_{i,k} = \mu_C\hat{C}_{i,k}$ where $\hat{A}_{i,k}$, $\hat{C}_{i,k}$ are deterministic matrices and $\mu_A, \mu_C \in \mathbb{R}$ are random variables with zero means and unit variances. Consider any agent $i \in \mathscr{V}$ and its neighbors $j \in \mathcal{N}_i$. Given a prescribed disturbance attenuation level $\gamma > 0$, weighted matrices $\{Q_{i,k}\}_{k\in\mathcal{K}}$, $\{T_{i,k}\}_{k\in\mathcal{K}}$, S_{1i}, S_{2i}, and scalar sequences $\alpha_{i,k} \in \mathfrak{I}_{q_i}$, $\alpha_{j,k} \in \mathfrak{I}_{q_j}$. The finite-horizon distributed H_∞-consensus control problem is solvable if there exist matrix sequence $\{\bar{K}_{i,k}\}_{k\in\mathcal{K}}$, positive definite matrix sequences $\{P^1_{i,k}\}_{k\in\mathcal{K}\cup\{n\}}$ and $\{P^2_{i,k}\}_{k\in\mathcal{K}\cup\{n\}}$ with initial conditions $P^1_{i,0} \le \gamma^2 S_{1i}$ and $P^2_{i,0} \le \gamma^2 S_{2i}$ such that the following conditions hold:*

$$\begin{bmatrix} \Omega_{i,k}^{11} & * & * & * & * & * & * & * \\ \Omega_{i,k}^{21} & \Omega_{i,k}^{22} & * & * & * & * & * & * \\ \Omega_{i,k}^{31} & \Omega_{i,k}^{32} & -P_{i,k+1} & * & * & * & * & * \\ \Omega_{i,k}^{41} & 0 & 0 & -P_{i,k+1} & * & * & * & * \\ 0 & \Omega_{i,k}^{52} & 0 & 0 & -\Omega_{i,k}^{55} & * & * & * \\ 0 & 0 & \Omega_{i,k}^{63} & 0 & 0 & -\gamma^2 T_{i,k} & * & * \\ \Omega_{i,k}^{71} & 0 & 0 & 0 & 0 & 0 & -P_{i,k+1} & * \\ 0 & \Omega_{i,k}^{82} & 0 & 0 & 0 & 0 & 0 & -\Omega_{i,k}^{88} \end{bmatrix} \leq 0 \tag{9.31}$$

where

$$\begin{aligned}
\Omega_{i,k}^{11} &= -\frac{1+q_i(1-\alpha_{i,k})}{1+q_i}P_{i,k} + \hat{A}_{i,k}^T P_{i,k+1}\hat{A}_{i,k} + L_{i,k}^T Q_{i,k} L_{i,k}, \\
\Omega_{i,k}^{21} &= \left[-\frac{1}{p_i} L_{j_{i_1},k}^T Q_{i,k} L_{i,k} \quad \cdots \quad -\frac{1}{p_i} L_{j_{ip_i},k}^T Q_{i,k} L_{i,k} \right]^T, \\
\Omega_{i,k}^{22} &= \mathrm{diag}\left\{ -\frac{\alpha_{j_{i_1},k}}{1+q_{j_{i_1}}} P_{j_{i_1},k}, \ldots, -\frac{\alpha_{j_{ip_i},k}}{1+q_{j_{ip_i}}} P_{j_{ip_i},k} \right\} + \Upsilon_{i,k}, \\
\Upsilon_{i,k} &= \frac{1}{p_i^2}\left[L_{j_{i_1},k}^T \quad \cdots \quad L_{j_{ip_i},k}^T \right] Q_{i,k} \left[L_{j_{i_1},k}^T \quad \cdots \quad L_{j_{ip_i},k}^T \right]^T, \\
\Omega_{i,k}^{31} &= P_{i,k+1}\bar{A}_{i,k} - \bar{\gamma}_i p_i B_{i,k}\bar{K}_{i,k}\bar{C}_{i,k}, \quad \Omega_{i,k}^{32} = \left[\bar{\gamma}_{j_{i_1}} B_{i,k}\bar{K}_{i,k}\bar{C}_{j_{i_1},k} \quad \cdots \quad \bar{\gamma}_{j_{p_i}} B_{i,k}\bar{K}_{i,k}\bar{C}_{j_{ip_i},k} \right], \\
\Omega_{i,k}^{41} &= p_i\sigma_i B_{i,k}\bar{K}_{i,k}\bar{C}_{i,k}, \quad \Omega_{i,k}^{52} = \mathrm{diag}\left\{ \sigma_{j_{i_1}} B_{i,k}\bar{K}_{i,k}\bar{C}_{j_{i_1},k}, \ldots, \sigma_{j_{ip_i}} B_{i,k}\bar{K}_{i,k}\bar{C}_{j_{ip_i},k} \right\}, \\
\Omega_{i,k}^{55} &= \mathrm{diag}_{p_i}\left\{ P_{i,k+1} \right\}, \quad \sigma_i = \sqrt{\bar{\gamma}_i(1-\bar{\gamma}_i)}, \quad \Omega_{i,k}^{71} = p_i\sqrt{\bar{\gamma}_i} B_{i,k}\bar{K}_{i,k}\hat{C}_{i,k}, \\
\Omega_{i,k}^{63} &= \left[P_{i,k+1}E_{i,k} \quad -p_i B_{i,k}\bar{K}_{i,k}D_{i,k} \quad B_{i,k}\bar{K}_{i,k}D_{j_{i_1},k} \quad \cdots \quad B_{i,k}\bar{K}_{i,k}D_{j_{ip_i},k} \right]^T, \\
\Omega_{i,k}^{82} &= \mathrm{diag}\left\{ \sqrt{\bar{\gamma}_{j_{i_1}}} B_{i,k}\bar{K}_{i,k}\hat{C}_{j_{i_1},k}, \ldots, \sqrt{\bar{\gamma}_{j_{ip_i}}} B_{i,k}\bar{K}_{i,k}\hat{C}_{j_{ip_i},k} \right\}, \\
\Omega_{i,k}^{88} &= \mathrm{diag}_{p_i}\left\{ P_{i,k+1} \right\}, \quad P_{i,k+1} = (U_{i,k}^1)^T P_{i,k+1}^1 U_{i,k}^1 + (U_{i,k}^2)^T P_{i,k+1}^2 U_{i,k}^2.
\end{aligned}$$

Moreover, the distributed controller gains $K_{i,k} = V_{i,k}\sum_{i,k}(P_{i,k+1}^1)^{-1}\sum_{i,k}^{-1} V_{i,k}^T \bar{K}_{i,k}$.

Remark 9.3 Firstly, sufficient conditions for the distributed H_∞-consensus controllers are still useful, especially when the corresponding necessary and sufficient conditions are intractable or NP-hard [18]. Secondly, an important issue is the solvability of the proposed sufficient design conditions. Obviously, there exists a trade-off between the solvability of the sufficient conditions and the performance requirement (9.6). In order to enhance the solvability of sufficient conditions, some optional variables in (9.6) can be adjusted to enlarge the bound of the right side of (9.6). The simplest method is to improve the disturbance rejection level γ. Another method is to adjust some optional variables, such as the weighted matrices $Q_{i,k}$, $T_{i,k}$, S_{1i}, and S_{2i} ($i \in \mathscr{V}$).

Remark 9.4 This chapter investigates the finite-horizon H_∞-consensus output feedback control problem of the multi-agent systems with both random parameters and missing measurements. The obtained design conditions for every agent in Theorem 9.3 contain all the neighboring agent information including the agent system parameters, the missing measurement probabilities as well as the statistical information of the random parameters. The main novelties lie in the following aspects: (1) the WCM formula has been proposed in term of the Kroneck product and the Hadamard product, which can be employed to deal with the random parameter systems; (2) a locally dynamic H_∞-consensus performance constraint has been presented, which characterizes the behaviors of every agent based only on the neighboring agents; and (3) by using the vector dissipativity theory, the locally design algorithm for each agent is only related to the known information of the neighbors, which ensure that it can be implemented on each agent in a distributed manner.

9.3 An Illustrative Example

In this section, a simulation example is presented to illustrate the effectiveness of the proposed distributed controller design algorithms for the multi-agent system with both random parameters and missing measurements.

Consider a multi-agent system which has six agents. Its topology is represented by a digraph $\mathscr{G} = (\mathscr{V}, \mathscr{E}, \mathscr{A})$ with the set of nodes $\mathscr{V} = \{1, 2, 3, 4, 5, 6\}$, the set of edges $\mathscr{E} = \{(2, 1), (2, 4), (3, 2), (4, 3), (4, 6), (5, 4), (6, 5), (6, 2), (1, 6)\}$, as shown in Fig. 9.1.

The time-varying dynamics of the agents are described by (9.1) with parameters as

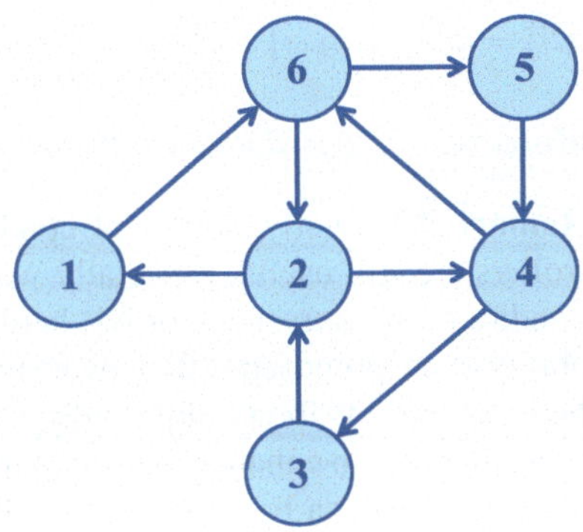

Fig. 9.1 Topology of multi-agent system

$$\bar{A}_{i,k} = \begin{bmatrix} 0.3 + 0.7\sin(k) & 0.4 \\ 0.35 & -0.65 \end{bmatrix}, \quad \tilde{A}_{i,k} = \mu_{i,k} \begin{bmatrix} 0.04 & 0 \\ 0 & 0.06 \end{bmatrix},$$
$$B_{i,k} = [1\ 0.5]^T, \quad L_{i,k} = [0.1\ 0.1], \quad D_{i,k} = 0.2,$$
$$\bar{C}_{i,k} = [0.75\ 0.75], \quad \tilde{C}_{i,k} = \varepsilon_{i,k}[0.02\ 0.03]$$

where $\mu_{i,k}$ and $\varepsilon_{i,k}$ ($i = 1, 2, 3, 4, 5, 6$) are mutually independent zero mean Gaussian noise sequences with unity covariances. Also, it is easily verified that

$$\hat{A}_{i,k} \equiv \text{diag}\{0.04, 0.06\}, \quad \hat{C}_{i,k} \equiv \begin{bmatrix} 0.02\ 0.03 \end{bmatrix},$$
$$B_{i,k} = \begin{bmatrix} 0.8944 & -0.4472 \\ 0.4472 & 0.8944 \end{bmatrix} \times \begin{bmatrix} 1.118 \\ 0 \end{bmatrix} \times 1.$$

The disturbance attenuation level γ^2 and time horizon $n - 1$ are taken as 0.1 and 50, respectively. In addition, the probabilities of missing measurement of each agent, the scalar sequence $\alpha_{i,k}$, the performance weighted matrices $Q_{i,k}$, $T_{i,k}$, the initial weighted matrix $P_{i,0}$ are given as $\bar{\gamma}_i = 0.9$, $\alpha_{i,k} \equiv 0.9$, $Q_{i,k} \equiv 1$, $T_{i,k} \equiv \text{diag}\{1, 1, 1, \cdots, 1\}$, $P_{i,0} = 30I$ ($i = 1, 2, 3, 4, 5, 6$). By using MATLAB (with the YALMIP 3.0), we can obtain all the desired controller gains of the six agents.

In the simulation, the initial conditions are set as $x_{1,0} = [-0.3\ \ 0.5]^T$, $x_{2,0} = [-0.1\ 0.2]^T$, $x_{3,0} = [-0.2\ 0.3]^T$, $x_{4,0} = [0.4\ \ -0.5]^T$, $x_{5,0} = [0.5\ \ -0.6]^T$, $x_{6,0} = [0.6\ \ -0.4]^T$. Simulation results are shown in Figs. 9.2, 9.3 and 9.4, where Figs. 9.2 and 9.3 depict the states of every agent, and Fig. 9.4 plots the consensus errors $z_{i,k} - z^*_{i,k}$, respectively. In these figures, $x^{(j)}_{i,k}$ ($j = 1, 2$) are the jth element of the system state $x_{i,k}$. From these figures, the closed-loop system achieves the preferable consensus effects by utilizing the designed algorithm. Consequently, it has been confirmed that the proposed distributed controller design method in this chapter is indeed effective.

9.4 Summary

In this chapter, the finite-horizon distributed H_∞-consensus control problem for a class of discrete time-varying multi-agent systems with random parameters and missing measurements has been addressed via the local performance analysis method stemming from the vector dissipativity theory. The WCM formula of random matrix and a novel dynamic local H_∞-consensus performance index have been proposed, respectively. A set of locally coupled sufficient conditions for every agent has been obtained such that the proposed locally dynamic H_∞-consensus performance constraints are satisfied in the presence of the external disturbances. In addition, the explicit expressions of the desired controller gains have been obtained. Finally, a simulation example has been employed to demonstrate the effectiveness of the proposed control scheme. When the local performance analysis method is applied to

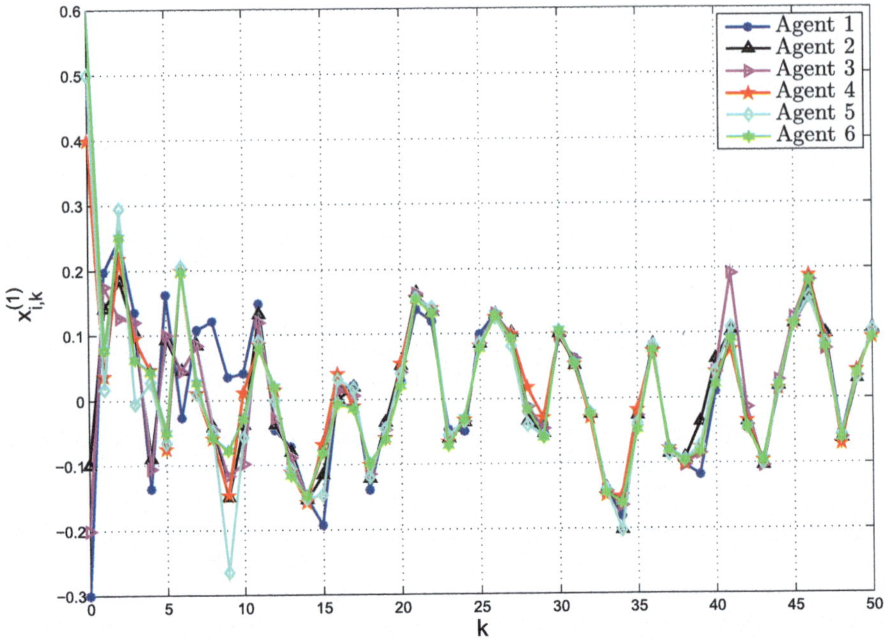

Fig. 9.2 The first element of agent state $x_{i,k}^{(1)}$

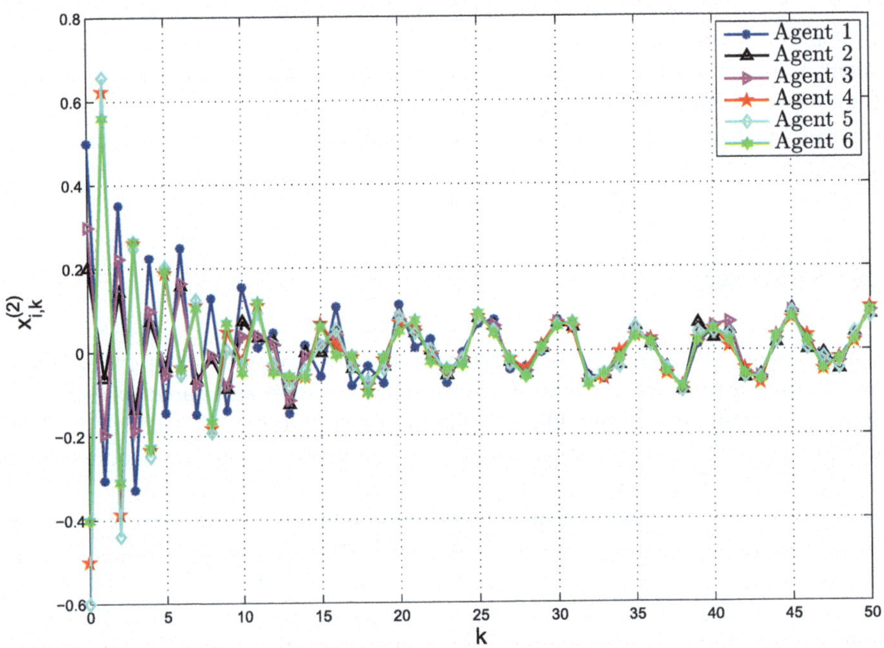

Fig. 9.3 The second element of agent state $x_{i,k}^{(2)}$

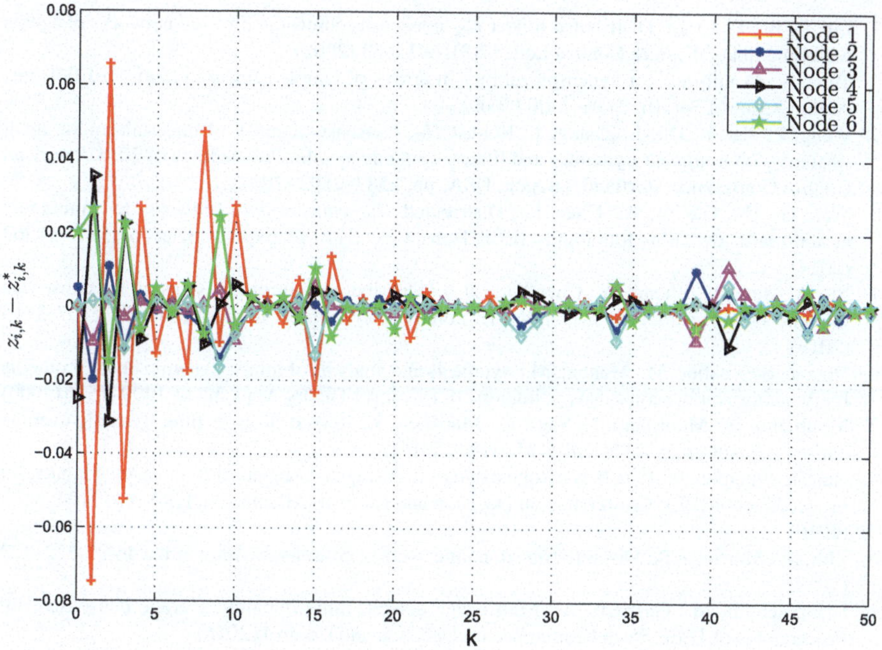

Fig. 9.4 The consensus errors $z_{i,k} - z^*_{i,k}$

solve the distributed consensus control problem, the main challenge comes from the modification of the traditional performance index. Such modification will bring convenience for establishing the vector dissipation inequality within the framework of the local performance analysis method. In the following chapter, as an important attempt, the local performance analysis method will be utilized to deal with the synchronization and filtering problems in the complex networks.

References

1. Ren, W., Beard, R.: Consensus seeking in multiagent systems under dynamically changing interaction topologies. IEEE Trans. Autom. Control **50**(5), 655–661 (2005)
2. Olfati-Saber, R., Murry, R.: Consensus problem in networks of agents with switching topology and time-delays. IEEE Trans. Autom. Control **49**(9), 1520–1533 (2004)
3. Spanos, D., Reza, O., Murray, R.: Dynamic consensus on mobile networks. In: Proceedings of the 2005 IFAC World Congress (2005)
4. Wang, Z., Ding, D., Dong, H., Shu, H.: H_∞ consensus control for multi-agent systems with missing measurements: the finite-horizon case. Syst. & Control Lett. **62**(10), 827–836 (2013)
5. Liu, Q., Wang, Z., He, X., Zhou, D.: Event-Based recursive distributed filtering over wireless sensor networks. IEEE Trans. Autom. Control **60**(9), 2470–2475 (2015)
6. Liu, Y., Jia, Y.: Consensus problem of high-order multi-agent systems with external disturbances: an H_∞ analysis approach. Int. J. Robust Nonlinear Control **20**(14), 1579–1593 (2010)

7. Lin, P., Jia, Y., Li, L.: Distributed robust H_∞ consensus control in directed networks of agents with time-delay. Syst. & Control Lett. **57**(8), 643–653 (2008)
8. Tuna, S.: Conditions for synchronizability in arrays of coupled linear systems. IEEE Trans. Autom. Control **54**(10), 2416–2420 (2009)
9. Wang, P., Jia, Y., Du, J., Zhang, J.: Robust H_∞ consensus control of uncertain multi-agent systems with nonlinear dynamics and time-varying delays. In: Proceedings of 2014 American Control Conference, Portland, Oregon, USA, pp. 2587–2592 (2014)
10. Wen, G., Hu, G., Yu, W., Chen, G.: Distributed H_∞ consensus of higher order multiagent systems with switching topologies. IEEE Trans. Circ. Syst.-II: Express Briefs **61**(5), 359–363 (2014)
11. Yu, W., Ren, W., Zheng, W., Chen, G., Lü, J.: Distributed control gains design for consensus in multi-agent systems with second-order nonlinear dynamics. Automatica **49**(7), 2107–2115 (2013)
12. Nosrati, S., Shafiee, M., Menhaj, M.: Synthesis and analysis of robust dynamic linear protocols for dynamic average consensus estimators. IET Control Theory Appl. **3**(11), 1499–1516 (2009)
13. Montijano, E., Montijano, J., Sags, C., Martłnez, S.: Robust discrete time dynamic average consensus. Automatica **50**(12), 3131–3138 (2014)
14. Bai, H., Freeman, R., Lynch, K.: Robust dynamic average consensus of time-varying inputs. In: Proceedings of IEEE Conference on Decision and Control, Atlanta, Ga, USA, pp. 3104–3109 (2010)
15. Zhu, M., Martinez, S.: Discrete-time dynamic average consensus. Automatica **46**(2), 322–329 (2010)
16. Franceschelli, M., Gasparri, A.: Multi-stage discrete time dynamic average consensus, In: Proceedings of IEEE 55-th Conference on Decision and Control (2016)
17. Haddad, W., Hui, Q., Chellaboina, V., Nersesov, S.: Vector dissipativity theory for discrete-time large-scale nonlinear dynamical systems. Adv. Diff. Equ. **2004**, 37–66 (2004)
18. Ugrinovskii, V.: Distributed robust estimation over random switching networks using H_∞ consensus. Automatica **49**(1), 160–168 (2013)
19. Löfberg, J.: YALMIP: A toolbox for modeling and optimization in MATLAB. In: Proceedings of the CACSD Conference, Taipei, Taiwan (2004)

Chapter 10
Bounded H_∞ Synchronization and H_∞ Filtering for Discrete-Time Complex Networks

In sensor network and multi-agent systems, the couplings among different nodes stem from the distributed filter or distributed controller, which utilizes the information from neighboring nodes. Different from the mentioned couplings, complex network itself has the outer-couplings among different nodes and the inner-couplings among different elements of a node. From the previous chapters, the local performance analysis method has been proved to be a powerful tool to solve the complicated couplings. Therefore, it is an interesting problem to apply this method to solve the different coupling problems pertaining to complex networks.

This chapter deals with the finite-horizon bounded H_∞ synchronization and H_∞ filteringH_∞ filtering for the discrete-time complex networks with missing measurements based on the local performance analysis method, respectively. First and foremost, a new local description of the bounded H_∞ synchronization performance index is proposed, which considers only the synchronization errors among neighboring nodes. In addition, a more general sector-bounded condition is presented, where the parameter matrices are different for different nodes. Next, by establishing the vector dissipativity-like for the complex network dynamics, the H_∞ synchronization criterion is derived in terms of the locally coupled conditions for each node. These conditions implemented in a cooperative manner can judge whether the complex network reach synchronization or not. Similarly, the existence conditions for the filter on each node are obtained, and then the filter parameters are designed via the recursive linear matrix inequalities. The distinctive features of the algorithms are low complexity, scalability, and distributed execution. At last, two numerical examples are utilized to verify the effectiveness and applicability of the proposed algorithms.

F. Han et al., *Distributed Filtering, Control and Synchronization*,
Studies in Systems, Decision and Control 428,
https://doi.org/10.1007/978-3-030-97075-8_10

10.1 Problem Formulation of Bounded H_∞ Synchronization

Consider the following array of discrete time-varying complex networks consisting of N coupled nodes defined on $k \in \mathscr{K} \triangleq \{0, 1, \ldots, n-1\}$:

$$\begin{cases} x_{i,k+1} = f(x_{i,k}) + \sum_{j=1}^{N} w_{ij}\Gamma x_{j,k} + L_{i,k}v_{i,k}, \\ z_{i,k} = M_k x_{i,k}, \\ x_{i,0} = \psi_i \end{cases} \tag{10.1}$$

where $x_{i,k} \in \mathbb{R}^{n_x}$ and $z_{i,k} \in \mathbb{R}^{n_z}$ are the state and the interested output of node i, respectively. ψ_i is an initial value of node i. $v_{i,k} \in \mathbb{R}^{n_{v_i}}$ is the disturbance input belonging to $l_2[0, n-1]$. $L_{i,k}$ and M_k are time-varying matrices with appropriate dimensions. $\Gamma = \mathrm{diag}\{r_1, r_2, \ldots, r_{n_x}\}$ is the matrix linking the j-th state variable if $r_j \neq 0$. $W = [w_{ij}]_{N\times N}$ is the coupled configuration matrix of the network with $w_{ij} \geq 0 (i \neq j)$ and $w_{ii} < 0$ $(i \in \mathscr{V})$ but not all zeros and satisfying $\sum_{j=1}^{N} w_{ij} = 0$.

Next, a digraph $\mathscr{G}$ is employed to describe the topology of the complex network (10.1). More specifically, $\mathscr{G} = \{\mathscr{V}, \mathscr{E}, W\}$, where $\mathscr{V} = \{1, 2, \ldots, N\}$, $\mathscr{E} = \{(i, j)|i, j \in \mathscr{V}\}$, and $W = [w_{ij}]_{N\times N}$. Here, $(i, j) \in \mathscr{E}$ implies that there exists a connection from node j to node i. If $w_{ij} > 0$, then node j is said to be a neighbor of node i. The in-degree of node i is p_i defined as $p_i = \sum_{j=1}^{N} \mathbb{I}_{w_{ij}>0}$, and the out-degree of node i is q_i defined as $q_i = \sum_{j=1}^{N} \mathbb{I}_{w_{ji}>0}$, where $\mathbb{I}_{w_{ij}>0}$ is an indicator function of $w_{ij} > 0$. And then, the set of neighbors of node i can be denoted as $\mathscr{N}_i = \{j_{i_1}, j_{i_2}, \ldots, j_{i_{p_i}}\}$. In this chapter, it is assumed that the topology of complex network is weakly connected.

The continuous nonlinear vector-valued function f: $\mathbb{R}^{n_x} \mapsto \mathbb{R}^{n_x}$ is assumed to satisfy $f(0) = 0$ and the sector-bounded condition

$$\left[f(x_{i,k}) - \Phi_{i,k}^1 x_{i,k}\right]^T \left[f(x_{i,k}) - \Phi_{i,k}^2 x_{i,k}\right] \leq 0 \tag{10.2}$$

for all $x_{i,k} \in \mathbb{R}^{n_x}$, where $\Phi_{i,k}^1$ and $\Phi_{i,k}^2$ are known real matrices with appropriate dimensions. Actually, (10.2) is a more general assumption than that in the existing references [1–5], since the parameter matrices $\Phi_{i,k}^1$ and $\Phi_{i,k}^2$ are different for different nodes.

As is well known, every node in the complex networks communicates only with its neighboring nodes according to the given digraph. That is, if j is a neighbor of node i, then the state information $z_{j,k}$ from node j is available for node i. Thus, node i can calculate the synchronization error with respect to $z_{i,k}$ and $z_{j,k}$, which is employed to describe a novel local synchronization, i.e., $z_{i,k} = z_{j,k}$, $\forall i \in \mathscr{V}$, $j \in \mathscr{N}_i$. Obviously, this description is weaker than the traditional global synchronization for all nodes, i.e., $z_{i,k} = z_{j,k}$, $\forall i, j \in \mathscr{V}$. Next, an interesting relationship between these two synchronization descriptions will be presented under the assumption of weakly connected topology.

Proposition 10.1 *For a given complex network with weakly connected network topology* $\mathscr{G}$, $z_{i,k} = z_{j,k}$ *for* $\forall\, i \in \mathscr{V}$, $j \in \mathscr{N}_i$ *if and only if* $z_{i,k} = z_{j,k}$ *for* $\forall\, i, j \in \mathscr{V}$.

Proof Necessity is obvious, and thus we only prove sufficiency. If $z_{i,k} = z_{j,k}$ holds for $\forall\, i \in \mathscr{V}$, $j \in \mathscr{N}_i$, then it can be rewritten as the following equation group

$$\begin{cases} z_{1,k} = z_{j,k},\ j \in \mathscr{N}_1 \\ \quad\vdots \\ z_{i,k} = z_{j,k},\ j \in \mathscr{N}_i \\ \quad\vdots \\ z_{N,k} = z_{j,k},\ j \in \mathscr{N}_N \end{cases}. \tag{10.3}$$

For any two nodes $i, j \in \mathscr{V}$, according to the assumption of weakly connection, there always exists at least one undirected path from node i to node j. More specifically, let this undirected path be: $i \leftrightarrow i_1 \leftrightarrow l \leftrightarrow \cdots \leftrightarrow m \leftrightarrow j_1 \leftrightarrow j$ $(i_1, l, m, j_1 \in \mathscr{V})$. Without loss of generality, this corresponding direct path between node i and j can be assumed as follows: $i \leftarrow i_1 \rightarrow l \leftarrow \cdots \leftarrow m \rightarrow j_1 \rightarrow j$. Since i_1 is a neighbor of node i, then $z_{i,k} = z_{i_1,k}$ according to (10.3). In a similar way, one also obtains $z_{i_1,k} = z_{l,k}, \cdots, z_{m,k} = z_{j_1,k}$, and $z_{j_{1,k}} = z_{j,k}$. Obviously, one has $z_{i,k} = z_{j,k}$. Therefore, the proof is complete.

Proposition 10.1 shows that the local description of the synchronization is equivalent to the global description for a weak connected topology. It is noteworthy that, this conclusion also holds for other connected topology, such as fully connected, strongly connected. In the following, the bounded H_∞ synchronization is presented based on Proposition 10.1.

Definition 10.1 For given some initial weighted matrices $S_i > 0$, the discrete time-varying complex network (10.1) is said to be bounded H_∞ synchronized with a disturbance attention level $\gamma > 0$ over the finite horizon $\mathscr{K}$ if the following inequality holds

$$\frac{1}{N}\sum_{i=1}^{N}\sum_{k=0}^{n-1}\sum_{j\in\mathscr{N}_i}\|z_{i,k} - z_{j,k}\|^2 \le \frac{\gamma^2}{N}\sum_{i=1}^{N}\left(\eta_{i,0}^T S_i \eta_{i,0} + \sum_{k=0}^{n-1}\|v_{i,k}\|^2\right). \tag{10.4}$$

Remark 10.1 In [2], the bounded H_∞ synchronization has been proposed, where the synchronized errors of any two nodes were considered, that is, $\sum_{1\le i<j\le N}\|z_{i,k} - z_{j,k}\|^2$. Actually, if j is not a neighbor of i, then j do not know the state of i due to the local communication constraint. Namely, it is unpracticable to calculate the synchronization error between any two nodes. Taking it into account, Definition 10.1 proposes a kind of new synchronization descriptions only including the synchronization errors among neighboring nodes. In addition, it is proved that this novel

synchronization description is equivalent to the traditional synchronization description under the assumption of weakly connected from Proposition 10.1.

10.2 Bounded H_∞ Synchronization for Discrete Time-Varying Complex Networks

In this section, we are ready to address the bounded H_∞ synchronization problem for the discrete time-varying complex network (10.1). First of all, the definition of the vector dissipativity-like over the finite horizon is presented for the complex network (10.1) according to Definition 10.1. For this purpose, in view of the inequality constraint (10.4), choose the following supply rate function of node i:

$$S_i(z_k, v_{i,k}) \triangleq \gamma^2 \|v_{i,k}\|^2 - \sum_{j \in \mathscr{N}_i} \|z_{i,k} - z_{j,k}\|^2. \tag{10.5}$$

For convenience, denote $x_k \triangleq [x_{1,k}^T \ \cdots \ x_{N,k}^T]^T$, $z_k \triangleq [z_{1,k}^T \ \cdots \ z_{N,k}^T]^T$, and $v_k \triangleq [v_{1,k}^T \ \cdots \ v_{N,k}^T]^T$.

Definition 10.2 The discrete time-varying complex network (10.1) is said to be vector dissipative-like over the finite horizon $\mathscr{K}$ with respect to a vector of supply rate functions $\mathbf{S}(z_k, v_k) \triangleq [S_1(z_k, v_{1,k}) \ \cdots \ S_N(z_k, v_{N,k})]^T$, if there exists a vector of continuous nonnegative definite storage functions $\mathbf{V}(x_k) \triangleq [V_1(x_{1,k}) \cdots V_N(x_{N,k})]^T$ with $\mathbf{V}(0) = 0$, and a sequence of nonsingular column substochastic dissipation matrices $W_k \in \mathbb{R}^{N\times N}$ such that the vector dissipation inequality

$$\mathbf{V}(x_{k+1}) \leq\leq W_k \mathbf{V}(x_k) + \mathbf{S}(z_k, v_k) \tag{10.6}$$

is satisfied for any $k \in \mathscr{K}$.

Next, one begins to establish the vector dissipativity for the complex networks (10.1). The critical point is to establish the vector dissipation inequality (10.6), where the candidate of the storage function is the Lyapunov function (i.e., $V_i(x_{i,k}) = x_{i,k}^T P_{i,k} x_{i,k}$ with $P_{i,k} > 0$) and the dissipation matrix needs to be constructed. For constructing a desirable dissipation matrix, define an interval-valued function with respect to out-degree of node i as follows:

$$\mathfrak{I}_{q_i} = \begin{cases} (0, \frac{1+q_i}{2q_i}), & \text{if } q_i \neq 0; \\ (0, 1], & \text{if } q_i = 0. \end{cases} \tag{10.7}$$

In addition, denote

$$W_k(\alpha_{1,k},\ldots,\alpha_{N,k}) \triangleq \begin{bmatrix} \frac{1+q_1(1-\alpha_{1,k})}{1+q_1} & a_{12}\frac{\alpha_{2,k}}{1+q_2} & \cdots & a_{1N}\frac{\alpha_{N,k}}{1+q_N} \\ a_{21}\frac{\alpha_{1,k}}{1+q_1} & \frac{1+q_2(1-\alpha_{2,k})}{1+q_2} & \cdots & a_{2N}\frac{\alpha_{N,k}}{1+q_N} \\ \vdots & \vdots & \ddots & \vdots \\ a_{N1}\frac{\alpha_{1,k}}{1+q_1} & a_{N2}\frac{\alpha_{2,k}}{1+q_2} & \cdots & \frac{1+q_N(1-\alpha_{N,k})}{1+q_N} \end{bmatrix} \tag{10.8}$$

where $\alpha_{i,k} \in \mathbb{J}_{q_i}$ $(i \in \mathscr{V})$ are known constants. For simplicity, $W_k(\alpha_{1,k},\ldots,\alpha_{N,k})$ is denoted as W_k.

Lemma 10.1 *For any node $i \in \mathscr{V}$ and its neighboring nodes $j \in \mathscr{N}_i$, let a positive scalar γ, two scalar sequences $\{\alpha_{i,k}\}_{k\in\mathscr{K}} \in \mathbb{J}_{q_i}$ and $\{\alpha_{j,k}\}_{k\in\mathscr{K}} \in \mathbb{J}_{q_j}$ be given. If there exist p_i+1 sequences of positive definite matrices $\{P_{i,k}\}_{k\in\mathscr{K}\cup\{n\}}$ and $\{P_{j,k}\}_{k\in\mathscr{K}\cup\{n\}}$ and a sequence of positive scalars $\{\lambda_{i,k}\}_{k\in\mathscr{K}}$ satisfying the following RLMIs for any $k\in\mathscr{K}$*

$$\begin{bmatrix} -\Xi^i_{i,k} & * & \cdots & * & * & * & * \\ -M_k^TM_k & -\Xi^{j_1}_{i,k} & \cdots & * & * & * & * \\ \vdots & \vdots & \ddots & \vdots & \vdots & \vdots & \vdots \\ -M_k^TM_k & 0 & \cdots & -\Xi^{j_{p_i}}_{i,k} & * & * & * \\ \frac{\lambda_{i,k}}{2}[\Phi^1_{i,k}+\Phi^2_{i,k}] & 0 & \cdots & 0 & -\lambda_{i,k}I & * & * \\ w_{ii}P_{i,k+1}\Gamma & w_{ij_1}P_{i,k+1}\Gamma & \cdots & w_{ij_{p_i}}P_{i,k+1}\Gamma & P_{i,k+1} & -P_{i,k+1} & * \\ 0 & 0 & \cdots & 0 & 0 & L^T_{i,k}P_{i,k+1} & -\gamma^2 I \end{bmatrix} < 0 \tag{10.9}$$

where

$$\Xi^i_{i,k} = \frac{1+q_i(1-\alpha_{i,k})}{1+q_i}P_{i,k} - p_iM_k^TM_k + \frac{\lambda_{i,k}}{2}\left((\Phi^1_{i,k})^T\Phi^2_{i,k} + (\Phi^2_{i,k})^T\Phi^1_{i,k}\right),$$

$$\Xi^{j_{i_1}}_{i,k} = \frac{\alpha_{j_{i_1},k}}{1+q_{j_{i_1}}}P_{j_{i_1},k} - M_k^TM_k,\quad \Xi^{j_{p_i}}_{i,k} = \frac{\alpha_{j_{i p_i},k}}{1+q_{j_{i p_i}}}P_{j_{i p_i},k} - M_k^TM_k.$$

Here, the subscripts $j_{i_1},\ldots,j_{i_{p_i}}$ are for neighbors of node i. Then, the complex network (10.1) is vector dissipative-like over the finite horizon $\mathscr{K}$ with respect to the vector supply the rate function $\mathbf{S}(z_k, v_k)$, the vector storage function $\mathbf{V}(x_k)$, and the dissipation matrix sequence W_k (10.8).

Proof First of all, by calculating the storage function of node i along the trajectory of the complex network (10.1), one can obtain that

$$\begin{aligned} & V_i(x_{i,k+1}) \\ &= x^T_{i,k+1}P_{i,k+1}x_{i,k+1} \\ &= 2f^T(x_{i,k})P_{i,k+1}\sum_{j\in\mathscr{N}_i\cup\{i\}} w_{ij}\Gamma x_{j,k} \end{aligned}$$

$$\begin{aligned}&+ v_{i,k}^T(L_{i,k}^T P_{i,k+1} L_{i,k} - \gamma^2 I)v_{i,k} + \gamma^2 v_{i,k}^T v_{i,k}\\&+ 2v_{i,k}^T L_{i,k}^T P_{i,k+1} f(x_{i,k}) + 2v_{i,k}^T L_{i,k}^T P_{i,k+1} \sum_{j\in\mathcal{N}_i\cup\{i\}} w_{ij}\Gamma x_{j,k}\\&+ f^T(x_{i,k})P_{i,k+1} f(x_{i,k}) + \sum_{j\in\mathcal{N}_i\cup\{i\}} x_j^T w_{ij}\Gamma P_{i,k+1} \sum_{j\in\mathcal{N}_i\cup\{i\}} w_{ij}\Gamma x_{j,k}. \end{aligned}\tag{10.10}$$

Obviously, it follows from (10.9) that the inequality $\Pi_{i,k} \triangleq \gamma^2 I - L_{i,k}^T P_{i,k+1} L_{i,k} > 0$. Via the completing square technique, the worst disturbance $v_{i,k}^* \triangleq \Pi_{i,k}^{-1} L_{i,k}^T P_{i,k+1} f(x_{i,k}) + \Pi_{i,k}^{-1} L_{i,k}^T P_{i,k+1} \sum_{j\in\mathcal{N}_i\cup\{i\}} w_{ij}\Gamma x_{j,k}$. Thus, one has

$$\begin{aligned}&v_{i,k}^T(L_{i,k}^T P_{i,k+1} L_{i,k} - \gamma^2 I)v_{i,k} + 2v_{i,k}^T L_{i,k}^T P_{i,k+1} f(x_{i,k}) + 2v_{i,k}^T L_{i,k}^T P_{i,k+1} \sum_{j\in\mathcal{N}_i\cup\{i\}} w_{ij}\Gamma x_{j,k}\\&\le f^T(x_{i,k})P_{i,k+1} L_{i,k}\Pi_{i,k}^{-1} L_{i,k}^T P_{i,k+1} f(x_{i,k})\\&\quad + 2f^T(x_{i,k})P_{i,k+1} L_{i,k}\Pi_{i,k}^{-1} L_{i,k}^T P_{i,k+1} \sum_{j\in\mathcal{N}_i\cup\{i\}} w_{ij}\Gamma x_{j,k}\\&\quad + \sum_{j\in\mathcal{N}_i\cup\{i\}} x_{j,k}^T w_{ij}\Gamma P_{i,k+1} L_{i,k}\Pi_{i,k}^{-1} L_{i,k}^T P_{i,k+1} \sum_{j\in\mathcal{N}_i\cup\{i\}} w_{ij}\Gamma x_{j,k}.\end{aligned}$$

Further, by utilizing the matrix inverse formula, we have

$$\begin{aligned}&P_{i,k+1} L_{i,k}\left(\gamma^2 I - L_{i,k}^T P_{i,k+1} L_{i,k}\right)^{-1} L_{i,k}^T P_{i,k+1} + P_{i,k+1}\\&= \left(P_{i,k+1}^{-1} - \frac{1}{\gamma^2} L_{i,k} L_{i,k}^T\right)^{-1}\\&\triangleq \Psi_{i,k+1} > 0.\end{aligned}$$

Thereby, one obtains a more compact expression

$$\begin{aligned}&\sum_{j\in\mathcal{N}_i\cup\{i\}} x_{j,k}^T w_{ij}\Gamma P_{i,k+1} \sum_{j\in\mathcal{N}_i\cup\{i\}} w_{ij}\Gamma x_{j,k} + 2f^T(x_{i,k})P_{i,k+1} \sum_{j\in\mathcal{N}_i\cup\{i\}} w_{ij}\Gamma x_{j,k}\\&+ f^T(x_{i,k})P_{i,k+1} L_{i,k}\Pi_{i,k}^{-1} L_{i,k}^T P_{i,k+1} f(x_{i,k}) + f^T(x_{i,k})P_{i,k+1} f(x_{i,k})\\&+ \sum_{j\in\mathcal{N}_i\cup\{i\}} x_{j,k}^T w_{ij}\Gamma P_{i,k+1} L_{i,k}\Pi_{i,k}^{-1} L_{i,k}^T P_{i,k+1} \sum_{j\in\mathcal{N}_i\cup\{i\}} w_{ij}\Gamma x_{j,k}\\&+ 2f^T(x_{i,k})P_{i,k+1} L_{i,k}\Pi_{i,k}^{-1} L_{i,k}^T P_{i,k+1} \sum_{j\in\mathcal{N}_i\cup\{i\}} w_{ij}\Gamma x_{j,k}\\&= f^T(x_{i,k})\Psi_{i,k+1}^{-1} f(x_{i,k}) + \sum_{j\in\mathcal{N}_i\cup\{i\}} x_{j,k}^T w_{ij}\Gamma\Psi_{i,k+1}^{-1} \sum_{j\in\mathcal{N}_i\cup\{i\}} w_{ij}\Gamma x_{j,k}\\&+ 2f^T(x_{i,k})\Psi_{i,k+1}^{-1} \sum_{j\in\mathcal{N}_i\cup\{i\}} w_{ij}\Gamma x_{j,k}\\&\triangleq \Upsilon_{i,k}.\end{aligned}$$

Thus, it follows from (10.10) that

$$V_i(x_{i,k+1}) \leq \Upsilon_{i,k} + \gamma^2 v_{i,k}^T v_{i,k}.$$

Also, by (10.2), one has

$$V_i(x_{i,k+1}) \leq \Upsilon_{i,k} + \gamma^2 v_{i,k}^T v_{i,k} - \lambda_{i,k} \left(f(x_{i,k}) - \Phi_{i,k}^1 x_{i,k}\right)^T \left(f(x_{i,k}) - \Phi_{i,k}^2 x_{i,k}\right)$$

where scalar $\lambda_{i,k} > 0$.

According to (10.9) and the Schur Complement Lemma, one acquires

$$\begin{aligned}
&\Upsilon_{i,k} + \sum_{j \in \mathscr{N}_i} \|z_{i,k} - z_{j,k}\|^2 - \lambda_{i,k} \left(f(x_{i,k}) - \Phi_{i,k}^1 x_{i,k}\right)^T \left(f(x_{i,k}) - \Phi_{i,k}^2 x_{i,k}\right) \\
&- \sum_{j \in \mathscr{N}_i} \frac{\alpha_{j,k}}{1+q_j} x_{j,k}^T P_{j,k} x_{j,k} - \frac{1+q_i(1-\alpha_{i,k})}{1+q_i} x_{i,k}^T P_{i,k} x_{i,k} \\
&= \xi_{i,k}^T \begin{bmatrix} -\Xi_{i,k}^i & * & \cdots & * & * \\ -M_k^T M_k & -\Xi_{i,k}^{j_1} & \cdots & * & * \\ \vdots & \vdots & \ddots & \vdots & \vdots \\ -M_k^T M_k & 0 & \cdots & -\Xi_{i,k}^{j_{p_i}} & * \\ \frac{\lambda_{i,k}}{2}[\Phi_{i,k}^1 + \Phi_{i,k}^2] & 0 & \cdots & 0 & -\lambda_{i,k} I \end{bmatrix} \xi_{i,k} \\
&+ \xi_{i,k}^T \begin{bmatrix} w_{ii}\Gamma \\ w_{ij_1}\Gamma \\ \cdots \\ w_{ij_{p_i}}\Gamma \\ I \end{bmatrix} \Psi_{i,k+1}^{-1} \begin{bmatrix} w_{ii}\Gamma \\ w_{ij_1}\Gamma \\ \cdots \\ w_{ij_{p_i}}\Gamma \\ I \end{bmatrix}^T \xi_{i,k} \\
&< 0
\end{aligned} \tag{10.11}$$

where $\xi_{i,k} = [x_{i,k}^T \quad x_{j_{i_1},k}^T \quad \cdots \quad x_{j_{ip_i},k}^T \quad f^T(x_{i,k})]^T$.

Thus, it follows from (10.5) and (10.11) that

$$\begin{aligned}
&V_i(x_{i,k+1}) \\
&\leq \Upsilon_{i,k} - \lambda_{i,k} \left(f_{i,k} - \Phi_{i,k}^1 x_{i,k}\right)^T \left(f_{i,k} - \Phi_{i,k}^2 x_{i,k}\right) \\
&\quad + \sum_{j \in \mathscr{N}_i} \|z_{i,k} - z_{j,k}\|^2 + \gamma^2 v_{i,k}^T v_{i,k} - \sum_{j \in \mathscr{N}_i} \|z_{i,k} - z_{j,k}\|^2 \\
&\leq S_i(z_k, v_{i,k}) + \sum_{j \in \mathscr{N}_i} \frac{\alpha_{j,k}}{1+q_j} x_{j,k}^T P_{j,k} x_{j,k} + \frac{1+q_i(1-\alpha_{i,k})}{1+q_i} x_{i,k}^T P_{i,k} x_{i,k}.
\end{aligned} \tag{10.12}$$

Then, according to this set of coefficients $\frac{a_{i1}\alpha_{1,k}}{1+q_1}, \ldots, \frac{1+q_i(1-\alpha_{i,k})}{1+q_i}, \ldots,$ and $\frac{a_{iN}\alpha_{N,k}}{1+q_N}$, matrix W_k of form (10.8) can be constructed. It is easily verified that 1) W_k is strictly diagonally dominant since $\alpha_{i,k} \in \mathfrak{I}_{q_i}$; 2) $\mathbf{1}^T W_k \leq\leq \mathbf{1}^T$. As a result, W_k is the desired dissipation matrix. Via the constructed W_k, (10.12) becomes

$$V_i(x_{i,k+1}) \leq [W_k \mathbf{V}(x_k)]_i + S_i(z_k, v_{i,k}) \tag{10.13}$$

where $[W_k \mathbf{V}(x_k)]_i$ is the i-th element of vector $W_k \mathbf{V}(x_k)$. Since (10.13) holds for $\forall\, i \in \mathscr{V}$, it is rewritten as a vector form:

$$\mathbf{V}(x_{k+1}) \leq\leq W_k \mathbf{V}(x_k) + \mathbf{S}(z_k, v_k). \tag{10.14}$$

Finally, from Definition 10.2, the proof of Lemma 10.1 is complete.

Remark 10.2 Actually, for establishing the vector dissipation inequality, it is of challenge to construct a nonsingular column substochastic dissipation matrix (10.8). Firstly, it is well-known that a strictly diagonally dominant matrix is nonsingular. Obviously, matrix (10.8) is strictly diagonally dominant in the light of $\alpha_{i,k} \in \mathfrak{I}_{q_i}$. Secondly, a substochastic matrix is nonnegative. Matrix (10.8) is nonnegative from $\alpha_{i,k} \in \mathfrak{I}_{q_i}$. Thirdly, the column sums of a substochastic matrix are all less than or equal to 1. Actually, no matter what $\alpha_{i,k}$ is, this can always be guaranteed by the structure of matrix (10.8).

In the following, the obtained vector dissipation inequality (10.14) needs to be transformed into the desirable bounded H_∞ synchronization performance constraint (10.4).

Theorem 10.1 *For any node $i \in \mathscr{V}$ and its neighbors $j \in \mathscr{N}_i$, let the scalar sequences $\alpha_{i,k} \in \mathfrak{I}_{q_i}$, $\alpha_{j,k} \in \mathfrak{I}_{q_j}$, initial weighted matrices $S_i > 0$ and $S_j > 0$, and estimator parameters $\{K_{i,k}\}_{k \in \mathscr{K}}$ be given. The complex network (10.1) is said to be bounded H_∞ synchronized over the finite horizon $\mathscr{K}$ with the disturbance attention $\gamma > 0$, if there exist a sequence of positive scalars $\{\lambda_{i,k}\}_{k \in \mathscr{K}}$ and $p_i + 1$ sequences of positive definite matrices $\{P_{i,k}\}_{k \in \mathscr{K} \cup \{n\}}$, $\{P_{j,k}\}_{k \in \mathscr{K} \cup \{n\}}$ with the initial condition $P_{i,0} \leq \gamma^2 S_i$, $P_{j,0} \leq \gamma^2 S_j$ satisfying the RLMIs (10.9) for any $k \in \mathscr{K}$.*

Proof First of all, from the definition of the substochastic matrix, it is pre-multiplied by $\mathbf{1}^T$ on both sides of the vector dissipation inequality (10.14), then one has

$$\mathbf{1}^T \mathbf{V}(x_{k+1}) \leq \mathbf{1}^T \mathbf{V}(x_k) + \mathbf{1}^T \mathbf{S}(z_k, v_k).$$

Obviously, $\mathbf{1}^T \mathbf{V}(x_k) = \sum_{i=1}^{N} V_i(x_{i,k})$ and $\mathbf{1}^T \mathbf{S}(z_k, v_k) = \sum_{i=1}^{N} S_i(z_k, v_{i,k})$. Then, one has

$$\sum_{i=1}^{N} V_i(x_{i,k+1}) \leq \sum_{i=1}^{N} V_i(x_{i,k}) + \sum_{i=1}^{N} S_i(z_k, v_{i,k}). \tag{10.15}$$

It follows from (10.15) that

$$\sum_{i=1}^{N}[V_i(x_{i,k+1}) - V_i(x_{i,k})] + \sum_{i=1}^{N}\sum_{j\in\mathcal{N}_i} \|z_{i,k} - z_{j,k}\|^2 \leq \gamma^2 \sum_{i=1}^{N} v_{i,k}^T v_{i,k}, \quad (10.16)$$

which implies that

$$\sum_{i=1}^{N}[V_i(x_{i,n}) - V_i(x_{i,0})] + \sum_{k=0}^{n-1}\sum_{i=1}^{N}\sum_{j\in\mathcal{N}_i} \|z_{i,k} - z_{j,k}\|^2 \leq \gamma^2 \sum_{k=0}^{n-1}\sum_{i=1}^{N} v_{i,k}^T v_{i,k}. \quad (10.17)$$

That is,

$$\frac{1}{N}\sum_{k=0}^{n-1}\sum_{i=1}^{N}\sum_{j\in\mathcal{N}_i} \|z_{i,k} - z_{j,k}\|^2 \leq \frac{1}{N}\sum_{i=1}^{N}\left(x_{i,0}^T P_{i,0} x_{i,0} + \gamma^2 \sum_{k=0}^{n-1}\sum_{i=1}^{N} v_{i,k}^T v_{i,k}\right).$$

Finally, in view of the initial condition $P_{i,0} \leq \gamma^2 S_i$, the bounded H_∞ synchronization performance constraint (10.4) is satisfied. Therefore, the proof is complete.

Remark 10.3 Owing to the local performance analysis method, the developed algorithm in this chapter is composed of sub-algorithms corresponding to each node in the complex network. Furthermore, the desirable performance is achieved by the cooperation of these sub-algorithms among neighboring nodes. Such algorithm structure can yield the following merits: 1) low complexity, that is, the computational burden is dispersed among massive nodes in the complex networks; 2) scalability, this design method is convenient to adapt the variation of the network topology, such as, some nodes lost their connections from other nodes, or new nodes join the network; and 3) distributed execution, every sub-algorithm can be executed on the corresponding node by receiving neighbor information, which is an indeed distributed calculations.

Remark 10.4 Theorem 10.1 indicates the locally coupled conditions of the synchronization problem, where the synchronization criteria can be verified in a cooperative manner by all nodes. Next, one analyzes the scheme complexity on node i ($i \in \mathcal{V}$). As is well known, the polynomial-time complexity of a standard LMI system is bounded by $O(\mathcal{M}\mathcal{N}^3 \log(\mathcal{V}/\varepsilon))$, where $\mathcal{M}$ is the total row size of the LMI system, $\mathcal{N}$ is the total number of scalar decision variables, $\mathcal{V}$ is a data-dependent scaling factor, and ε is the relative accuracy set for the scheme. From (10.9), it obtains that $\mathcal{M} = (p_i + 2)n_x + n_{v_i}$ and $\mathcal{N} = \frac{(1+n_x)n_x}{2} + 1$. When it is executed recursively n steps, the computational complexity for node i can be represented as $O(np_i n_x^7)$, that is, it is related not to the network size N but to the number of neighbors p_i. No matter how large the size of the network is, the sub-algorithm burden on each node only depends on the in-degree p_i. Nevertheless, the burden of the criteria based on the globally augmented method in [2, 3, 5–7] depends polynomially on the network size N. Therefore, the complexity of our scheme is significantly reduced.

10.3 Problem Formulation of Finite-Horizon H_∞ Filtering

In this chapter, the measurement output of node i with missing measurements is modeled by

$$y_{i,k} = \gamma_{i,k} C_{i,k} x_{i,k} + D_{i,k} w_{i,k} \tag{10.18}$$

where $y_{i,k} \in \mathbb{R}^{n_y}$ is the measurement output of node i, $w_{i,k}$ is the disturbance input belonging to $l_2[0, n-1]$. $C_{i,k}$ and $D_{i,k}$ are known matrices with appropriate dimensions. Additionally, $\gamma_{i,k}$ is a Bernoulli distributed white sequence of taking values of 1 or 0 with $\text{Prob}\{\gamma_{i,k} = 1\} = \bar{\gamma}_i$ and $\text{Prob}\{\gamma_{i,k} = 0\} = 1 - \bar{\gamma}_i$. Here, $\bar{\gamma}_i \in [0, 1]$ is a known constant. In addition, it is assumed that all stochastic variables $\gamma_{i,k}$ are independent in k and i.

The filter structure of node i is given as follows:

$$\begin{cases} \hat{x}_{i,k+1} = f(\hat{x}_{i,k}) + \sum_{j=1}^{N} w_{ij} \Gamma \hat{x}_{j,k} + K_{i,k}(y_{i,k} - \bar{\gamma}_i C_{i,k} \hat{x}_{i,k}), \\ \hat{z}_{i,k} = M_k \hat{x}_{i,k} \end{cases} \tag{10.19}$$

where $K_{i,k}$ is the filter parameter to be calculated later.

By setting $e_{i,k+1} \triangleq x_{i,k+1} - \hat{x}_{i,k+1}$ and $\tilde{z}_{i,k} \triangleq z_{i,k} - \hat{z}_{i,k}$, the filtering error dynamics of node i is obtained from (10.1), (10.18), and (10.19) as follows:

$$\begin{cases} e_{i,k+1} = \tilde{f}_{i,k} + \sum_{j=1}^{N} w_{ij} \Gamma e_{j,k} + L_{i,k} v_{i,k} - K_{i,k} D_{i,k} w_{i,k} \\ \qquad - \bar{\gamma}_i K_{i,k} C_{i,k} e_{i,k} - (\gamma_{i,k} - \bar{\gamma}_i) K_{i,k} C_{i,k} x_{i,k}, \\ \tilde{z}_{i,k} = M_k e_{i,k} \end{cases} \tag{10.20}$$

where

$$\tilde{f}_{i,k} = f(x_{i,k}) - f(\hat{x}_{i,k}).$$

Further, denote $\eta_{i,k} \triangleq [x_{i,k}^T \ e_{i,k}^T]^T$ and $\tilde{v}_{i,k} \triangleq [v_{i,k}^T \ w_{i,k}^T]^T$. One obtains the following locally augmented dynamics as follows:

$$\begin{cases} \eta_{i,k+1} = f_{i,k} + \mathscr{L}_{i,k} \tilde{v}_{i,k} - \bar{\gamma}_i \mathscr{K}_{i,k} \eta_{i,k} - (\gamma_{i,k} - \bar{\gamma}_i) \mathscr{C}_{i,k} \eta_{i,k} + \sum_{j=1}^{N} \mathscr{W}_{ij} \eta_{j,k}, \\ \tilde{z}_{i,k} = \mathscr{M}_k \eta_{i,k} \end{cases} \tag{10.21}$$

where

$$f_{i,k} = \left[f^T(x_{i,k}) \;\; \tilde{f}_{i,k}^T \right]^T, \quad \mathscr{M}_k = [0 \;\; M_k],$$
$$\mathscr{W}_{ij} = \text{diag}\left\{ w_{ij}\Gamma, w_{ij}\Gamma \right\}, \quad \mathscr{K}_{i,k} = \text{diag}\left\{ 0, K_{i,k}C_{i,k} \right\},$$
$$\mathscr{L}_{i,k} = \begin{bmatrix} L_{i,k} & 0 \\ L_{i,k} & -K_{i,k}D_{i,k} \end{bmatrix}, \quad \mathscr{C}_{i,k} = \begin{bmatrix} 0 & 0 \\ K_{i,k}C_{i,k} & 0 \end{bmatrix}.$$

From (10.2), one has

$$\left(f_{i,k} - \bar{\Phi}_{i,k}^1 \eta_{i,k} \right)^T \left(f_{i,k} - \bar{\Phi}_{i,k}^2 \eta_{i,k} \right) \leq 0 \tag{10.22}$$

where $\bar{\Phi}_{i,k}^1 = \text{diag}\left\{ \Phi_{i,k}^1, \Phi_{i,k}^1 \right\}$ and $\bar{\Phi}_{i,k}^2 = \text{diag}\left\{ \Phi_{i,k}^2, \Phi_{i,k}^2 \right\}$.

Definition 10.3 Given a set of weighted matrices $S_i > 0$, the locally augmented system (10.21) is said to satisfy the H_∞ performance constraint over the finite horizon $\mathscr{K}$ with the disturbance attention level $\gamma > 0$ if the following inequality holds

$$\frac{1}{N} \sum_{i=1}^{N} \sum_{k=0}^{n-1} \mathbb{E}\|\tilde{z}_{i,k}\|^2 \leq \frac{\gamma^2}{N} \sum_{i=1}^{N} \left(\eta_{i,0}^T S_i \eta_{i,0} + \sum_{k=0}^{n-1} \|\tilde{v}_{i,k}\|^2 \right). \tag{10.23}$$

In the following, one figures out to design a set of filters of the form (10.19) on each node for the complex network (10.1) with the measurement model (10.18). In other words, we are going to design the filter parameters $K_{i,k}$ such that the locally augmented dynamics (10.21) satisfies the H_∞ performance constraint (10.23) over the finite horizon $\mathscr{K}$.

10.4 Finite-Horizon H_∞ Filtering for Complex Networks

Now, we are in a position to address the finite-horizon H_∞ filteringH_∞ filtering along the similar line of the above synchronization problem. Thus, the detailed proofs of following Lemma 10.2, Theorems 10.2 and 10.3 are all omitted. First of all, the definition of stochastic vector dissipativity-like over the finite horizon is presented for the locally augmented dynamics (10.21). From the H_∞ performance constraint (10.23) in Definition 10.3, take the supply rate function of node i as

$$S_i(\tilde{z}_{i,k}, \tilde{v}_{i,k}) = \gamma^2 \tilde{v}_{i,k}^T \tilde{v}_{i,k} - \tilde{z}_{i,k}^T \tilde{z}_{i,k}. \tag{10.24}$$

For simplicity, let

$$\tilde{z}_k \triangleq \left[\tilde{z}_{1,k}^T \;\; \cdots \;\; \tilde{z}_{N,k}^T \right]^T,$$
$$\tilde{v}_k \triangleq \left[\tilde{v}_{1,k}^T \;\; \cdots \;\; \tilde{v}_{N,k}^T \right]^T,$$
$$\eta_k \triangleq \left[\eta_{1,k}^T \;\; \cdots \;\; \eta_{N,k}^T \right]^T.$$

Definition 10.4 The locally augmented system (10.21) is said to be stochastic vector dissipative-like over the finite-horizon $\mathscr{K}$ regarding a vector of supply rate functions $\mathbf{S}(\tilde{z}_k, \tilde{v}_k) \triangleq [S_1(\tilde{z}_{1,k}, \tilde{v}_{1,k}) \ \cdots \ S_N(\tilde{z}_{N,k}, \tilde{v}_{N,k})]^T$ if there exists a vector of continuous nonnegative definite storage functions $\mathbf{V}(\eta_k) = [V_1(\eta_{1,k}) \ \cdots \ V_N(\eta_{N,k})]^T$ with $\mathbf{V}(0) = 0$, and a sequence of nonsingular column substochastic matrices $W_k \in \mathbb{R}^{N \times N}$ such that the vector dissipation inequality is satisfied for any $k \in \mathscr{K}$:

$$\mathbb{E}\{\mathbf{V}(\eta_{k+1})\} \leq\leq W_k \mathbb{E}\{\mathbf{V}(\eta_k)\} + \mathbb{E}\{\mathbf{S}(\tilde{z}_k, \tilde{v}_k)\}. \tag{10.25}$$

For further analysis, the Lyapunov function $V_i(x_{i,k}) = x_{i,k}^T P_{i,k} x_{i,k}$ ($P_{i,k} > 0$ is unknown) is chosen as the storage function of node i.

Lemma 10.2 *For any node $i \in \mathscr{V}$ and its neighbors $j \in \mathscr{N}_i$, let the scalar $\gamma > 0$, scalar sequences $\{\alpha_{i,k}\}_{k \in \mathscr{K}} \in \mathfrak{I}_{q_i}$, $\{\alpha_{j,k}\}_{k \in \mathscr{K}} \in \mathfrak{I}_{q_j}$, and estimator parameter sequence $\{K_{i,k}\}_{k \in \mathscr{K}}$ be given. If there exists a sequence of positive scalars $\{\mu_{i,k}\}_{k \in \mathscr{K}}$, $p_i + 1$ sequences of positive definite matrices $\{P_{i,k}\}_{k \in \mathscr{K} \cup \{n\}}$ and $\{P_{j,k}\}_{k \in \mathscr{K} \cup \{n\}}$ such that the following matrix inequalities hold for any $k \in \mathscr{K}$*

$$\begin{bmatrix} \Upsilon_{i,k}^{i} & * & \cdots & * & * \\ \mathscr{W}_{ij_1}^T \Psi_{i,k}^{-1} (\mathscr{W}_{ii} - \bar{\gamma}_i \mathscr{K}_{i,k}) & -\Upsilon_{i,k}^{j_1} & \cdots & * & * \\ \vdots & \vdots & \ddots & \vdots & \vdots \\ \mathscr{W}_{ij_{p_i}}^T \Psi_{i,k}^{-1} (\mathscr{W}_{ii} - \bar{\gamma}_i \mathscr{K}_{i,k}) & \mathscr{W}_{ij_{p_i}}^T \Psi_{i,k}^{-1} \mathscr{W}_{ij_1} & \cdots & -\Upsilon_{i,k}^{j_{p_i}} & * \\ \frac{\mu_{i,k}}{2} [\bar{\Phi}_{i,k}^1 + \bar{\Phi}_{i,k}^2] + \Psi_{i,k}^{-1} (\mathscr{W}_{ii} - \bar{\gamma}_i \mathscr{K}_{i,k}) & \Psi_{i,k}^{-1} \mathscr{W}_{ij_1} & \cdots & \Psi_{i,k}^{-1} \mathscr{W}_{ij_{p_i}} & -\mu_{i,k} I + \Psi_{i,k}^{-1} \end{bmatrix} < 0 \tag{10.26}$$

where

$$\Psi_{i,k} = P_{i,k+1}^{-1} - \frac{1}{\gamma^2} \mathscr{L}_{i,k} \mathscr{L}_{i,k}^T > 0, \quad \Upsilon_{i,k}^{j_1} = \frac{\alpha_{j_{i_1},k}}{1 + q_{j_{i_1}}} P_{j_{i_1},k} + \mathscr{W}_{ij_1}^T \Psi_{i,k}^{-1} \mathscr{W}_{ij_1},$$

$$\Upsilon_{i,k}^{j_{i p_i}} = \frac{\alpha_{j_{i p_i},k}}{1 + q_{j_{i p_i}}} P_{j_{i p_i},k} + \mathscr{W}_{ij_{p_i}}^T \Psi_{i,k}^{-1} \mathscr{W}_{ij_{p_i}},$$

$$\begin{aligned} \Upsilon_{i,k}^{i} = {} & \mathscr{M}_k^T \mathscr{M}_k + \bar{\gamma}_i (1 - \bar{\gamma}_i) \mathscr{C}_{i,k}^T P_{i,k+1} \mathscr{C}_{i,k} + (\mathscr{W}_{ii} - \bar{\gamma}_i \mathscr{K}_{i,k})^T \Psi_{i,k}^{-1} (\mathscr{W}_{ii} - \bar{\gamma}_i \mathscr{K}_{i,k}) \\ & - \frac{1 + q_i (1 - \alpha_{i,k})}{1 + q_i} P_{i,k} - \frac{\mu_{i,k}}{2} \left((\bar{\Phi}_{i,k}^1)^T \bar{\Phi}_{i,k}^2 + (\bar{\Phi}_{i,k}^2)^T \bar{\Phi}_{i,k}^1 \right). \end{aligned}$$

Here, the subscripts $j_1, \cdots, j_{p_i}$ are for neighbors of node i. Then, the local augment dynamics (10.21) is stochastic vector dissipative-like over the finite-horizon $\mathscr{K}$ regarding to the vector supply rate function $\mathbf{S}(\tilde{z}_k, \tilde{v}_k)$, the vector storage function $\mathbf{V}(\eta_k)$, and the dissipation matrix W_k (10.8).

Theorem 10.2 *For any node $i \in \mathscr{V}$ and its neighbors $j \in \mathscr{N}_i$, let scalar sequences $\{\alpha_{i,k}\}_{k\in\mathscr{K}} \in \mathfrak{I}_{q_i}$, $\{\alpha_{j,k}\}_{k\in\mathscr{K}} \in \mathfrak{I}_{q_j}$, initial weight matrices $S_i > 0$, $S_j > 0$, and estimator parameter sequence $\{K_{i,k}\}_{k\in\mathscr{K}}$ be given. If there exists a sequence of positive scalars $\{\mu_{i,k}\}_{k\in\mathscr{K}}$, $p_i + 1$ sequences of positive definite matrices $\{P_{i,k}\}_{k\in\mathscr{K}\cup\{n\}}$, $\{P_{j,k}\}_{k\in\mathscr{K}\cup\{n\}}$ with the initial condition $P_{i,0} \le \gamma^2 S_i$, $P_{j,0} \le \gamma^2 S_j$ such that the matrix inequalities (10.26) hold for any $k \in \mathscr{K}$. Then, the local augment dynamics (10.21) satisfies the H_∞ performance constraint (10.23) with the disturbance attenuation level $\gamma > 0$.*

Theorem 10.3 *For any node $i \in \mathscr{V}$ and its neighbors $j \in \mathscr{N}_i$, let the disturbance attenuation level $\gamma > 0$, positive definite matrices S_{1i}, S_{2i}, S_{1j}, S_{2j}, and scalar sequences $\{\alpha_{i,k}\}_{k\in\mathscr{K}} \in \mathfrak{I}_{q_i}$, $\{\alpha_{j,k}\}_{k\in\mathscr{K}} \in \mathfrak{I}_{q_j}$ be given. The finite-horizon H_∞ estimate problem for the complex network (10.1) is solvable, if there exists a set of positive scalars $\{\mu_{i,k}\}_{k\in\mathscr{K}}$, a sequence of real matrices $\{E_{i,k}\}_{k\in\mathscr{K}}$, $2(p_i + 1)$ sequences of positive definite matrices $\{P^1_{i,k}\}_{k\in\mathscr{K}\cup\{n\}}$, $\{P^2_{i,k}\}_{k\in\mathscr{K}\cup\{n\}}$, $\{P^1_{j,k}\}_{k\in\mathscr{K}\cup\{n\}}$ and $\{P^2_{j,k}\}_{k\in\mathscr{K}\cup\{n\}}$ with the initial conditions $P^1_{i,0} \le \gamma^2 S_{1i}$, $P^2_{i,0} \le \gamma^2 S_{2i}$, $P^1_{j,0} \le \gamma^2 S_{1j}$ and $P^2_{j,0} \le \gamma^2 S_{2j}$ such that the following RLMIs hold for any $k \in \mathscr{K}$*

$$\begin{bmatrix} -\bar{\Upsilon}^i_{i,k} & * & * & * & * & * \\ 0 & -\Phi^{22}_{i,k} & * & * & * & * \\ \Phi^{31}_{i,k} & 0 & -\mu_{i,k} I & * & * & * \\ \Phi^{41}_{i,k} & \Phi^{42}_{i,k} & I & -P_{i,k+1} & * & * \\ \sigma_i P_{i,k+1}\mathscr{C}_{i,k} & 0 & 0 & 0 & -P_{i,k+1} & * \\ 0 & 0 & 0 & \mathscr{L}^T_{i,k} P_{i,k+1} & 0 & -\gamma^2 \end{bmatrix} < 0 \tag{10.27}$$

where

$$\bar{\Upsilon}^i_{i,k} = \frac{1 + q_i(1 - \alpha_{i,k})}{1 + q_i} P_{i,k} - \mathscr{M}_k^T \mathscr{M}_k + \frac{\mu_{i,k}}{2}\left[(\bar{\Phi}^1_{i,k})^T \bar{\Phi}^2_{i,k} + (\bar{\Phi}^2_{i,k})^T \bar{\Phi}^1_{i,k}\right],$$

$$\Phi^{22}_{i,k} = \mathrm{diag}\left\{\frac{\alpha_{j_{i_1},k}}{1 + q_{j_{i_1}}} P_{j_{i_1},k}, \ldots, \frac{\alpha_{j_{i_{p_i}},k}}{1 + q_{j_{i_{p_i}}}} P_{j_{i_{p_i}},k}\right\}, \quad \Phi^{31}_{i,k} = \frac{\mu_{i,k}}{2}\left[\bar{\Phi}^1_{i,k} + \bar{\Phi}^2_{i,k}\right],$$

$$\Phi^{41}_{i,k} = P_{i,k+1}(\mathscr{W}_{ii} - \bar{\gamma}_i \mathscr{K}_{i,k}), \quad \Phi^{42}_{i,k} = P_{i,k+1}\mathscr{W}_{\mathscr{N}_i}, \quad \sigma_i = \sqrt{\bar{\gamma}_i(1 - \bar{\gamma}_i)},$$

$$P_{i,k+1} = \mathrm{diag}\left\{P^1_{i,k+1}, P^2_{i,k+1}\right\}, \quad E_{i,k} = P^2_{i,k+1} K_{i,k}, \quad \mathscr{W}_{\mathscr{N}_i} = \begin{bmatrix}\mathscr{W}_{ij_1} & \cdots & \mathscr{W}_{ij_{p_i}}\end{bmatrix}.$$

Moreover, the filter parameters $K_{i,k}$ are given as follows:

$$K_{i,k} = (P^2_{i,k+1})^{-1} E_{i,k}. \tag{10.28}$$

The proofs of Lemma 10.2, Theorems 10.2 and 10.3 are all omitted due to the similarity to Lemma 10.1 and Theorem 10.1.

10.5 Illustrative Examples

In this section, two simulation examples are presented to illustrate the effectiveness of the synchronization criteria and filter design method developed in this chapter.

Consider a CN which has 10 nodes. Its topology is represented by a digraph $\mathscr{G} = \{\mathscr{V}, \mathscr{E}, W\}$ with the set of nodes $\mathscr{V} = \{1, 2, 3, 4, 5, 6, 7, 8, 9, 10\}$, the set of edges $\mathscr{E} = \{(2, 1), (3, 2), (4, 3), (5, 4), (6, 5), (7, 6), (8, 7), (9, 8), (10, 9), (1, 10)\}$.

The coupling configuration matrix $W = [W_{ij}]$ is given as follows: $W_{ii} = -0.1$, $W_{1,10} = W_{21} = W_{32} = W_{43} = W_{54} = W_{65} = W_{76} = W_{87} = W_{98} = W_{10,9} = 0.1$, otherwise $W_{ij} = 0$.

The inner-coupling matrix and the output matrix are given as $\Gamma = \text{diag}\{0.1, 0.1\}$ and $M_k = [0.1\ 0.1]$. The disturbance matrices are presented as follows:

$$
\begin{aligned}
&L_{1,k} = \begin{bmatrix} 0.01\ 0.01 \end{bmatrix}, \quad L_{2,k} = \begin{bmatrix} 0\ 0.01 \end{bmatrix}, \quad L_{3,k} = \begin{bmatrix} 0.1\ 0 \end{bmatrix}, \quad L_{4,k} = \begin{bmatrix} 0.015\ 0.01 \end{bmatrix}, \\
&L_{5,k} = \begin{bmatrix} 0.02\ 0.01 \end{bmatrix}, \quad L_{6,k} = \begin{bmatrix} 0.01\ 0.02 \end{bmatrix}, \quad L_{7,k} = \begin{bmatrix} 0.012\ -0.015 \end{bmatrix}, \\
&L_{8,k} = \begin{bmatrix} -0.01\ 0.01 \end{bmatrix}, \quad L_{9,k} = \begin{bmatrix} 0.01\ -0.01 \end{bmatrix}, \quad L_{10,k} = \begin{bmatrix} 0.013\ 0.012 \end{bmatrix}.
\end{aligned}
$$

The nonlinear vector-valued function $f(x_{i,k})$ is chosen as

$$
f(x_{i,k}) = \begin{bmatrix} x_{i,k}^{(1)} + 0.3x_{i,k}^{(2)} - \tanh(0.3x_{i,k}^{(1)}) \\ -x_{i,k}^{(2)} + 0.4x_{i,k}^{(1)} + \tanh(0.2x_{i,k}^{(2)}) \end{bmatrix}
$$

where $x_{i,k}^{(j)} (j = 1, 2)$ is the j-th element of $x_{i,k}$. Then, it is easy to see that the constraint (10.2) is met with

$$
\Phi_{i,k}^1 = \begin{bmatrix} 0.7 & 0.3 \\ 0.4 & -1 \end{bmatrix}, \quad \Phi_{i,k}^2 = \begin{bmatrix} 1 & 0.3 \\ 0.4 & -0.8 \end{bmatrix}.
$$

In addition, set the external disturbance as $v_{i,k} = 0.1 * \cos(5k)$, the time horizon $n - 1$ as 50, and the initial values of the complex network (10.1) as

$$
\begin{aligned}
&\psi_1 = [-0.3\ 0.5]^T, \quad \psi_2 = [-0.1\ 0.4]^T, \quad \psi_3 = [-0.2\ -0.5]^T, \quad \psi_4 = [-0.4\ 0.6]^T, \\
&\psi_5 = [-0.5\ 0.7]^T, \quad \psi_6 = [-0.6\ -0.5]^T, \quad \psi_7 = [-0.3\ -0.4]^T, \\
&\psi_8 = [-0.4\ 0.5]^T, \quad \psi_9 = [-0.2\ 0.5]^T, \quad \psi_{10} = [-0.1\ 0.8]^T.
\end{aligned}
$$

An example for the bounded H_∞ synchronization criterion:
First of all, the disturbance attenuation level γ, the scalar parameter $\alpha_{i,k}$, the performance weighted matrix $Q_{i,k}$, the initial weighted matrix S_i and $P_{i,0}$ are given as $\gamma = 0.01, \alpha_{i,k} \equiv 0.8, Q_{i,k} \equiv I, S_i = P_{i,0} = I\ (i \in \mathscr{V})$. Then, according to Theorem 10.1, we can conclude that the complex network (10.1) is bounded H_∞ synchronized by using the YALMIP toolbox. Finally, by utilizing the initial values of the

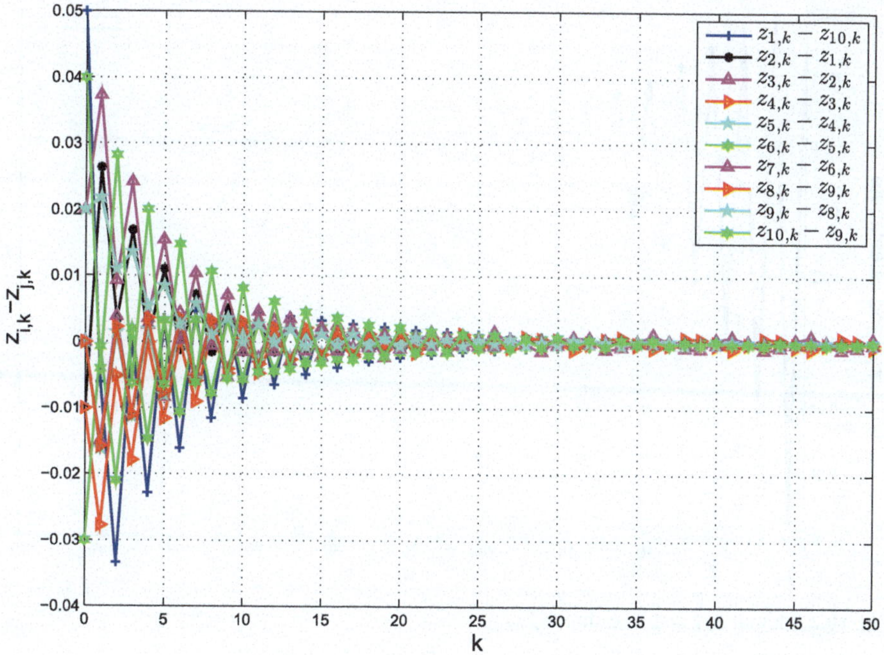

Fig. 10.1 Synchronization errors $z_{i,k} - z_{j,k}$ $(i = 1, 2, \cdots, 10, j \in \mathcal{N}_i)$

complex network, the synchronization errors between every node and its neighbors are depicted in Fig. 10.1. From this figure, the synchronization of different nodes has a satisfactory presentation. As a result, it has been confirmed that the proposed synchronization criteria in this chapter is indeed effective.

An example for the H_∞ filtering algorithm.
First and foremost, the missing measurement probability $\bar{\gamma}_i$, the disturbance attenuation level γ, the scalar parameter $\alpha_{i,k}$, the performance weight matrix $Q_{i,k}$, the initial weighted matrix S_i and $P_{i,0}$ are taken as $\bar{\gamma}_i = 0.8$, $\gamma = \sqrt{0.05}$, $\alpha_{i,k} \equiv 0.6$, $Q_{i,k} \equiv I$, $S_i = P_{i,0} = 3I$ $(i \in \mathcal{V})$, respectively. Then, by using the YALMIP toolbox, we can obtain all estimator parameters $K_{i,k}$ $(i \in \mathcal{V})$ from Theorem 10.3. Next, set the initial values of the estimates as

$$\hat{x}_{1,0} = [-0.2\ 0.4]^T,\ \hat{x}_{2,0} = [-0.2\ 0.3]^T,\ \hat{x}_{3,0} = [-0.25\ 0.6]^T,\ \hat{x}_{4,0} = [-0.3\ 0.5]^T,$$
$$\hat{x}_{5,0} = [-0.4\ 0.6]^T,\ \hat{x}_{6,0} = [-0.5\ 0.6]^T,\ \hat{x}_{7,0} = [-0.4\ 0.5]^T,\ \hat{x}_{8,0} = [-0.5\ 0.6]^T,$$
$$\hat{x}_{9,0} = [-0.3\ 0.4]^T,\ \hat{x}_{10,0} = [-0.2\ 0.7]^T.$$

Simulation results are shown in Figs. 10.1, 10.2, 10.3, 10.4, 10.5, 10.6, 10.7, 10.8, 10.9, 10.10 and 10.11, which depict the output and its estimate of node i $(i = 1, 2, \ldots, 10)$, respectively. From these figures, it can be concluded that the

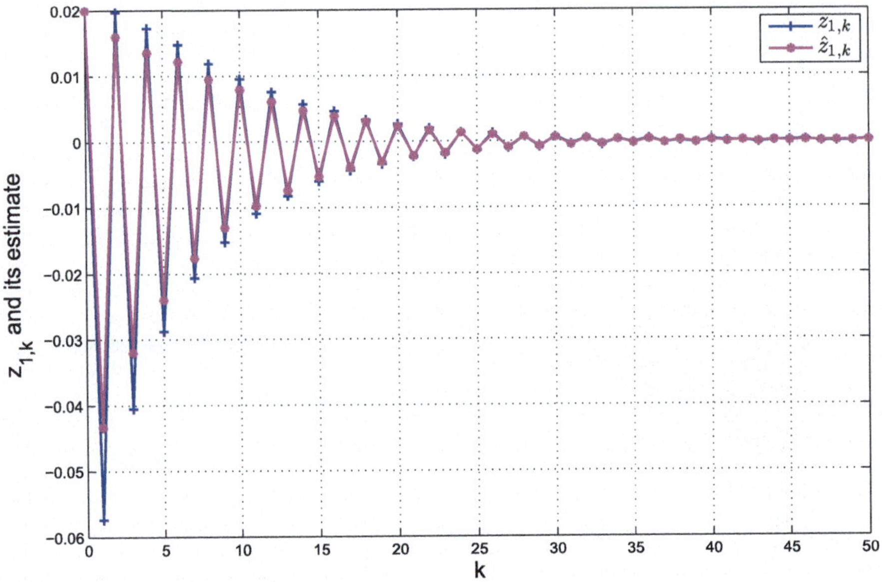

Fig. 10.2 Output $z_{1,k}$ and its estimate $\hat{z}_{1,k}$

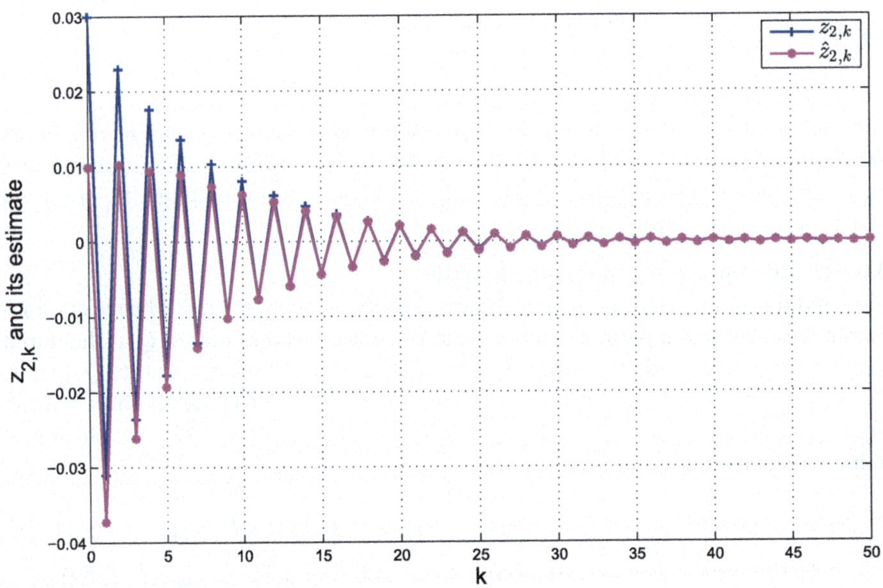

Fig. 10.3 Output $z_{2,k}$ and its estimate $\hat{z}_{2,k}$

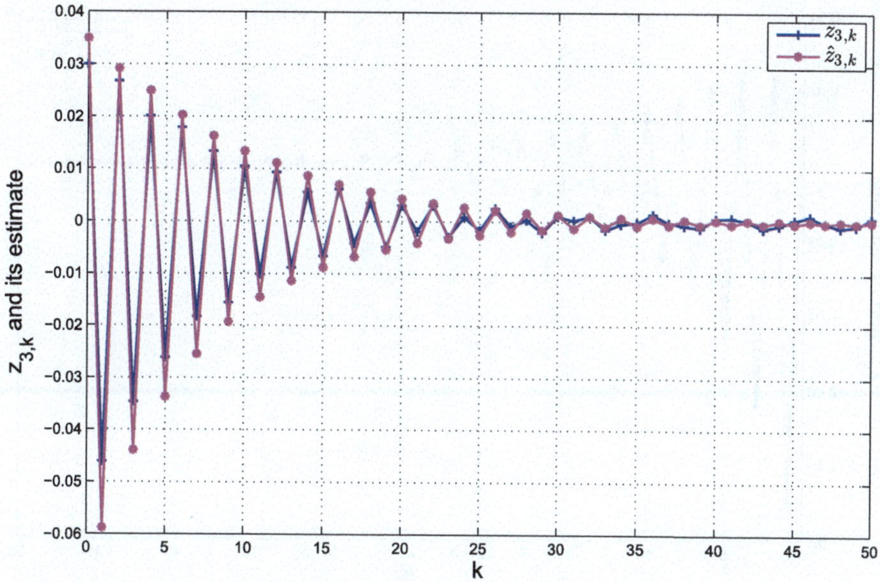

Fig. 10.4 Output $z_{3,k}$ and its estimate $\hat{z}_{3,k}$

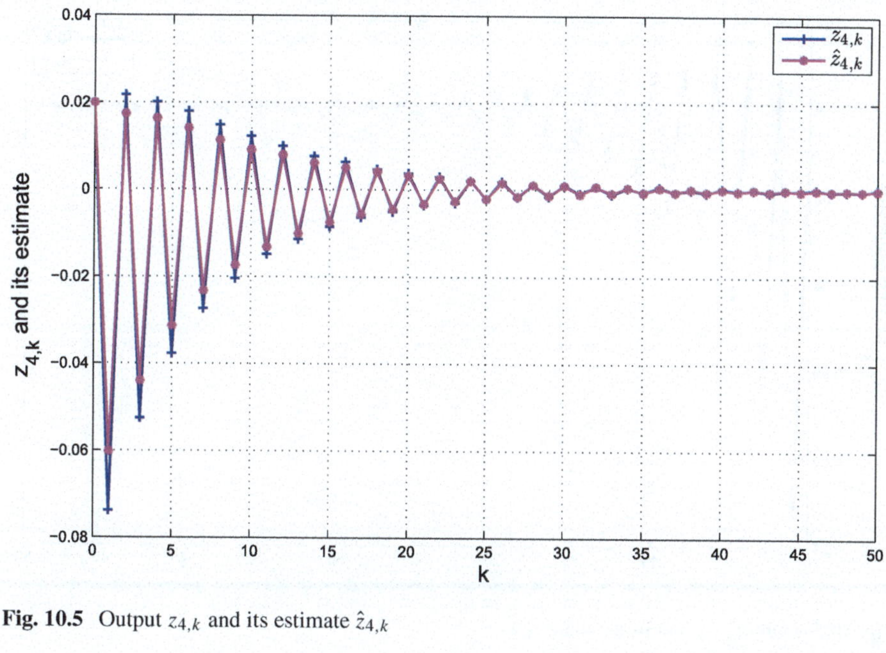

Fig. 10.5 Output $z_{4,k}$ and its estimate $\hat{z}_{4,k}$

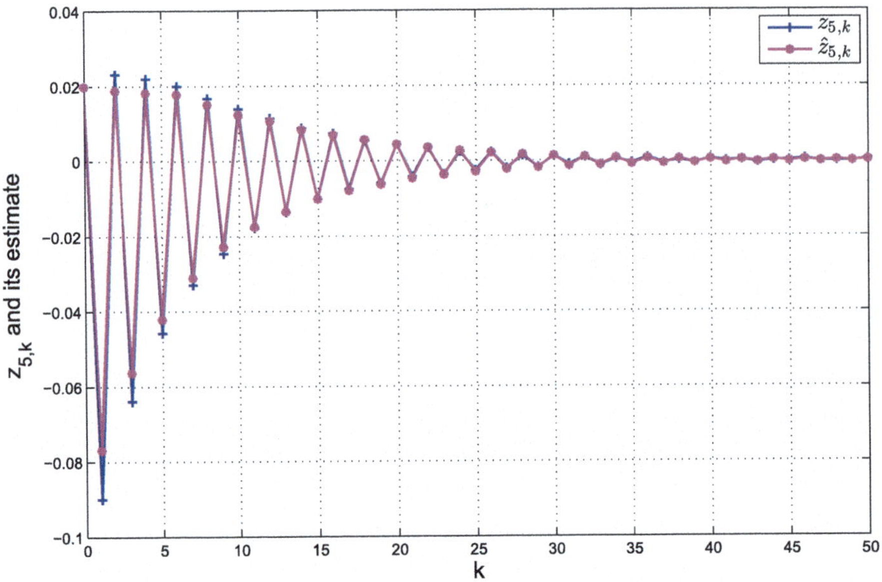

Fig. 10.6 Output $z_{5,k}$ and its estimate $\hat{z}_{5,k}$

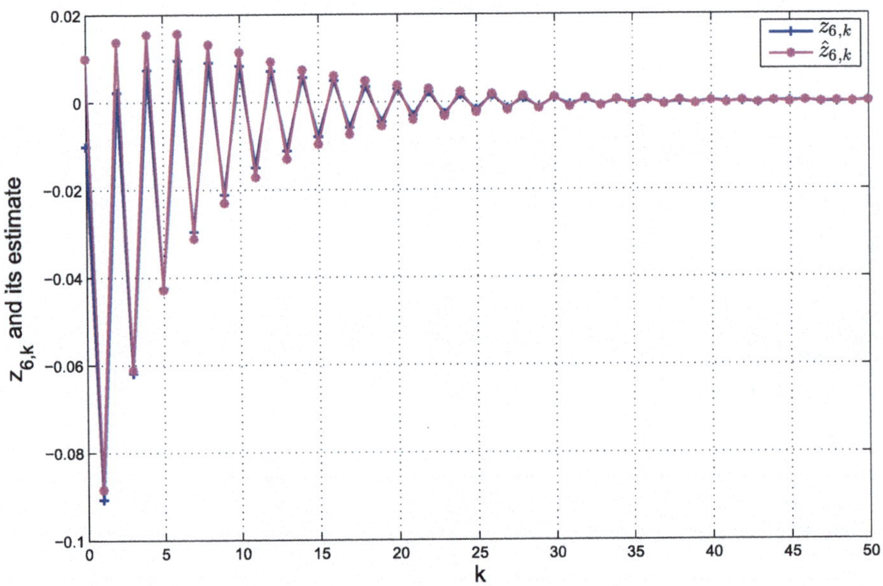

Fig. 10.7 Output $z_{6,k}$ and its estimate $\hat{z}_{6,k}$

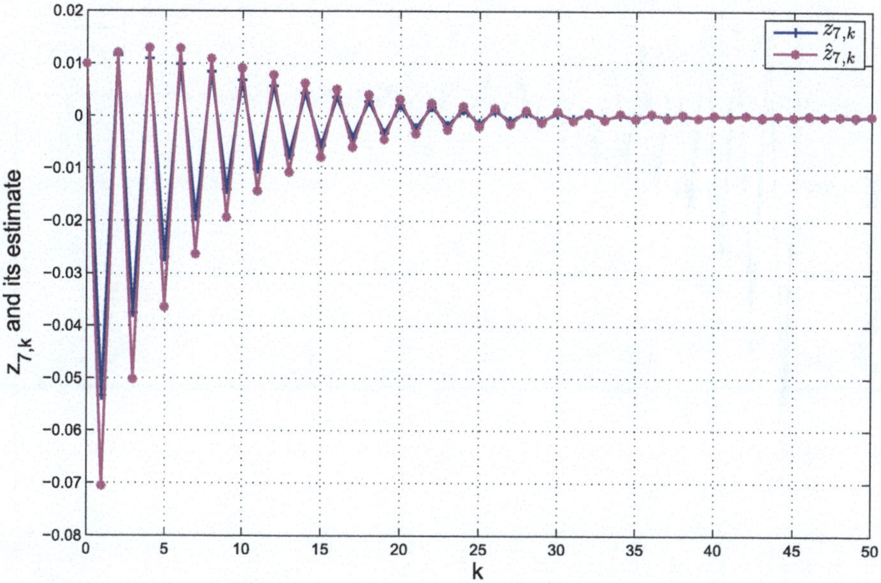

Fig. 10.8 Output $z_{7,k}$ and its estimate $\hat{z}_{7,k}$

Table 10.1 The running times of several main functions in the global criterion

Function name	Calls	Total time	Self time
Global criterion	1	831.442	0.439
Solvesdp	50	826.243	0.113
Calllmilab	50	821.843	0.279
Feasp	50	819.284	0.060
Feaslv (MEX-file)	50	818.402	818.402

estimate can track the original output for each node. This is to say, this simulation has verified that the filtering scheme is indeed effective for the addressed class of complex networks with missing measurements.

The comparison between the local criterion and the global criterion.
With the same parameters, the running times of the criteria of the local performance analysis method are compared to that of the global augmented method. According to Profile Summary in MATLAB, the running times of several main functions in criteria are presented in Tables 10.1 and 10.2, where Calls is the number of times that this function is called, Total Time is the total time spent in this function, and Self Time is the time spent in a function excluding the time spent in its child functions. Additionally, the time unit is second.

From the comparison of Tables 10.1 and 10.2, it can be seen that the running times of functions in the local criteria can be significantly reduced compared with that in

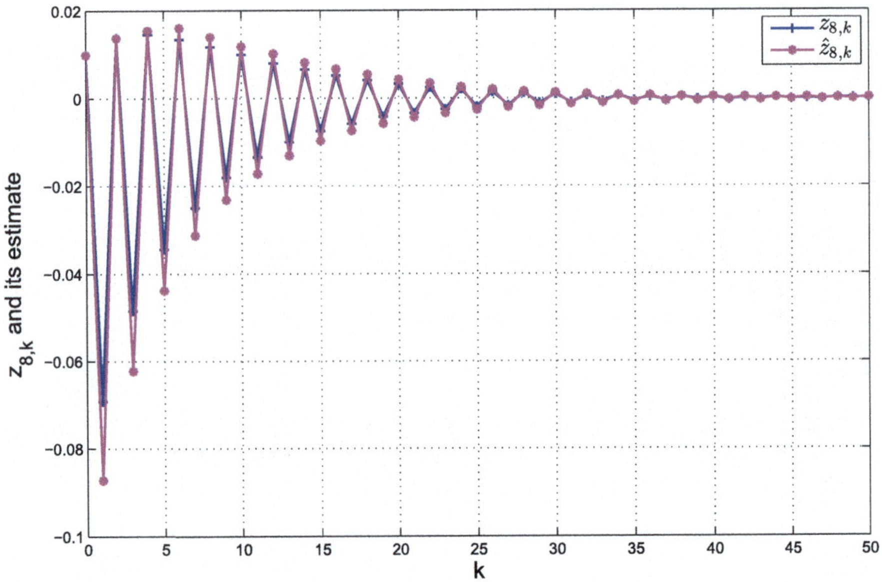

Fig. 10.9 Output $z_{8,k}$ and its estimate $\hat{z}_{8,k}$

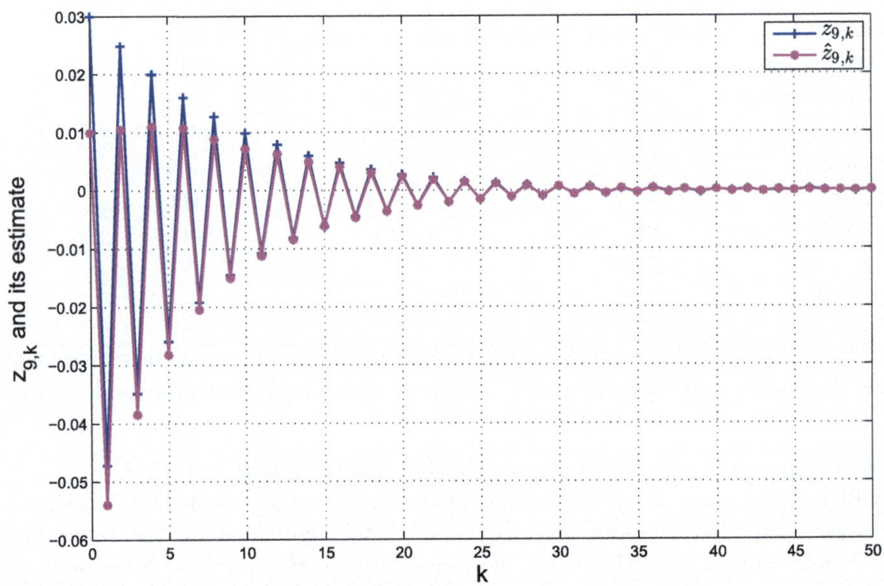

Fig. 10.10 Output $z_{9,k}$ and its estimate $\hat{z}_{9,k}$

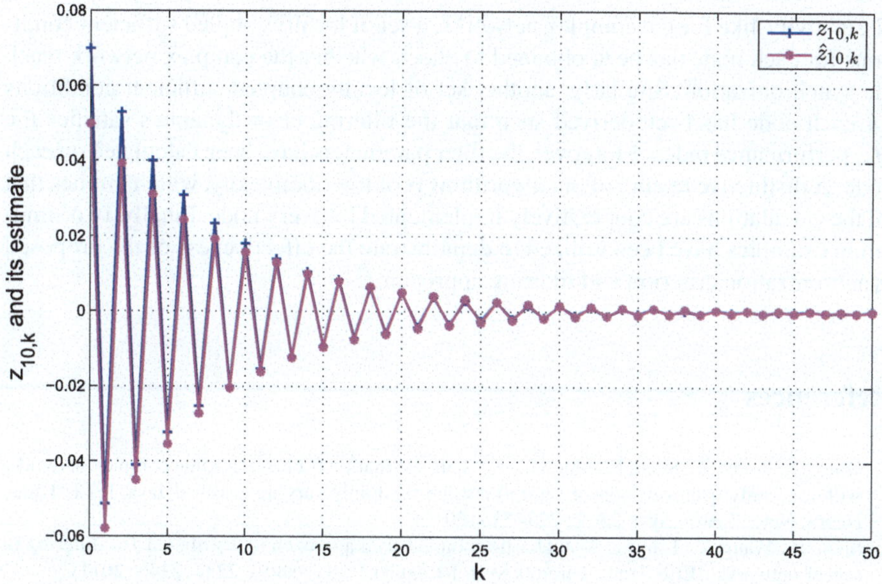

Fig. 10.11 Output $z_{10,k}$ and its estimate $\hat{z}_{10,k}$

Table 10.2 The running times of several main functions in the local criterion

Function name	Calls	Total time	Self time
Local criterion	1	40.696	1.377
Solvesdp	50	22.840	0.098
Calllmilab	50	18.226	0.845
Feasp	50	12.376	11.348
Feaslv (MEX-file)	50	11.348	4.002

the global criteria, which is consistent with the time-complexity analysis in Remark 10.4. Therefore, the distinguished advantages of the local criteria proposed in this chapter are quite obvious.

10.6 Summary

In this chapter, based on the vector dissipativity theory, the local performance analysis method has been developed to address the bounded H_∞ synchronization and H_∞ filtering problem for a class of discrete time-varying complex networks. A novel local description of bounded H_∞ synchronization has been presented only considering the synchronization errors among neighboring nodes. By establishing the vector dissipativity-like for the complex networks, a set of locally coupled sufficient conditions for each node has been obtained to check whether the complex network reach the synchronization. Similarly, another set of locally coupled sufficient conditions for each node has been derived such that the filtering error dynamics satisfies the H_∞ performance index. Moreover, the filter parameters have been calculated on each node. A distinctive feature of our algorithms is of low complexity, which implies that all the calculations are cooperatively implemented by every node. Finally, two simulation examples have been utilized to demonstrate the effectiveness of the proposed synchronization criterion and filtering approach.

References

1. Ding, D., Wang, Z., Shen, B., Shu, H.: H_∞ state estimation for discrete-time complex networks with randomly occurring sensor saturations and randomly varying sensor delays. IEEE Trans. Neural Netw. Learn. Syst. **23**(5), 725–736 (2012)
2. Shen, B., Wang, Z., Liu, X.: A stochastic sampled-data approach to distributed H_∞ filtering in sensor networks. IEEE Trans. Circuits Syst. I-Regular Papers **58**(9), 2237–2246 (2011)
3. Shen, B., Wang, Z., Ding, D., Shu, H.: H_∞ state estimation for complex networks with uncertain inner coupling and incomplete measurements. IEEE Trans. Neural Netw. Learn. Syst. **24**(12), 2027–2037 (2013)
4. Liang, J., Wang, Z., Liu, X.: Exponential synchronization of stochastic delayed discrete-time complex networks. Nonlinear Dyn. **53**(1–2), 153–165 (2008)
5. Liang, J., Wang, Z., Liu, X.: State estimation for coupled uncertain stochastic networks with missing measurements and time-varying delays: the discrete-time case. IEEE Trans. Neural Netw. **20**(5), 781–793 (2009)
6. Liu, Y., Wang, Z., Liang, J., Liu, X.: Synchronization and state estimation for discrete-time complex networks with distributed delays. IEEE Trans. Syst. Man Cybern. B Cybern. **38**(5), 1314–1325 (2008)
7. Wang, Z., Wang, Y., Liu, Y.: Global synchronization for discrete-time stochastic complex networks with randomly occurred nonlinearities and mixed time delays. IEEE Trans. Neural Netw. **21**(1), 11–25 (2010)

Chapter 11
Conclusions and Future Work

This chapter draws some conclusions on this book and points out some possible research directions related to the work done in this book.

11.1 Conclusions

The focus of the book has been placed on the distributed filtering, control and synchronization for discrete-time nonlinear stochastic systems. Specifically, several research problems have been investigated in detail.

- In Chap. 2, the distributed filtering problem has been dealt with for a class of discrete-time piecewise linear systems over SNs. First of all, the modes of the system and the filter of each node have been described by the available information, respectively. Next, two vectors of mode have been defined for the globally augmented plant and filter. Then, the performance index has been presented for the augmented filtering error system. Next, by means of Lyapunov stability theory and convex optimization method, sufficient conditions have been derived such that the distributed PWL filtering error dynamics is globally asymptotically stable while achieving H_∞ performance constraint. In the following, distributed H_∞ filter parameters are designed in term of a set of LMIs. Finally, an illustrative simulation has been provided to demonstrate the effectiveness of proposed distributed filter approach. It is noted that the computational burden is related with the size of sensor network. To address this problem, the local performance analysis method has been developed in the following chapters.
- In Chap. 3, we have investigated the distributed H_∞-consensus filtering problem for a class of discrete time-varying systems with stochastic nonlinearities and missing measurements in SNs. The statistical properties of the stochastic nonlinearity

F. Han et al., *Distributed Filtering, Control and Synchronization*,
Studies in Systems, Decision and Control 428,
https://doi.org/10.1007/978-3-030-97075-8_11

function have been presented. A set of Bernoulli distributed random variables has been employed to describe the occurrence of the missing measurements. The available information has been employed to construct the distributed filter for each node, where the traditional Kalman filter structure and the consensus error from its neighboring nodes are both considered. A novel H_∞-consensus performance constraint has been proposed to simultaneously consider the consensus errors and the filtering errors. By using the vector dissipativity theory, a set of sufficient conditions has been established for each node such that the locally augmented dynamics satisfies the proposed H_∞-consensus performance constraint. Furthermore, the corresponding filter gains have been calculated on each node.

- In Chap. 4, the local performance analysis method has been employed to deal with the distributed filtering problem for a class of discrete time-varying systems with the event-triggering protocol and random parameters. An information matrix has been proposed to formulate the statistical property of a random matrix. Next, this matrix has been employed to calculate the weighted-covariance matrix by means of the Kronecker product and the Hadamard product. A novel event-triggering protocol has been introduced to the distributed filtering problem in order to prolong the life of sensor node. A set of locally coupled design sufficient conditions for each node has been derived such that the filtering error dynamics satisfies the vector dissipativity over the finite-horizon, as well as the proposed H_∞-consensus constraint. In addition, the explicit expressions of the desired filter gains have been obtained. Finally, a simulation example has been employed to demonstrate the effectiveness of the proposed filtering algorithms.
- In Chap. 5, the distributed H_∞-consensus filtering problem has been investigated for a class of discrete time-varying systems with randomly occurring norm-bounded uncertainties and sensor saturations with the RRP over SNs. Typical norm-bounded uncertainties have been considered to enter into the network in a random manner. Sensor saturations have been involved to reflect the limited measure capacity of nodes. In order to reduce the usage of network resources and sensor energy, the RRP has been employed to allow each node to receive information only from one neighboring node at each time instant. Then, to establish the vector dissipativity of the filtering error dynamics, the supply rate functions have been chosen according to the desirable performance index. Next, the vector dissipation inequality has been transformed to the desirable H_∞-consensus performance index by using properties of the dissipation matrix. Sufficient conditions have been established for the filtering error dynamics to achieve certain pre-specified disturbance attenuation criterion in the H_∞ measure. Two special cases have been examined to show the flexibility and scalability of the designed scheme. Finally, a simulation example has been shown to illustrate the effectiveness of the proposed distributed filtering scheme.
- In Chap. 6, we have investigated the scalable distributed H_∞-consensus filtering problem for a class of discrete time-varying systems with multiplicative noises and censored measurements over SNs where only partial nodes have the ability to conduct the measure task. A set of deterministic indicative variables has been employed to characterize the censored measurements. A part of nodes in

sensor network can measure the information of the plant, while another part of nodes cannot. Then, based on the available information, the distributed filter has been constructed for each node based on the available information. A novel H_∞-consensus performance index has been established to reflect the impacts of the censored measurements on the filtering error dynamics. By using the local performance analysis method, a sufficient condition has been established so as to meet the prescribed H_∞-consensus performance requirement. Furthermore, the desired filter gains have been recursively calculated in a distributed manner. Finally, the effectiveness of the main results has been illustrated by a simulation example.

- In Chap. 7, we have investigated the local design problem of the distributed H_∞-consensus filter for the discrete time-varying system with multiplicative noises and deception attacks over SNs. Two sequences of mutually uncorrelated Gaussian white noises with known statistical information have been employed to model the multiplicative noises. A Bernoulli distributed random variable has been utilized to formulate whether the deception attacks are successful or not. Next, the sector-like bounded condition has been utilized to formulate the unknown malicious signal involved in the deception attacks . Then, the desirable H_∞-consensus performance constraint has been established to tolerate a certain degree of deception attacks . In view of the stochastic vector dissipativity theory, sufficient condition has been established in terms of a set of matrix inequalities for each node to guarantee the prescribed H_∞-consensus performance. Furthermore, the corresponding filter gains have been designed by recursively solving certain matrix inequalities. Finally, a simulation example has also been given to verify the effectiveness of the proposed filter algorithm.
- In Chap. 8, we have studied the distributed resilient filtering problem for a class of nonlinear time-delayed systems subject to stochastic perturbations. Two Gaussian white stochastic processes with known statistical information have been utilized to describe the system noises and the measurement noises, respectively. Also, a resilient filter has been designed for each node by means of the parameter uncertainties and Bernoulli-distributed random variables. Then, a novel exponential-bounded performance index has been put forward to measure the disturbance rejection level of the distributed filtering against the external disturbances and the influence of the initial values. A new vector dissipation definition including multiple vectors of energy storage functions has been established to deal with the time-delay filtering error dynamics. Within the framework of local performance analysis inspired by this new definition of vector dissipation, sufficient conditions in terms of recursive linear matrix inequalities have been constructed for each node to guarantee the desirable performance index. Next, a local optimization problem subject to a set of recursive linear matrix inequalities has been presented for each node to minimize the upper bound in the performance index, where the calculations can be conducted on every node in a distributed manner and the filter gains have been also calculated. Finally, an illustrative simulation example has been provided to verify the applicability of the proposed estimators.
- In Chap. 9, the finite-horizon distributed H_∞-consensus control problem has been addressed for a class of discrete time-varying multi-agent systems with random

parameters and missing measurements via the local performance analysis method. With the aid of the assumption that the considered topology has a spanning tree, it has been proved that the local consensus based on the information among neighboring agents is equivalent to the global consensus described by any two agents. Then, a novel dynamic local H_∞-consensus performance index have been proposed just considering the consensus error among neighboring nodes to cater for the requirement of local performance method. The WCM formula of random matrix involving the covariance information between any two random variables has been established via the Kronecker product and the Hadamard product. A set of locally coupled sufficient conditions for every agent has been obtained such that the proposed locally dynamically H_∞-consensus performance constraints are satisfied in the presence of the external disturbances. In addition, the explicit expressions of the desired controller gains have been obtained. Finally, a simulation example has been employed to demonstrate the effectiveness of the proposed control scheme.

- In Chap. 10, the local performance analysis method has been employed to address the bounded H_∞ synchronization and H_∞ filtering problem for a class of discrete time-varying CNs. For this purpose, a novel local description of bounded H_∞ synchronization has been presented only considering the synchronization errors among neighboring nodes. Moreover, it has been proved that this description is equivalent to the traditional global synchronization, which considers the synchronization errors of all nodes. By establishing the vector dissipativity-like for the complex networks, a set of locally coupled sufficient conditions for each node has been obtained to check whether the complex network reach the synchronization. Similarly, another set of locally coupled sufficient conditions for each node has been derived such that the filtering error dynamics satisfies the H_∞ filtering performance index. Moreover, the filter parameters have been calculated on each node. A distinctive feature of our algorithms is of low complexity, which implies that all the calculations are cooperatively implemented by every node. Finally, two simulation examples have been utilized to demonstrate the effectiveness of the proposed synchronization criterion and state estimation approach.

11.2 Future Work

Related topics for the future research work are listed below:

- The local performance method would be applied to solve the distributed filtering with various complicated system dynamics and network-induced phenomena, which result in new challenges for the local performance method.
- Another urgent work is to extend the local performance method to the different networks with different characteristics, for example, the neural networks and the gene regulatory network.

- The parameter optimization problems need to be considered under the local performance method. Via optimizing some given parameters, it is possible to achieve a better performance behavior.
- An important work is to realize the truly distributed calculations by means of the parallel computation software and protocols.
- Obviously, the utilization of the linear matrix inequalities will yield conservatism since only a feasible solution is available. Meanwhile, it is impossible to acquire the specific relationship among parameters. As a result, it is desirable to find some new tools to address distributed filtering problem within the framework of local performance method.
- The distributed filtering with dynamic topology would be intensively investigated via the local performance method.

Index

F. Han et al., *Distributed Filtering, Control and Synchronization*, Studies in Systems, Decision and Control 428,
https://doi.org/10.1007/978-3-030-97075-8

The manufacturer's authorised representative in the EU is Springer Nature Customer Service Centre GmbH, Europaplatz 3, 69115 Heidelberg, Germany. If you have any concerns regarding our products, please contact ProductSafety@springernature.com

Printed and bound by CPI Group (UK) Ltd, Croydon, CR0 4YY
22/09/2026
02220961-0005